NEW DIRECTIONS IN THE SKELETAL BIOLOGY OF GREECE

Hesperia Supplements

The *Hesperia* Supplement series (ISSN 1064–1173) presents book-length studies in the fields of Greek archaeology, art, language, and history. Founded in 1937, the series was originally designed to accommodate extended essays too long for inclusion in the journal *Hesperia.* Since that date the Supplements have established a strong identity of their own, featuring single-author monographs, excavation reports, and edited collections on topics of interest to researchers in classics, archaeology, art history, and Hellenic studies.

Hesperia Supplements are electronically archived in JSTOR (www.jstor.org) where all but the most recent titles may be found. For order information and a complete list of titles, see the ASCSA website (www.ascsa.edu.gr).

Occasional Wiener Laboratory Series (OWLS)

OWLS is a subseries of *Hesperia* Supplements under the general editorship of Sherry C. Fox, director of the Wiener Laboratory at the American School of Classical Studies at Athens. Although the topics covered are scientific in nature, reflecting the interests of the Wiener Laboratory, it is hoped that the books will appeal to a wide archaeological audience. Authors are therefore encouraged to avoid overly technical discussions, and clearly indicate why their work is important to researchers outside their immediate field.

1 L. A. Schepartz, S. C. Fox, and C. Bourbou, *New Directions in the Skeletal Biology of Greece* (2009)

Hesperia Supplement 43

NEW DIRECTIONS IN THE SKELETAL BIOLOGY OF GREECE

EDITED BY
LYNNE A. SCHEPARTZ, SHERRY C. FOX,
AND CHRYSSI BOURBOU

The American School of Classical Studies at Athens
2009

Cover illustration: Skull from Theopetra Cave, Mesolithic period. Radiograph showing pseudopathological lesions that mimic the "hair-on-end" pattern, but are actually due to postmortem erosion. Radiograph S. Fox

Library of Congress Cataloging-in-Publication Data

New directions in the skeletal biology of Greece / edited by Lynne A. Schepartz, Sherry C. Fox, and Chryssi Bourbou.
p. cm. — (Hesperia supplement ; 43)
Includes bibliographical references and index.
ISBN 978-0-87661-543-0 (alk. paper)
1. Physical anthropology—Greece. 2. Anthropometry—Greece. 3. Human remains (Archaeology)—Greece. 4. Human skeleton—Analysis. 5. Greece—Antiquities. I. Schepartz, Lynne Alison. II. Fox, Sherry C. III. Bourbou, Chryssi.
GN50.45.G8N48 2009
599.909495—dc22 2008042038

This volume is dedicated to the memory of J. Lawrence Angel, a pioneer in skeletal biology research in the eastern Mediterranean.

CONTENTS

ILLLUSTRATIONS

TABLES

INTRODUCTION: NEW DIRECTIONS IN THE SKELETAL BIOLOGY OF GREECE

by Lynne A. Schepartz, Sherry C. Fox, and Chryssi Bourbou

Come, tell me how you live. This is more than the title of Agatha Christie's mesmerizing account of life with her archaeologist husband Max Mallowan and her curiosity about other cultures and past times.[1] It is also the question skeletal biologists ask of every individual and population they study. Such inquiries invariably lead to investigations of health and diet, stress and violence, physical activity, social status, and how we cared for and venerated people in the past. These are things that most often only a skeletal biologist can determine, although they must work closely with archaeological data to render a complete answer. Like Agatha Christie, skeletal biologists are mistakenly thought to be focused on death, yet it is clearly life and behavior that is the driving motivation behind our work.

WHY GREECE?

Greece is the home of pioneering efforts in skeletal biology and archaeology that still constitute the "backbone" of research today. It has an outstanding array of national scholars and international institutes that continue these traditions and expand the boundaries of skeletal biology research. Greece played a key role in the development of significant initiatives in skeletal biology through J. Lawrence Angel's attention to paleodemography and populational health, and it remains a forerunner in skeletal biology via the integrative directions of the Wiener Laboratory of the American School of Classical Studies at Athens, as discussed further in Chapter 1.[2]

There are numerous reasons why the academic climate in Greece is so conducive to innovative skeletal biology research: There is an extensive archaeological record, with a well-known chronology spanning early prehistory to the near present. Protection of cultural heritage is a high priority, and part of a national ethic that values the past and its lessons for the present. There is a well-developed research infrastructure that includes national, district, and local museums; governmental research groups; analytical laboratories; and a large cohort of well-trained professionals from the region and beyond. Important collections are available for comparative study and reanalysis using newer methods and techniques, and the region is a leader in archaeological science.

1. Christie Mallowan 1946.
2. For an excellent discussion of the historical development of skeletal biology in Greece, readers should consult Roberts et al. 2005.

But resources alone do not really explain why Greece became, and continues to be, a leading area for skeletal biology. The subject has also developed because Greece is a region where there are challenging questions of population affinities and interactions across space and time, where the range of natural environments led populations to selectively exploit local and nonlocal resources, and where the development of social complexity involved many polities and had far-reaching effects throughout the Mediterranean and even more distant regions.

In short, while one can argue that conditions for skeletal biology research are never perfect—the bone preservation is not ideal, the time and resources many archaeological projects allot to skeletal research is still less than desired, the number of positions for skeletal biologists is considerably less than the number of highly qualified scholars—a multitude of historical and institutional factors ensure that the potential for inventive skeletal biology research in Greece far exceeds that of almost any other world region. Proof of this is the exciting scholarship by a new and burgeoning generation of researchers in this volume.

SCOPE OF THE VOLUME

The impetus for *New Directions in the Skeletal Biology of Greece* was a colloquium organized by Lynne A. Schepartz and Sherry C. Fox for the 104th Annual Meeting of the Archaeological Institute of America in 2003. The following year, Chryssi Bourbou organized a session at the 15th European Meetings of the Paleopathology Association on "Bioarchaeology in Greece." Both sessions highlighted the flourishing of skeletal biology research in Greece and worked toward building a community of scholars interested in similar questions. The result is this volume, the first in a new supplement series of *Hesperia* under the rubric *OWLS (Occasional Wiener Laboratory Series)* and dedicated to archaeological science in Greece and the Greek world.

The contributors to this volume have backgrounds in many disciplines, including anthropology, archaeology, bioarchaeology, medicine, dentistry, genetics, chemistry, and paleoanthropology. They truly represent the perspective that our knowledge of the human skeleton and human health and behavior is dependent upon collaboration among these diverse fields.

Each paper makes a unique contribution to our understanding of Greek populations. At the same time, there are areas of overlap so that the reader can appreciate both the differences and commonalities that constitute our current understanding of ancient Greece. Table 1 provides a summary of the regions and themes covered in this book, arranged in order of the relative age of the samples under discussion.

Temporal Range

The skeletal biology of Greece begins with its first known inhabitants, represented by Middle Pleistocene populations. Harvati's detailed study of the Petralona cranium in Chapter 2, using a three-dimensional geometric

TABLE 1. CONTRIBUTED PAPERS ARRANGED BY RELATIVE AGE OF STUDY SAMPLES

Author	*Region*	*Sample*	*Theme*
Harvati (Ch. 2)	Macedonia	Middle Paleolithic	Population origins, paleoanthropology
Stravopodi, Manolis, Kousoulakos, Aleporou, and Schultz (Ch. 16)	Various	Mesolithic–Bronze Age	Paleopathology
Lorentz (Ch. 5)	Euboia (Central)	Neolithic	Cranial variability, bioarchaeology
Papathanasiou, Zachou, and Richards (Ch. 13)	Lokris (Central)	Neolithic–Bronze Age	Dietary reconstruction, bioarchaeology
Kirkpatrick Smith (Ch. 6)	Attica	Bronze Age	Paleopathology, bioarchaeology
Schepartz, Miller-Antonio, and Murphy (Ch. 10)	Peloponnese	Bronze Age	Women's health, dentition
Iezzi (Ch. 11)	Lokris (Central)	Bronze Age	Population variability, mobility
Petroutsa, Richards, Kolonas, and Manolis (Ch. 14)	Peloponnese	Bronze Age	Dietary reconstruction
Liston and Preston Day (Ch. 4)	Crete	Geometric	Paleopathology, bioarchaeology
Charlier, Poupon, Goubard, and Descamps (Ch. 3)	Attica	Classical	Cremation, mortuary behavior
Hillson (Ch. 9)	Dodecanese	Classical	Infant osteology, population variation, mortuary behavior
Bourbou and Tsilipakou (Ch. 8)	Macedonia	Byzantine	Health, paleopathology
Papageorgopoulou and Xirotiris (Ch. 12)	Epirus	Byzantine	Population variation, dietary reconstruction, health
Bourbou (Ch. 7)	Crete	Byzantine	Paleopathology
Garvie-Lok (Ch. 15)	Peloponnese	Frankish and Ottoman	Population mobility, dietary reconstruction
Georgiou, Zouganelis, Spiliopoulou, and Koutselinis (Ch. 17)	Crete	Historical	mt-DNA, genetic identity

morphometric technique, illustrates its morphological distance from later Neanderthals as well as its affinities with the early European Sima de los Huesos 5 and the African Kabwe specimens. Thus, we see that the early peopling of Greece took place during a time of significant European population expansions and complex biological changes. From the founding populations we move forward in time to the next periods with somewhat larger skeletal samples from the Mesolithic and the Neolithic cultures. Although many papers deal with prehistoric populations, with particular emphasis on the Bronze Age, fully half of the contributions examine later societies. Several time periods that were until recently largely neglected, such as the Byzantine and post-Byzantine, are also represented.

Geographic Range

In addition to assembling papers covering most of Greece's long temporal history, we also chose contributions that reflect the diversity of Greece's landscape and the impact this has on human health and subsistence. There are papers on better-known areas, such as the Peloponnese and Crete, and studies of less investigated regions (e.g., Epirus and Macedonia). Map 1 shows the major sites discussed in this volume.

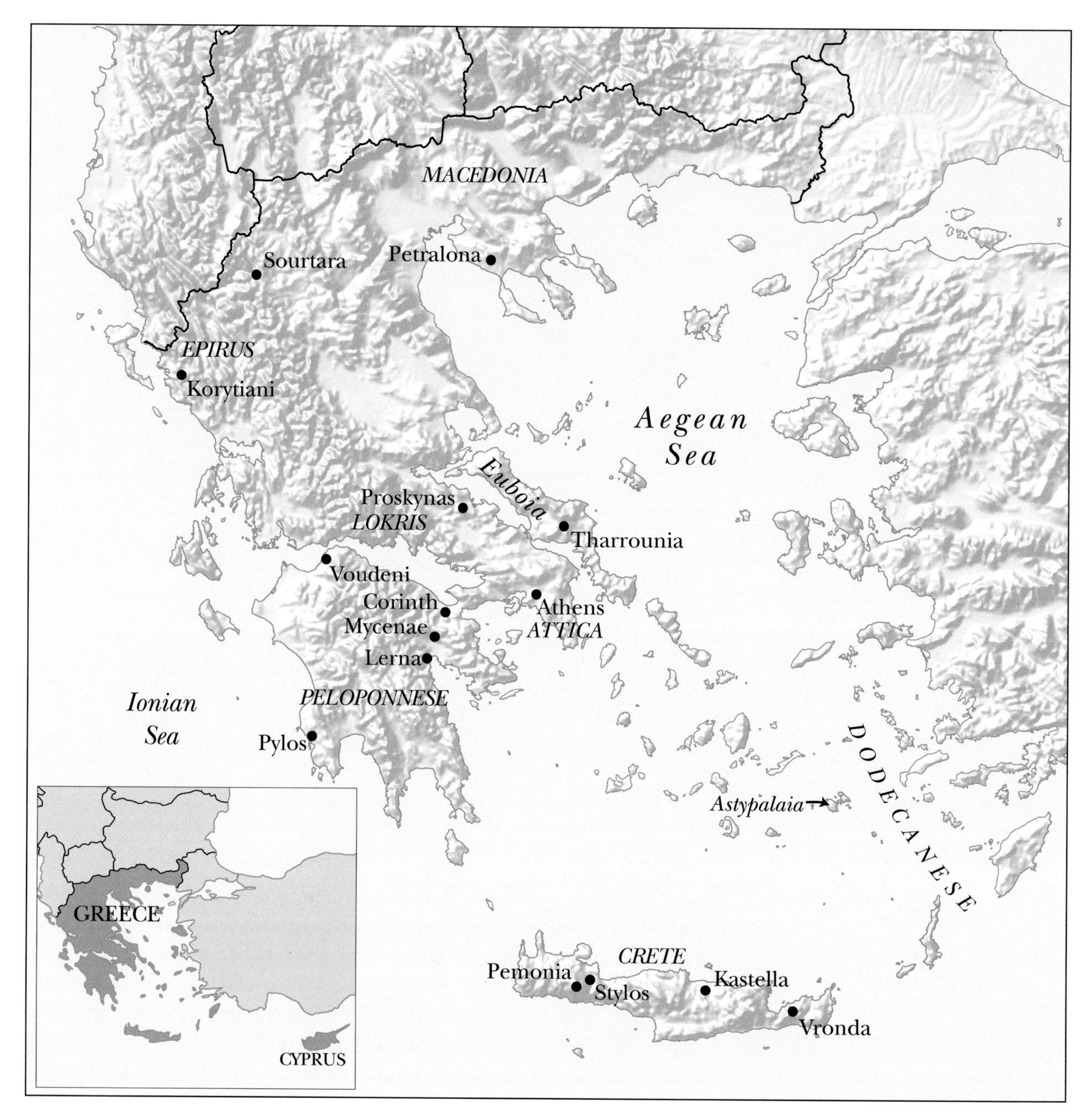

Map 1. Greece, showing locations of main sites discussed in the volume. Mapping Specialists, Ltd.

Research Questions

It is impossible to easily categorize each of these papers by its major theme or research question, although we provide some keywords in Table 1. Almost every contribution touches upon several themes that we feel are seminal to an appreciation of the new directions in the skeletal biology of Greece.

Discussions on the synergy of biology, social context, and mortuary practices are found in Chapters 3, 4, 5, and 10. The violence of the past is examined in Chapter 6 by Kirkpatrick Smith (for the militaristic Mycenaeans of Athens) and in Chapter 7 by Bourbou (in the lives of medieval urban dwellers in Crete). The dynamic of changing society and health is the focus of the studies presented in Chapters 8, 9, and 10.

Many of the papers exemplify the value of advances in archaeological science for addressing new or persistent skeletal biology questions. Dietary reconstruction through isotopic analyses is the objective of Chapters 12, 13, and 14. The development of this research area in Greece surpasses efforts in any other region[3] and it continues to provide us with exciting insights about dietary diversity. Issues of population mobility are examined in Chapter 11 by Iezzi (using limb morphology) and in Chapter 15 by Garvie-Lok (using chemical analyses). In Chapter 16, the authors take the study of porotic hyperostosis and malarial interactions to a fundamentally new level with their histological and comparative investigations. The potential identification of an important historical religious leader is resolved by the mt-DNA analysis of skeletal samples from Crete, presented in Chapter 17.

Together, these papers exemplify the compelling and challenging questions of skeletal biology in Greece today and in the future. Whether it is placing population relationships into the context of mobility, migration, and economy rather than merely populational affinities; or shifting studies of behavior toward understanding human impacts on the landscape and the consequences for human health; or redefining our knowledge of diet and dietary signatures within archaeological and ecological frameworks; or examining the construction of social and gender identity through the lens of human health, the future of skeletal biology research in Greece appears brighter than ever.

ACKNOWLEDGMENTS

Collectively, the editors would like to thank Malcolm H. Wiener for his generous and continuous support throughout all these years since the establishment of the Wiener Laboratory of the American School of Classical Studies at Athens in 1992. His pioneering vision of a laboratory that would promote archaeological science in Greece is now a reality, well reflected in the work of numerous researchers who carried out their projects within the Laboratory. Special thanks to the Princeton Publication Office for housing the publication of OWLS, and particularly Charles Watkinson, Tracey Cullen, and Carol A. Stein for dealing with all the issues that arose during the preparation of this volume.

Lynne A. Schepartz would like to acknowledge the people responsible for her transformation into a scholar of ancient Greek populations: Sharon Stocker and Jack L. Davis of the University of Cincinnati. This volume would not have been conceived without their intellectual curiosity, encouragement, friendship, and generosity. Funds from Florida State University were used to support the cost of volume illustrations. Sherry C. Fox greatly appreciates the steadfast support of Malcolm H. Wiener as well as the Wiener Laboratory community at the American School of Classical Studies at Athens. Chryssi Bourbou would like to thank Charlotte Roberts for financial support and useful comments during the preparation of her session in Durham, and all the colleagues and friends whose valuable help and continuous support made the publication of this volume possible.

3. M. Richards (pers. comm.).

REFERENCES

Christie Mallowan, A. 1946. *Come, Tell Me How You Live*, New York.

Roberts, C., C. Bourbou, S. Triantaphyllou, A. Lagia, and A. Tsaliki. 2005. "Health and Disease in Greece: Past, Present, and Future," in *Health in Antiquity,* ed. H. King, London, pp. 32–58.

CHAPTER I

Bioarchaeological Approaches to Aegean Archaeology

by Jane Buikstra and Anna Lagia

A remarkable range of questions of significance in interpreting Aegean pasts is amenable to bioarchaeological study. These include issues ranging from genetic relatedness to diet and health.

Is there a sharp shift in diets from the Mesolithic to the Neolithic in the eastern Mediterranean as is witnessed in other parts of Europe?[1] What was the contribution of agricultural and animal products to human diet and economy across time?[2] What was the participation of women as reflected in their activity levels in different periods?[3] Are the Mesara tholoi the interment facilities of multiple families?[4] Did state formation during Minoan times lead to diet and health differences between social classes? Is the settlement hierarchy and regional political organization suggested for the Bronze Age reflected in regional health, dietary, and activity patterns?[5] What marriage patterns linked the elite across the Aegean world?[6] Is the end of the Bronze Age associated with in-migration or perhaps with changes in warfare?[7] How should we explain the apparent preponderance of males within certain ancient Aegean cemeteries,[8] while in others the sex ratio favors females?[9] Do the ethnic boundaries reported for Archaic Lesbos, modeled on material culture, also reflect discontinuities in gene flow?[10] To what extent do didactic texts written by upper-class male citizens reflect the conditions under which most people lived in classical antiquity?[11] Was the place of women, as depicted in the mortuary record, marginal?[12] How were individuals with severe disorders and disabilities treated at death in different times?[13] How were infants and children accepted by society in different time periods?[14] What disease caused the ca. 430 B.C. plague in Athens?[15] When did tuberculosis become epidemic within the Mediterranean?[16] What caused the prevalence and co-occurrence of certain forms of genetic anemias in certain regions in Greece from antiquity to modern times?[17]

The answers to these and other questions critical to the interpretation of ancient Aegean peoples are accessible to bioarchaeologists who carefully consider human remains within archaeological contexts. This requires, however, that cemetery analyses be embedded within problem-oriented research designs.

1. Richards and Hedges 1999; Richards, Price, and Koch 2003; Richards, Schulting, and Hedges 2003.
2. Dalby 1996; Wilkins, Harvey, and Dobson 1995.
3. Moore and Scott 1997; Grauer and Stuart-Macadam 1998; Laffineur and Betancourt 1997.
4. Branigan 1988, 1993.
5. Branigan 2000; Driessen, Schoep, and Laffineur 2002.
6. Renfrew and Cherry 1986.
7. Drews 1993; Laffineur 1999.
8. Angel 1959, 1971a, 1973; Charles 1963; Lagia, Petroutsa, and Manolis 2007.
9. Angel 1970, 1971b; Triantaphyllou 2001.
10. Spencer 1995b; Lambrianides 1995.
11. Jameson 1977–1978; Morris 1994a.
12. Cohen 1989; Stears 1995; Scheidel 1995; Johnstone 1998.
13. Garland 1995; Dettwyler 1991; Hawkey 1998; Vlahogiannis 2005; Lagia 2007b.
14. Golden 1988, 1990, 1997; Garnsey 1989; Morris 1987, 1992; Neils and Oakley 2003; Lagia 2007a.
15. Thucydides III, 87; Grmek 1983; Holden 1996.
16. Grmek 1989; Roberts and Buisktra 2003.
17. Angel 1964, 1966a, 1967; Grmek 1989; Lagia, Eliopoulos, and Manolis 2007.

Figure 1. J. Lawrence Angel in front of the Kourion Museum, Cyprus, around 1950. Courtesy National Anthropological Archives, Smithsonian Institution (Papers of J. Lawrence Angel, Box 157)

Our overview of Aegean bioarchaeology, that is, the study of ancient cemeteries and their contents,[18] identifies three apparent trends. The first and dominant of these is exquisite, detailed osteological description, typical of appendices in many site reports. There is good news in that such data sets—their long lists of measurements and other observations—can contain the raw materials for further inferences about ancient lives. On the other hand, because such information frequently results from antiquated bioanthropological rituals—measurements by rote rather than by problem orientation—such data sets may *not* contain the most useful clues for decoding the ancient past. Newly developed methods are seldom reflected and the relevance of such copious data sets to archaeological or historical issues may *not* be obvious.

A second trend is for skeletal biologists within the Aegean region to follow data collection protocols and problem sets as defined by J. Lawrence Angel (Fig. 1), an American biological anthropologist who championed an orientation he called "social biology."[19] By this term he emphasized what we might today call a life history approach, which has much to recommend it.[20] Angel's social biology was problem-oriented and concerned with context. He addressed significant archaeological and historical questions of his day, focusing upon issues of demography, health, and disease.

Social biology, as developed by Angel in the 1940s, was clearly ahead of its time, perhaps by as many as two decades. In its unmodified form, however, it is now somewhat outdated and requires critical review. There are new ways of investigating the past unknown at the time of Angel's monumental efforts. Quantities of data require statistical testing not commonly conducted during the mid-20th century. And, of course, we *have* learned that pelvic "parturition pits" cannot be used to estimate fertility rates—a stimulating yet overly optimistic suggestion championed by Angel.[21] As we shall see later, however, there are other means of generating this vital rate through innovations unknown in Angel's time. Given his proclivity for promoting new techniques, for example, trace element analysis,[22] it is

18. Buikstra 1977.
19. Angel 1946; Buikstra and Hoshower 1990; Ortner and Kelley 1990; Jacobsen and Cullen 1990.
20. Wright and Yoder 2003.
21. Angel 1971a; critique in Stewart 1970; Suchey et al. 1979.
22. Angel and Bisel 1986.

obvious that Angel would have embraced such advances. His basic tenet, the use of skeletal biological data to enhance our knowledge of ancient lifeways, is therefore to be endorsed. To let our purview be circumscribed by the research designs common to Angel's era is, however, unduly constraining.

The third group of skeletal biological papers, an admitted minority, are those that—like Angel—explicitly focus upon archaeological or historical questions. They are distinguished from Angel's work, however, by the application of quantitative methods and tests of alternative models. Most commonly, these have been oriented to issues of migrations or other expectations concerning genetic relationships using population-based approaches.[23] Bioarchaeological study is, however, suitable for a much wider range of problems, including questions commonly asked in Aegean contexts. These include issues that range from social complexity to the history of epidemic disease. As will be seen in the following pages, bioarchaeological research in the Aegean has decisively turned to this direction during the recent years.[24]

THEORIES OF MORTUARY BEHAVIOR

Before we turn to topics explicitly grounded in skeletal biology, however, we must recognize the influence of alternative theoretical perspectives within bioarchaeological research. While most of us dealing with skeletal biology are grounded in evolutionary theory, archaeologists typically adhere to theories developed in social science and historical studies. Cemetery investigations within the past few decades have been a battleground for archaeological theoretical debates between proponents of two contrastive schools. The first of these, developed through the influence of Lewis Binford,[25] Arthur Saxe,[26] and James Brown,[27] is identified with the "new" or processual archaeology of the 1960s and 1970s.[28] In the Saxe-Binford-Brown approach, emphasis is placed upon the grave or tomb, with variation in treatment of the dead, grave wealth, and tomb elaboration assumed to reflect social status of the deceased. More recently, Hodder and colleagues—under the aegis of contextual or interpretive archaeology—have reminded us that it is the *living* who bury the dead and that funerary ritual often serves as an arena wherein the living engage in political, social, and economic competition.[29] For example, as the state assumes dominance, it may take control of the ancestors and their tombs to legitimize its power. Alternatively, in times of conflict, the desecration of the ancients commonly empowers the living protagonists.

While the postprocessual critique has much to recommend it, there is an aspect of the Saxe-Binford-Brown approach that continues to hold potent explanatory power. As Saxe[30] and Lynne Goldstein[31]—and more recently Ian Morris,[32] working within the Classical world—emphasize, placement of the ancestors is frequently a conscious choice designed to establish legitimacy over scarce and restricted resources such as land or water. In this way, placement of the dead within cemeteries and the location of cemeteries themselves can be explained. These and other implications

23. For example, Musgrave and Evans 1980.
24. See also Roberts et al. 2005 and Macchinon 2007 for compiled bibliography.
25. Binford 1971.
26. Saxe 1970.
27. Brown 1971.
28. Brown 1995.
29. Hodder 1980, 1982; Parker Pearson 1982.
30. Saxe 1970.
31. Goldstein 1981.
32. Morris 1991.

of alternative theoretical perspectives developed within archaeological investigations of mortuary behavior should be understood by any who engage in the study of cemetery sites.

In recent years, there has been a strong urge toward a more anthropological approach to classical archaeology and the application of methodological and conceptual developments of other fields to classics.[33] Attempts to apply archaeological theory and to integrate anthropological and archaeological data include the work of Liston,[34] Fox Leonard,[35] Triantaphyllou,[36] Little and Papadopoulos,[37] Rotroff, Little, and Snyder,[38] Papathanasiou, Larsen, and Norr,[39] Papathanasiou,[40] Garvie-Lok,[41] Tsaliki,[42] Karali and Tsaliki,[43] Hillson,[44] Bessios and Triantaphyllou,[45] Ingvarsson-Sundström,[46] Bourbou,[47] Liston and Papadopoulos,[48] Grammenos and Triantaphyllou,[49] Lagia,[50] Tritsaroli,[51] Vika,[52] Nafplioti,[53] Bourbou and Richards,[54] Petroutsa,[55] Ubelaker and Rife,[56] Rife et al.,[57] Kavvadias and Lagia,[58] Lagia and Richards,[59] and Papathanasiou and Richards.[60]

GENETIC RELATIONSHIPS

Many issues of past and current interest in Aegean studies involve hypothesized migrations and population genetic relationships. These may be on a global scale, such as propositions relating waves of Egyptian or Anatolian migration to Greek contexts, or they may reflect more subtle micro-level concerns about residence and marriage patterns. By choosing appropriate methods, answers to each of these questions are accessible.

Recently developed ancient DNA (aDNA) methods, as discussed in chapter 17 in this volume by Georgiou et al., can address numerous issues, ranging from migration patterns to residence and even sex identifications. Mitochondrial DNA (mt-DNA) appears most appropriate for global issues, while nuclear variation such as in HLA (Human Leukocyte Antigen) systems facilitates regional studies. Another form of nuclear variation, micro-satellite or VNTR (Variable Number of Tandem Repeats) loci, is useful in identifying kin-based cemeteries and inferring residential rules.[61] Nuclear DNA studies have also been successful in distinguishing males from females and thus hold promise for the intractable problem of estimating sex in juvenile skeletal remains.[62]

33. Morris 1994b, 2000; Spencer 1995a; Dyson 1998; Davies 2001; Ober et al. 2007.
34. Liston 1993, 2007.
35. Fox Leonard 1997, 2005.
36. Triantaphyllou 1998, 1999, 2001.
37. Little and Papadopoulos 1998.
38. Rotroff, Little, and Snyder 1999.
39. Papathanasiou, Larsen, and Norr 2000.
40. Papathanasiou 2001, 2003, 2005.
41. Garvie-Lok 2001.
42. Tsaliki 2001, 2005, 2008.
43. Karali and Tsaliki 2001.
44. Hillson 2002, this volume.
45. Bessios and Triantaphyllou 2002.
46. Ingvarsson-Sundström 2003.
47. Bourbou 2003, 2004.
48. Liston and Papadopoulos 2004.
49. Grammenos and Triantaphyllou 2004.
50. Lagia 2007a, 2007b, forthcoming.
51. Tritsaroli 2006.
52. Vika 2007.
53. Nafplioti 2007, 2008.
54. Bourbou and Richards 2007.
55. Petroutsa 2007.
56. Ubelaker and Rife 2007.
57. Rife et al. 2007.
58. Kavvadias and Lagia, forthcoming.
59. Lagia and Richards, in prep.
60. Papathanasiou and Richards, in prep.
61. Brown and Brown 1992; Haydon 1993; Hermann and Hummel 1994; Pääbo, Higuchi, and Wilson 1989; Merriwether, Rothhammer, and Ferrell 1994; Stone and Stoneking 1993; Tuross 1994; Stone 2000.
62. Stone et al. 1996.

As such, aDNA analysis can help address questions concerning, for example, female infanticide. It is noteworthy that in certain contexts such as the Roman site of Ashkelon,[63] and the Neolithic Tel Teo,[64] where infanticide of females has been suggested to explain the large numbers of infant remains in "unusual" contexts, limited DNA analysis found more males to be present in the samples. It is worth investigating similar questions that have arisen from exceptional contexts such as the Late Hellenistic well in the Athenian Agora,[65] or the infant amphora graves at the Gymnasion of Messene,[66] as well as in contexts and periods in which the presence of infants and young children in the mortuary record appears to be highly selective.[67]

Three pilot studies of aDNA analysis have taken place so far in Greece addressing questions of kinship and biological sex. All attest the difficulties of aDNA preservation from local ecosystems. Brown et al.[68] analyzed 22 samples from Grave Circle B at Mycenae and successfully determined the sex of eight individuals. Two of the results, however, conflicted with the morphological determination of sex and with that based on grave goods and raised questions bilaterally concerning the biochemical and macroscopic methods of analysis. Similarly, Evison et al.[69] and Evison[70] analyzed 43 specimens of teeth and bone from diverse time periods and contexts in Greece, ranging from the Upper Paleolithic to the Classical periods. Their analysis emphasizes the technical and practical difficulties of aDNA preservation and extraction in Greece and offers a protocol of analysis. Most recently, Chilvers et al.[71] attempted to isolate ancient DNA from eight Neolithic and Bronze Age sites from mainland Greece and Crete. Only two sites, which included 22 of the 88 specimens, provided evidence for mitochondrial and/or nuclear DNA. The authors emphasize the importance of considering the thermal history of a site in assessing the feasibility of recovering aDNA. Moreover, they call attention to the significance of endorsing a self-critical attitude, in addition to the adherence to stringent criteria of authenticity, in any attempt to isolate aDNA, as has been recommended by Cooper and Poinar,[72] Pääbo et al.,[73] and Gilbert et al.[74] The presence of the same mtDNA haplogroup in two individuals from Grave Circle B in Mycenae, believed by facial reconstruction to have shared family relatedness, was attributed by Chilvers et al. to a sibling relation rather than marriage.[75]

Even with the availability of aDNA, however, Aegean bioarchaeology should not lose sight of its traditional strength—the use of phenotypic variation to estimate relatedness. Observation of skeletal features as a means of estimating genetic affinity has a long history in Aegean studies.[76] This approach has primarily focused upon cranial measurements, dismissing other forms of variation as less genetically grounded. However useful such measurements may be, it is clear that neglecting other sources of information is premature. Especially appropriate in situations where imperfect preservation or cremation may render traditional cranial and dental measurements inapplicable are the so-called nonmetric or discontinuous skeletal features.[77] Experimental studies in controlled contexts have verified the genetic basis of nonmetric traits, identifying hyperostotic features such as ossicles and bony bridges as having heritabilities as high as measured dimensions.[78]

63. Smith and Kahila 1992.
64. Smith et al. 1999.
65. Angel 1945; Little 1999; Rotroff, Little, and Snyder 1999.
66. Themelis 2000.
67. For example, Morris 1987; Lagia 2007a.
68. Brown et al. 2000.
69. Evison et al. 2000.
70. Evison 2001.
71. Chilvers et al. 2008.
72. Cooper and Poinar 2000.
73. Pääbo et al. 2004.
74. Gilbert et al. 2005.
75. Bouwman et al. 2008.
76. For example, Manolis 1990; Stravopodi, Manolis, and Neroutsos 1997.
77. Berry and Berry 1967; Buikstra and Ubelaker 1994; Hauser and DeStefano 1989.
78. Cheverud and Buikstra 1982.

At present, there are several statistical techniques for analyzing inherited variations that have not been applied in the ancient Aegean world. These include the use of population genetic modeling, usually based in the estimation of within-group genetic variance (F_{st}). This standardized statistic is useful for comparison across populations and can also be used within cemeteries to estimate the degree of variance in each sex and thus infer residence patterning.[79] In the North American mid-continent, for example, we find clear evidence that male kin were buried together and presumably living together during the Mesolithic or Woodland periods. Residence patterns change, however, with the development of maize agriculture by Mississippian times and is apparently associated with the adoption of matrilocal residence rules.[80]

In contrast to model-bound techniques, such as those based on the estimation of F_{st}, model-free approaches have the advantage of being equally applicable to distributions of skeletal traits and artifact attributes. These can be used to investigate the degree of isomorphism between cultural and genetic boundaries, important in studies of ethnic variation.[81] Although many migration models speak of artifacts as if they had lives of their own, moving across the landscape at will, the most convincing way to document population movement or residence patterning requires the study of human remains. Acknowledging this void in Aegean cemetery studies, Nafplioti[82] applied metric and nonmetric cranial and dental analysis in three Bronze Age populations from the Argolid, Naxos, and Central Crete to investigate archaeological hypotheses that implicate population movements to interpret discontinuities in the material record. Intra-population variation and inter-population biodistance analysis by Nafplioti does not appear to support such theories.

On the other hand, analysis of Y chromosome haplotypes from living populations in Eastern Crete and the mainland offers support for the hypothesis of population movement from the mainland to Crete during the Late Bronze Age.[83] Analyses of mtDNA and Y chromosome markers, aimed at making inferences about migrations in the Neolithic and the Bronze Age, point to a substantial Middle Eastern, Central Asian, and European genetic input in eastern Crete during the Neolithic, while dismissing any significant North African affinity.[84] An interesting suggestion is that the genetic uniqueness of the Lasithi plateau appears to support former interpretations of the site as a mountain refugium.[85]

PALEODEMOGRAPHY

Paleodemography, a technique critically evaluated by Angel in his studies of ancient Aegean peoples, holds potential for the estimation of vital rates such as mortality and fertility. Under archaeologically controlled circumstances, it can also be used to develop models of population distributions and density.

Recent critiques of paleodemography, initially developed in Europe by Bocquet-Appel and Masset,[86] have focused upon two key issues: (1) accuracy and bias in age indicators, and (2) bias due to the age structure

79. Konigsberg and Buikstra 1995.
80. Konigsberg 1988.
81. Konigsberg and Buikstra 1995.
82. Nafplioti 2007.
83. King et al. 2008.
84. Martinez et al. 2008; King et al. 2008.
85. Martinez et al. 2007, 2008.
86. Bocquet-Appel and Masset 1982, 1985.

of reference populations. This critique has stimulated considerable debate and methodological advancement. Attention has focused upon development of age indicators accurate in older adult years, as well as the question of bias. In addition, statistical techniques for cross-population evaluations have proliferated.

Contemporary studies have verified the accuracy and lack of apparent bias in age estimates based upon the sternal extremity of the fourth rib and cemental annuli of the dentition.[87] Other means of estimating age at death among older individuals are being sought.[88] Given the population-specific nature of skeletal maturation and sexual dimorphism, the paleodemographic significance of the modern reference collection of the University of Athens that was founded at the Wiener Laboratory is immense.[89]

Statistical responses to recent paleodemographic concerns, including reference sample mimicry and violations of stable and stationarity assumptions, have taken two forms. Both eschew the development of life tables as ends in themselves. The least complex response focuses upon comparative models contrasting high-fertility/high-mortality traditional populations such as the Yanamamo with low-fertility/low-mortality peoples like the !Kung. The study sample is typically evaluated against these two ethnographic samples and then characterized in terms of high or low fertility and mortality.[90]

More statistically complex approaches also develop comparative perspectives, frequently based upon hazard models, in which a smooth mathematical function, the hazard of death, is fit to mortality data contained within a life table. Hazard models provide smooth summaries of age-at-death data and can be used in cross-population analyses.[91] Many parameters can be made to fit life table data, and not all are biologically amenable to interpretation.[92]

An alternative statistical approach employs maximum likelihood estimates (MLE).[93] MLEs are related to hazard models, but are more useful because they can estimate probabilities for specific parameters. In other words, the MLE approach can determine what the probability is that the crude birth rate was, for example, 0.04 and the crude death rate 0.02, given the observed set of values.[94]

Recent paleodemographic critiques have developed a series of studies evaluating our ability to estimate vital rates such as mortality and fertility from mean-age-at-death information, which is readily available from cemetery sites. Somewhat counterintuitive is the fact that fertility is more amenable to estimation than mortality.[95] Large numbers of infant deaths, however, are frequently associated with rapidly increasing populations rather than an overall change in adult mortality patterns. In North America, we have statistically developed paleodemographic models that link elevated fertility levels with early evidence of horticulture, increased dependence upon high carbohydrate seeds, and attendant changes in ceramic technology. We believe that the availability of a grain-based weaning food encouraged short birth intervals, larger families, and population increase. This inference is supported by the observation of caries or cavities in deciduous dentitions and the nature of microscopically visible dental wear, also on primary teeth.[96]

87. Charles et al. 1989; Loth and İşcan 1989.
88. Boldsen et al. 2002.
89. Pike 1997; Eliopoulos, Lagia, and Manolis 2007.
90. Milner, Humpf, and Harpending 1989.
91. Gage 1988, 1991; Wood et al. 1992a.
92. Konigsberg and Frankenberg 1994.
93. Paine 1989.
94. Konigsberg and Frankenberg 1994.
95. Sattenspiel and Harpending 1983.
96. Buikstra, Konigsberg, and Bullington 1986; Cook and Buikstra 1979.

As a footnote, we should add that recently published results, based on computer simulations, indicate that the effects of increased fertility and in-migration are statistically indistinguishable.[97] Other classes of evidence should, however, facilitate interpretative rigor.

A common theme in archaeological reports is to interpret "large numbers" of infant graves (or of individuals in a grave) as the result of an epidemic, famine, infanticide, murder, or sacrifice. The opposite, i.e., the scarcity of infants from cemeteries, is taken to reflect absence of interest/emotional involvement or of exclusion of infants and young children because they are not accepted as full members of the society. For this reason, an essential first step in a bioarchaeological analysis is to assess the composition of a cemetery sample by comparing it to model life tables.[98] Such (basic) comparisons launched by Angel[99] are now commonly applied in bioarchaeological studies.[100]

Furthermore, considerations of preservation and recovery bias are addressed and regional patterns in mortuary behavior are sought to differentiate between preservation versus burial customs. For example, it was shown that the inclusion of infants and young children in Athens during the Classical, Hellenistic, and Early Roman periods varied temporally, reflecting changes in the mortuary customs. These changes characterized several regions in Greece that shared similar burial rites.[101]

When assessing biases introduced in the course of excavations, it should be noted that improved recovery techniques and water sieving of cremated remains from the Early Iron Age Kavousi in Crete considerably increased the number of infants at the site and questioned previous assumptions on children not being cremated and of being excluded from formal burial during the Dark Age.[102] Taphonomic issues are also important in interpreting demographic patterning,[103] especially in the analysis of multiple burials where the burial program may include more than one phase.[104] The application of forensic techniques to understanding the procedures applied to secondary cremation burials at Early Iron Age Kavousi in Crete[105] and at the Early to Middle Roman site of Kenchreai near Corinth[106] has provided interesting information on the role that cremation played in mortuary practices. Interestingly, both studies claim that only a select percentage of the fully cremated remains was collected from the pyres and was used for burial in the graves.

HEALTH AND DISEASE

The study of ancient health involves several distinctive but related research trajectories. One of these is the investigation of general population stress as a measure of the relative quality of life. Most commonly, such studies have been initiated to assess the impact of agricultural intensification upon human groups.[107] The same techniques can, however, be adapted to many contexts, for example, estimating the biological cost of state formation by comparing skeletal features across social strata.

Among the indicators of relative health status are those that reflect acute stress and recovery, such as radiopaque lines of growth arrest, or

97. Paine 1997.
98. Buikstra and Mielke 1985; Milner, Wood, and Boldsen 2000.
99. Angel 1969.
100. Liston 1993; Wittwer-Backofen et al. 2000; Triantaphyllou 2001; Papathanasiou 2001, 2005; Ingvarsson-Sundström 2003; Lagia, forthcoming.
101. Lagia 2007a.
102. Liston 1993.
103. Duday 1978; Duday et al. 1990; Haglund and Sorg 2002.
104. As, for example, Lagia 2002; Bessios and Triantaphyllou 2002; Vlahogianni, Lagia, and Sabetai, forthcoming.
105. Liston 2007.
106. Ubelaker and Rife 2007.
107. Cohen and Armelagos 1984.

Harris lines, and linear enamel defects, commonly termed *linear enamel hypoplasias.*[108] While age at insult can be estimated through the study of both Harris lines and hypoplasias, the fact that Harris lines can be reabsorbed makes dental defects the preferred acute stress markers.

Chronic stress is most readily evaluated through comparisons of growth attainment to dental age.[109] Given that dental development is more robust than skeletal maturation in the face of dietary or disease stress, cross-population studies of stature attainment can usefully address relative quality of life. In parallel fashion, cortical thickness per dental age can also be used as an indicator of chronic stress.[110]

Recently, such paleoepidemiological approaches have been criticized as "paradoxical."[111] Three issues are raised, including the fact that many paleodemographic and paleoepidemiological models violate the assumption of stationarity, central in demographic studies, as Angel pointed out three decades ago.[112] Although recent investigations have demonstrated that nonstationarity is not a fatal flaw, the two remaining criticisms must be seriously considered. One of these—the explicit recognition that the death sample is not equivalent to a birth cohort—was also emphasized by Angel.[113] Using multiple lines of evidence and not equating living and archaeological samples can counter this concern. The remaining critique emphasizes that many factors, ranging from genetic predisposition to environmental effects, may render an individual uniquely vulnerable (fragile) to specific health-related stressors and therefore make summary statistics and comparative studies meaningless. This latter criticism has yet to be fully evaluated in contemporary epidemiology, and its impact upon studies of ancient remains is as yet unknown.

The application of stress indicators to population studies is the most widely applied strategy among bioarchaeologists working in diverse contexts in Greece.[114] Rigor in the integration of multiple lines of evidence remains a challenge here, as it does elsewhere.

The history of specific diseases can also be addressed in Aegean contexts. The fact that certain forms of hemolytic anemias advantage individuals in malarial environments is well known. Therefore, skeletal evidence for anemia has been cited in the Aegean to document the history of malaria and related landscape transformations conducive to this condition.[115] Such models must, however, engage in rigorous differential diagnoses to distinguish the impact of inherited anemias from nutritional conditions.[116]

That no genetic anemia has been identified in the large prehistoric and historic Greek cemetery samples remains a paleopathological enigma.[117] Furthermore, it is hard to explain in clinical terms why the extreme forms of porotic hyperostosis are only known from the New World where genetic anemias did not exist prior to contact. Could it be that the psychosocial aspects of these diseases, related to the physical appearance and debilitation of the patients, were associated with social exclusion and isolation, as in contemporary societies?[118] Perhaps those socially shunned were also excluded from community cemeteries.

Angel[119] was particularly interested in identifying thalassemia in the eastern Mediterranean. He systematically examined postcranial elements in search of pathognomonic traits[120] and considered it essential to "find a

108. Goodman and Rose 1991; McHenry 1968; Rose, Condon, and Goodman 1985.
109. Saunders 2000.
110. Goodman et al. 1984.
111. Wood et al. 1992b.
112. Angel 1969.
113. Angel 1969.
114. Liston 1993; Papathanasiou, Larsen, and Norr 2000; Fox Leonard 1997; Fox 2005; Manolis and Neroutsos 1997; Agelarakis 2000, 2002; Papathanasiou 2001, 2003, 2005; Triantaphyllou 1998, 1999, 2001, 2004; Stravopodi and Manolis 2000; Ingvarsson-Sundström 2003; Bourbou 2004; Lagia forthcoming; Tritsaroli 2006; Lagia, Petroutsa, and Manolis 2007.
115. Angel 1964, 1966a.
116. Hill and Armelagos 1990.
117. Lagia 1993; Lagia, Eliopoulos, and Manolis 2007.
118. Ratip and Modell 1996.
119. Angel 1964, 1966a, 1967, 1971a, 1978.
120. Angel 1964, 1967, 1978.

wide range of bone response to thalassemia since this term includes several different molecular changes in the α- and β-chains of the globin part of the hemoglobin molecule."[121] Today, with molecular biology and advanced knowledge of the genotypic and phenotypic variability in the various forms of the thalassemias, it may be possible to identify this group of diseases in the past. The recent analysis of the skeletal changes that are unique to this group of disorders based on a unique, documented case from the modern reference collection of the University of Athens will further advance our ability to identify this significant group of diseases.[122]

Careful chronologically controlled studies can also be used to investigate the history of other major diseases that continue to plague contemporary society. Among these are tuberculosis and the treponematoses, including venereal syphilis, both known in the ancient Old and New Worlds.

Tuberculosis, for example, has been documented in skeletons or mummies from Early Dynastic Egypt, Early Bronze Age Jordan, and Neolithic Italy.[123] There should, therefore, also be a skeletal history in ancient Aegean remains. Even within contexts with marginal preservation, the typically dense and ankylosed joints reflective of mycobacterial infection should be visible. As tuberculosis generally provides an indirect measure of population health, its presence and prevalence in ancient samples would enhance our appreciation both of Aegean health and the history of tuberculosis. Indirect evidence for the second major group of mycobacterial diseases, namely of leprosy, has been identified in a Paleochristian ossuary from the island of Kos.[124] The pencil-like appearance observed in a number of metatarsals, in addition to the ample presence of periosteal reactions among the commingled long bones, point strongly to the presence of leprosy in the population. Identification of the pathognomonic facial lesions is nevertheless pending.

Similarly, the skeletal symptoms of treponematosis involve sclerotic and dense osteological changes. While such modifications are typically linked to both venereal syphilis and nonvenereal forms such as bejel and endemic syphilis, age-related patterning can facilitate differential diagnoses. Pre–15th-century evidence of treponemal infection in the eastern Mediterranean would greatly assist global inferences concerning the history of that constellation of diseases.[125] For the moment the only clear case of syphilis in Greece has been diagnosed in a burial dating to the 18th century from Pylos in the Peloponnese.[126] The excellent condition of preservation of the skeleton allowed the observation of all pathognomonic features of *caries sicca* on the cranium as well as extensive osteosclerotic lesions bilaterally in the long bones.

While most disease diagnoses in archaeological samples are grounded in distributions of skeletal lesions within individuals and across populations, the identification of pathogens through aDNA analyses is also assuming prominence. For example, the PCR (polymerase chain reaction) amplification and subsequent sequencing of a 123-base-pair fragment of aDNA from a calcified lymph node recovered from a Peruvian mummy has established the presence of *Mycobacterium tuberculosis* in the Americas prior to Columbian contact.[127] Mycobacterial infection has also been documented in ancient Egyptian materials through the identification of acid-fast bacilli

121. Angel 1967, p. 387.
122. Lagia, Eliopoulos, and Manolis 2007.
123. Buikstra, Baker, and Cook 1993.
124. Lagia, Petroutsa, and Manolis 2002.
125. Baker and Armelagos 1988.
126. Lagia et al. 2003.
127. Salo et al. 1994.

and aDNA.[128] While care *must* be taken to guard against contamination, and we must realize that negative results does not mean the absence of infectious disease, aDNA studies hold potential for clarifying a range of issues, including the identification of various European plagues.

BEHAVIOR

The wear and tear of daily life also leaves a record within skeletal and dental structures. Angel, for example, championed elbow arthritis patterns as evidence for the use of the spear and spear thrower.[129] While Angel's characterization of "atlatl elbow" has been found wanting,[130] the use of arthritis to identify changes in repetitive activities remains sound. In North America, for example, degenerative changes in female remains clearly document the impact of maize agriculture and the effort required to produce and process grain.[131]

Developed areas of muscle attachment and the shape of long bones also reflect activity levels. A well-developed supinator crest at the elbow, for example, may identify an individual who frequently throws—but cannot distinguish a North American major league pitcher from a harpoon-wielding Inuit.[132] In parallel, biomechanical analyses document that bone cross sectional shape reflects types of mechanical loads, while cross sectional size is associated with levels of mechanical loads. Thus, agriculturalists tend to present a common femur shape whose cross sectional size may vary with the terrain across which the farmer moves.[133] For a review of these behavioral markers, see Jurmain.[134]

Triantaphyllou[135] observed a similar incidence of skeleto-muscular conditions between males and females in diverse samples from northern Greece and concluded that the two sexes participated equally in labor. She also used such evidence among adolescents and young adults to infer their moderate participation in occupational activities early on. Papathanasiou[136] used evidence of osteoarthritis and musculoskeletal stress markers to suggest increased physical activity and heavy workloads among the Neolithic people of Alepotrypa. Finally, Triantaphyllou[137] observed evidence of specialized occupational activities through the study of degenerative changes and dental trauma. Rigorous methods of quantification and evaluation of the significance of the observed skeletal patterns, however, are indispensable.

DIET

Although dental health has a long history as a basis for inferences concerning diet and nutritional balance,[138] more recent biochemical techniques have revolutionized our ability to investigate ancient cuisines. While the initial popularity of trace element analyses has waned as a result of documented propensities for postdepositional change or diagenesis,[139] it appears that elemental values for strontium and barium do vary systematically across ecosystems and within regional food chains.[140] They are also robust in the face of diagenetic change. Thus, although we have the technology for

128. Crubézy et al. 1998; Nerlich et al. 1997; Zink, Grabner, and Nerlich 2005.
129. Angel 1966b.
130. Bridges 1990.
131. Pickering 1984.
132. Kennedy 1989.
133. Ruff 1992.
134. Jurmain 1999.
135. Triantaphyllou 2001, 2004.
136. Papathanasiou 2001, 2005.
137. Triantaphyllou 2001.
138. Powell 1985; Hillson 1996, 2000.
139. Buikstra et al. 1989; Ezzo 1992; Lambert et al. 1982; Price et al. 1992.
140. Burton and Price 1990, 1991.

measuring a broad array of trace elements within archaeological materials, at the present time only strontium and barium appear to hold promise for environmental and dietary studies.

By contrast, cautious optimism continues to be expressed over certain light and heavy isotope methods. Ratios of stable carbon isotopes 13 to 12 in human bone collagen and especially bone apatite, transformed in relationship to a standard, document the presence of C^4 plants such as corn, millet, and sorghum in the diet. Standard carbon isotope measures, commonly expressed as $\delta^{13}C$, also reflect marine resource consumption. By using a bivariate plot of $\delta^{13}C$ and $\delta^{15}N$, researchers can isolate the effect of marine-based human diets from those reflecting C^4 plants.[141] In the Aegean context, it appears that distinguishing marine and terrestrial signals is of paramount importance.

Angel's inference[142] that "the Mycenaean aristocrats [from Grave Circle B] had a richer diet than the common people"—based on their "better growth, bigger size, and greater strength"—seems to have been confirmed by stable isotope analysis. Comparison of individuals from grave circles A and B with those from the chamber tombs of "middle class" people at Mycenae showed that the diet of at least some individuals from the latter group included more plants than animal protein. Also, the latter had a greater range of isotopic values.[143]

Many studies have now applied stable isotope data within the Aegean region, with analyses spanning periods from Neolithic to Early Modern.[144] Encouragingly, while a variety of laboratories in the UK, the USA, and Canada have been involved in these analyses, their results complement each other.[145] By far, the most complete and excellent treatise of ancient diets is that of Garvie-Lok,[146] who modeled diet by analyzing human, faunal, and food signatures from diverse sites of the medieval and early modern periods in Greece. The picture that so far emerges from the above-mentioned studies is a diet that relies heavily on C^3 terrestrial resources, mostly plants, and less on animal protein, i.e., meat or dairy products. A significant implication of Garvie-Lok's thesis is that in the sites she studied, dairy products were isotopically similar to meat, rendering questions concerning dietary differences by gender, economic, or ethnic group. Overall, a temporal trend in diets toward increasing contribution of animal protein is observed from prehistoric to historic times, albeit with regional variability. The highest values observed per region are those of the classical and medieval periods.[147]

In all time periods studied, the absence, or auxiliary significance, of marine diets is a conspicuous trend, although regional variability has been attested to reflect cultural differences as well as proximity and accessibility to marine resources. The greater inclusion of marine resources in the diet of the elite individuals buried in Grave Circles A and B at Mycenae has been cited as a marker of distinction from the occupants of the chamber tombs.[148] Marine food consumption was noted at the Middle Byzantine site of Kastella in Crete.[149] Garvie-Lok[150] analyzed local marine foods and noted a difficulty in differentiating purely terrestrial diets from diets that include small amounts of marine resources of low trophic level. C^4 plants,

141. Ambrose 1993; Sealy, Armstrong, and Schrire 1995; Schoeninger 1995.

142. Angel 1973, pp. 387–388.

143. Hedges and Richards 1999, pp. 214, 223, 227, 231; Richards and Hedges 2008.

144. Hedges and Richards 1999; Papathanasiou et al. 2000; Triantaphyllou 2001; Papathanasiou 2001; Garvie-Lok 2001; Lagia, Petroutsa, and Manolis 2007; Petroutsa 2007; Bourbou and Richards 2007; Richards and Hedges 2008; Richards and Vika 2008; Triantaphyllou et al. 2008; Lagia and Richards, in prep.; Papathanasiou and Richards, in prep.

145. Papathanasiou, in prep.

146. Garvie-Lok 2001.

147. Triantaphyllou 2001; Garvie-Lok 2001; Lagia and Richards, in prep.

148. Richards and Hedges 2008.

149. Bourbou and Richards 2007.

150. Garvie-Lok 2001.

possibly sorghum or millet, were found to be present in dietary signatures that date to the Neolithic from southern Greece,[151] whereas their reliance appears to increase during the Late Bronze and Early Iron Age in sites from northern Greece.[152] It is not clear, however, whether these rather minor signals originate from human or domesticated animal consumption. During the medieval period, it was clearly shown that C^4 grains—that now may also include maize—served directly as human food and supplemented a primary staple of C^3 grain.[153]

Also reflective of diet, but more useful in identifying the effects of population movement and even residence patterning is the assessment of isotope ratios for stable strontium $^{87}Sr/^{86}Sr$ and oxygen $^{18}O/^{16}O$. In that teeth crystallize the groundwater ratios of youth, bone values reflect more recent individual histories. Oxygen ratios tend to vary with latitude, while strontium isotope ratios differ between forms of bedrock and groundwater. Comparisons of strontium and oxygen isotope ratios in teeth and bones therefore can identify individuals who have changed residence across latitudinal or geologic zones.[154] This is another powerful way in which to document migrations and other changes in postmarital residence. Indeed, strontium isotope ratio analysis ($^{87}Sr/^{86}Sr$) of human bone and dental enamel was applied most recently by Nafplioti[155] to supplement her morphological analysis addressing the question of population movement and cultural discontinuity during the Late Bronze Age in the Aegean. Through negative evidence, Nafplioti supports the view that the introduction of novel cultural features in Crete need not have resulted from the actual movement of people from the mainland. The picture is still not clear, however, since Y chromosome analysis suggests that some of the "later Bronze Age changes in Crete were indeed underwritten by an incursion of mainland people."[156] It will be interesting to observe how these discrepancies are resolved as both strontium isotope ratio and Y chromosome analysis develop in scope and methodological vigor.

Population movement and ethnicity has also been addressed through C and N stable isotope variation by Garvie-Lok,[157] who identified greater diversity and more marine and C^4 plants inclusion in the diets of Muslim and Frankish groups versus local populations in the island of Mytilini and in Corinth, respectively. Similarly, Richards et al.[158] found great variability in the adult human C and N values from recent excavations at Neolithic Çatal Höyük and offered as possible explanations the great diversity of groups of people inhabiting the area, the great mobility of certain peoples, or differing herding practices at the site that resulted in a large spread in faunal isotope values. The same study questions former ideas about the site being an area of primarily cattle herding and consumption and found instead animal protein coming mainly from goat and sheep. They also recognized input of C^4 plants into human diets, likely through the consumption of animals that consumed C^4 plants. Finally, they found weaning to occur before the age of 1.5 years. Likewise, the two recent attempts to identify weaning ages at Middle Byzantine Kastella[159] and Middle Bronze Age Lerna[160] conclude that weaning in these sites took place at, or shortly before, the age of 2 to 2.5 years.

151. Papathanasiou 2001.
152. Triantaphyllou 2001.
153. Garvie-Lok 2001.
154. Price, Grupe, and Schrötter 1994; Price et al. 1994; Sealy et al. 1991; Schoeninger 1995.
155. Nafplioti 2007, 2008.
156. King et al. 2008, p. 212.
157. Garvie-Lok 2001.
158. Richards et al. 2003.
159. Bourbou and Richards 2007.
160. Triantaphyllou et al. 2008.

SUMMARY AND CONCLUSIONS

Thus, we have the tools for addressing many of the questions posed at the beginning of this brief review. Inherited nonmetric traits and perhaps aDNA analyses should be useful in establishing patterns of kinship within tholos tombs. Nonmetric features of 11th-century B.C. cremations can be compared with less fragmented materials to establish evidence for population replacement, while aDNA may be used to investigate unusual sex ratios noted within many Aegean skeletal collections. Model-free approaches to biological distance will permit direct comparisons of material culture styles and genetic boundaries over the long term, thus evaluating proposed migration patterns and the genetic reality of ethnic divisions. Paleoepidemiologic studies and the fortuitous preservation of ancient DNA may also assist in disease diagnosis, relevant both for the history of Old World diseases and in evaluating the health of ancient Aegeans. Appropriate biochemical and paleoepidemiological methods are already being used for assessing long-term changes in health patterns and diet in diverse contexts in the Aegean. Such methods can be further used, for example, in assessing the biological costs and benefits of state formation during Minoan times. Gender differences in activity levels are becoming apparent through the application of MSSM and degenerative changes in various contexts. We can anticipate that craft specialization, whether in 3rd-millennium B.C. Crete or elsewhere, should be reflected in bone shape and articular degeneration.

The existing bioarchaeological methods for addressing significant historical and archaeological questions through the study of ancient Aegean cemetery sites and their content offer obvious, desirable potential. They can be effective, however, only when the skeletal biologist is accepted as a full and active participant in research initiatives, beginning with the design of archaeological sampling strategies and carried through to interdisciplinary interpretations. Too many site reports are replete with descriptive appendices without integration of skeletal biological data within the body of the text. We should, therefore, follow J. Lawrence Angel's example by embracing problem-oriented approaches, but we should not be limited to the analytical techniques of his day. The most effective way to escape the descriptive appendices of archaeological site reports is to know the remarkable range of methods now available in bioarchaeological study, applying these within a problem-oriented framework. As this is happening, our appreciation of the distinguished Aegean past is immeasurably enriched.

REFERENCES

Agelarakis, A. 2000. "Aspects of Demography and Paleopathology among the Hellenistic Abderetes in Thrace, Greece," *Eulimene* 1, pp. 13–24.

———. 2002. "Investigations of Physical Anthropology and Paleopathology of the Ancient Necropolis of Thassos," in *Excavating Classical Culture: Recent Archaeological Discoveries in Greece* (*BAR-IS* 1031), ed. M. Stamatopoulou and M. Yeroulanou, Oxford, pp. 12–19.

Ambrose, S. H. 1993. "Isotopic Analysis: Methodological and Interpretive Considerations," in *Investigations of Ancient Human Tissue: Chemical Analyses in Anthropology,* ed. M. K. Sandford, Philadelphia, pp. 59–130.

Angel, J. L. 1945. "Skeletal Material from Attica," *Hesperia* 14, pp. 279–363.

———. 1946. "Social Biology of Greek Culture Growth," *American Anthropologist* 48, pp. 493–533.

———. 1959. "Early Helladic Skulls from Aghios Kosmas," in *Aghios Kosmas: An Early Bronze Age Settlement and Cemetery in Attica,* ed. G. E. Mylonas, Princeton, pp. 169–179.

———. 1964. "Osteoporosis: Thalassemia?" *American Journal of Physical Anthropology* 22, pp. 369–374.

———. 1966a. "Porotic Hyperostosis, Anemias, Malarias, and Marshes in the Prehistoric Eastern Mediterranean," *Science* 153, pp. 760–763.

———. 1966b. "Early Skeletons from Tranquility, California," *Smithsonian Contributions to Anthropology* 2, pp. 1–19.

———. 1967. "Porotic Hyperostosis or Osteoporosis Symmetrica," in *Diseases in Antiquity,* ed. D. R. Brothwell and A. T. Sandison, Springfield, Ill., pp. 378–389.

———. 1969. "The Bases of Paleodemography," *American Journal of Physical Anthropology* 30, pp. 427–437.

———. 1970. "Human Skeletal Remains at Karataş," *AJA* 74, pp. 253–259.

———. 1971a. *The People of Lerna: Analysis of a Prehistoric Aegean Population,* Princeton.

———. 1971b. "Early Neolithic Skeletons from Çatal Hüyük: Demography and Pathology," *AnatSt* 21, pp. 77–98.

———. 1973. "Human Skeletons from Grave Circles at Mycenae," in *Ο Ταφικός Κύκλος Β των Μυκηνών,* ed. G. E. Mylonas, Athens, pp. 379–397.

———. 1978. "Porotic Hyperostosis in the Eastern Mediterranean," *Medical College of Virginia Quarterly* 14, pp. 10–16.

Angel, J. L., and S. C. Bisel. 1986. "Health and Stress in an Early Bronze Age Population," in *Ancient Anatolia: Aspects of Change and Cultural Development. Essays in Honor of Machteld J. Mellink,* ed. J. V. Canby, Madison, pp. 12–30.

Baker, B. J., and G. J. Armelagos. 1988. "The Origin and Antiquity of Syphilis: Paleopathological Diagnosis and Interpretation," *CurrAnthr* 29, pp. 703–737.

Berry, A. C., and R. J. Berry. 1967. "Epigenetic Variation in the Human Cranium," *Journal of Anatomy* 101, pp. 361–379.

Bessios, M., and S. Triantaphyllou. 2002. "Ομαδικός τάφος από το Βόρειο Νεκροταφείο της Αρχαίας Πύδνας," *ΑΕΜΘ* 14 (2000), pp. 386–392.

Binford, L. R. 1971. "Mortuary Practices: Their Study and Their Potential," in *Approaches to the Social Dimensions of Mortuary Practices* (Memoirs of the Society for American Archaeology 25), ed. J. A. Brown, Washington D.C., pp. 6–29.

Bocquet-Appel, J. P., and C. Masset. 1982. "Farewell to Paleodemography," *Journal of Human Evolution* 11, pp. 321–333.

———. 1985. "Paleodemography: Resurrection or Ghost?" *Journal of Human Evolution* 14, pp. 107–111.

Boldsen, J. L., G. R. Milner, L. W. Konigsberg, and J. W. Wood. 2002. "Transition Analysis: A New Method for Estimating Age-Indicator Methods," in *Paleodemography: Age Distribution from Skeletal Samples,* ed. R. D. Hoppa and J. W. Vaupel, Cambridge, pp. 73–106.

Bourbou C. 2003. "Health Patterns of Proto-Byzantine Populations (6th–7th Centuries A.D.) in South Greece: The Cases of Eleutherna (Crete) and Messene (Peloponnese)," *International Journal of Osteoarchaeology* 13, pp. 303–313.

———. 2004. *The People of Early Byzantine Eleutherna and Messene (6th–7th Century A.D.): A Bioarchaeological Approach,* Athens.

Bourbou, C., and M. Richards. 2007. "The Middle Byzantine Menu: Paleodietary Information from Isotopic Analysis of Humans and Fauna from Kastella, Crete," *International Journal of Osteoarchaeology* 17, pp. 63–72.

Bouwman, A. S., K. A. Brown, A. J. N. W. Prag, and T. A. Brown. 2008. "Kinship between Burials from Grave Circle B at Mycenae Revealed by Ancient DNA Typing," *JAS* 35, pp. 2580–2584.

Branigan, K. 1988. *The Foundations of Palatial Crete: A Survey of Crete in the Early Bronze Age,* Amsterdam.

———. 1993. *Dancing with Death: Life and Death in Southern Crete c. 3000–2000 B.C.,* Amsterdam.

———. 2000. *Urbanism in the Aegean Bronze Age,* Sheffield.

Bridges, P. 1990. "Osteological Correlates of Weapon Use," in *A Life in Science: Papers in Honor of J. Lawrence Angel* (Center for American Archeology Scientific Papers 6), ed. J. E. Buikstra, Kampsville, Ill., pp. 87–98.

Brown, E. A., and K. A. Brown. 1992. "Ancient DNA and the Archaeologist," *Antiquity* 66, pp. 10–23.

Brown, J. A. 1971. "The Dimensions of Status in the Burials at Spiro," in *Approaches to the Social Dimensions of Mortuary Practices* (Memoirs of the Society for American Archaeology 25), ed. J. A. Brown, Washington, D.C., pp. 92–112.

———. 1995. "On Mortuary Analysis, with Special Reference to the Saxe-Binford Research Program," in *Regional Approaches to Mortuary Analysis,* ed. L. Beck, New York, pp. 3–26.

Brown, T. A., K. A. Brown, C. E. Flaherty, L. M. Little, and A. J. N. W. Prag. 2000. "DNA Analysis of Bones from Grave Circle B at Mycenae: A First Report," *BSA* 95, pp. 115–119.
Buikstra, J. E. 1977. "Biocultural Dimensions of Archeological Study: A Regional Perspective," in *Biocultural Adaptation in Prehistoric America* (Proceedings of the Southern Anthropological Society 11), ed. R. L. Blakely, Athens, Ga., pp. 67–84.
Buikstra, J. E., B. J. Baker, and D. C. Cook. 1993. "What Diseases Plagued Ancient Egyptians? A Century of Controversy Considered," in *Biological Anthropology and the Study of Ancient Egypt,* ed. W. V. Davies and R. Walker, London, pp. 24–53.
Buikstra, J. E., S. Frankenberg, J. Lambert, and L. Xue. 1989. "Multiple Elements, Multiple Expectations," in *Bone Biochemistry and Past Behavior* (School of American Research Seminar Series), ed. D. Price, Cambridge, pp. 155–210.
Buikstra, J. E., and L. M. Hoshower. 1990. "Introduction," in *A Life in Science: Papers in Honor of J. Lawrence Angel* (Center for American Archeology Scientific Papers 6), ed. J. E. Buikstra, Kampsville, Ill., pp. 1–16.
Buikstra, J. E., L. W. Konigsberg, and J. Bullington. 1986. "Fertility and the Development of Agriculture in the Prehistoric Midwest," *AmerAnt* 51, pp. 528–546.
Buikstra, J. E., and J. H. Mielke. 1985. "Demography, Diet, and Health," in *The Analysis of Prehistoric Diets,* ed. R. I. Gilbert and J. H. Mielke, Orlando, pp. 359–422.
Buikstra, J. E., and D. H. Ubelaker, eds. 1994. *Standards for Data Collection from Human Skeletal Remains,* Fayetteville, Ark.
Burton, J. H., and T. D. Price. 1990. "Ratio of Barium to Strontium as a Paleodietary Indicator of Consumption of Marine Resources," *JAS* 17, pp. 547–557.
———. 1991. "Paleodietary Applications of Barium Values in Bone," in *Proceedings of the 27th International Symposium on Archaeometry, Heidelberg, 1990,* ed. E. Pernicka and G. A. Wagner, Basel, pp. 787–795.
Charles, D. K., K. Condon, J. M. Cheverud, and J. E. Buikstra. 1989. "Estimating Age at Death from Growth Layer Groups in Cementum," in *Age Markers in the Human Skeleton,* ed. M. Y. İşcan, Springfield, Ill., pp. 277–301.
Charles, R. P. 1963. *Étude anthropologique des Nécropoles d'Argos* (Études péloponnésiens 3), Paris.
Cheverud, J. M., and J. E. Buikstra. 1982. "Quantitative Genetics of Skeletal Nonmetric Traits in the Rhesus Macaques on Cayo Santiago, III: Relative Heritability of Skeletal Nonmetric and Metric Traits," *American Journal of Physical Anthropology* 59, pp. 151–155.
Chilvers, E. R., A. S. Bouwman, K. A. Brown, R. G. Arnott, A. J. N. W. Prag, and T. A. Brown. 2008. "Ancient DNA in Human Bones from Neolithic and Bronze Age Sites in Greece and Crete," *JAS* 35, pp. 2707–2714.
Cohen, D. 1989. "Seclusion, Separation, and the Status of Women in Classical Athens," *GaR* 36, pp. 3–15.
Cohen, M. N., and G. J. Armelagos, eds. 1984. *Paleopathology at the Origins of Agriculture,* Orlando.
Cook, D. C., and J. E. Buikstra. 1979. "Health and Differential Survival in Prehistoric Populations: Prenatal Dental Defects," *American Journal of Physical Anthropology* 51, pp. 649–664.
Cooper, A. and H. N. Poinar. 2000. "Ancient DNA: Do It Right or Not at All," *Science* 289, p. 1139.
Crubézy, É., B. Ludes, J.-D. Poveda, J. Clayton, D. Crouau-Roy, and D. Montagnon. 1998. "Identification of *Mycobacterium* DNA in an Egyptian Pott's Disease of 5,400 Years Old," *Comptes rendus de l'Académie des sciences: Série III, Sciences de la vie* 321, pp. 941–951.
Dalby, A. 1996. *Siren Feasts: A History of Food and Gastronomy in Greece,* London.
Davis, J. L. 2001. "Classical Archaeology and Anthropological Archaeology in North America: A Meeting of Minds at the Millennium?" in *Archaeology at the Millennium,* ed. G. M. Feinman and T. D. Price, New York, pp. 415–437.

Dettwyler, K.A. 1991. "Can Paleopathology Provide Evidence for 'Compassion'?" *American Journal of Physical Anthropology* 84, pp. 375–384.

Drews, R. 1993. *The End of the Bronze Age: Changes in Warfare and the Catastrophe ca. 1200 B.C.*, Princeton.

Driessen, J., I. Schoep, and R. Laffineur. 2002. *Monuments of Minos: Rethinking the Minoan Palaces* (*Aegaeum* 23), Liège.

Duday, H. 1978. "Archéologie funéraire et anthropologie," *Cahiers d'anthropologie* 1, pp. 55–101.

Duday, H., P. Courtaud, E. Crubezy, P. Sellier, and A.-M. Tillier. 1990. "L'anthropologie 'de terrain': Reconnaissance et interprétation des gestes funéraires," *Bulletin et memoires de la Societé d'Anthropologie de Paris* 2, pp. 25–90.

Dyson, S. L. 1998. *Ancient Marbles to American Shores: Classical Archaeology in the United States,* Philadelphia.

Eliopoulos, C., A. Lagia, and S. Manolis. 2007. "A Modern, Documented Skeletal Collection from Greece," *HOMO: Journal of Comparative Human Biology* 58, pp. 221–228.

Evison, M. P. 2001. "Ancient DNA in Greece: Problems and Prospects," *Journal of Radioanalytical and Nuclear Chemistry* 247, pp. 673–678.

Evison, M. P., N. Kyparissi-Apostolika, E. Stravopodi, N. R. J. Fieller, and D. M. Smillie. 2000. "An Ancient HLA Type from a Paleolithic Skeleton from Theopetra Cave, Greece," in *Theopetra Cave: Twelve Years of Excavation and Research, 1987–1998. Proceedings of the International Conference, Trikala, 6–7 November 1998,* ed. N. Kyparissi-Apostolika, Athens, pp. 109–117.

Ezzo, J. A. 1992. "A Test of Diet versus Diagenesis at Ventana Cave, Arizona," *JAS* 19, pp. 23–37.

Fox, S. C. 2005. "Health in Hellenistic and Roman Times: The Case Studies of Paphos, Cyprus and Corinth, Greece," in *Health in Antiquity,* ed. H. King, London, pp. 59–82.

Fox Leonard, S. C. 1997. "Comparative Health from Paleopathological Analysis of the Human Skeletal Remains Dating to the Hellenistic and Roman Periods, from Paphos, Cyprus and Corinth, Greece" (diss. Univ. of Arizona).

Gage, T. B. 1988. "Mathematical Hazard Models of Mortality: An Alternative to Model Life Tables," *American Journal of Physical Anthropology* 76, pp. 429–441.

———1991. "Causes of Death and the Components of Mortality: Testing the Biological Interpretations of a Competing Hazards Model," *American Journal of Human Biology* 3, pp. 289–300.

Garland, R. 1995. *The Eye of the Beholder: Deformity and Disability in the Graeco-Roman World,* London.

Garnsey, P. 1989. "Infant Health and Upbringing in Antiquity," in *Food, Health and Culture in Classical Antiquity,* ed. P Garnsey, Cambridge.

Garvie-Lok, S. 2001. "Loaves and Fishes: A Stable Isotope Reconstruction of Diet in Medieval Greece" (diss. Univ. of Calgary).

Gilbert, M. T. P., H.-J. Bandelt, M. Hofreiter, and I. Barnes. 2005. "Assessing Ancient DNA Studies," *Trends in Ecology and Evolution* 20, pp. 541–44.

Golden, M. 1988. "Did the Ancients Care When Their Children Died?," *GaR* 35, pp. 152–63.

———. 1990. *Children and Childhood in Classical Athens,* Baltimore.

———. 1997. "Change or Continuity? Children and Childhood in Hellenistic Historiography," in *Inventing Ancient Culture: Historicism, Periodization and the Ancient World,* ed. M. Golden and P. Toohey, New York, pp. 176–190.

Goldstein, L. G. 1981. "One-Dimensional Archaeology and Multi-Dimensional People: Spatial Organization and Mortuary Analysis," in *Archaeology of Death,* ed. R. Chapman, I. Kinnes, and K. Randsborg, Cambridge, pp. 53–69.

Goodman, A. H., D. Martin, G. J. Armelagos, and G. Clark. 1984. "Indications of Stress from Bone and Teeth," in *Paleopathology at the Origins of Agriculture,* ed. M. N. Cohen and G. J. Armelagos, Orlando, pp. 13–49.

Goodman, A. H., and J. C. Rose. 1991. "Dental Enamel Hypoplasias as Indicators of Nutritional Status," in *Advances in Dental Anthropology,* ed. M. A. Kelley and C. S. Larsen, Orlando, pp. 279–294.

Grammenos, D. B., and S. Triantaphyllou, eds. 2004. *Ανθρωπολογικές Μελέτες από τη Βόρεια Ελλάδα* (Δημοσιεύματα του Αρχαιολογικού Ινστιτούτου Μακεδονικών και Θρακικών Σπουδών 5), Thessaloniki.

Grauer, A. L., and P. Stuart-Macadam, eds. 1998. *Sex and Gender in Paleopathological Perspective,* Cambridge.

Grmek, M. 1989. *Diseases in the Ancient Greek World,* Baltimore.

Haglund, W., and M. H. Sorg, eds. 2002. *Advances in Forensic Taphonomy: Method, Theory, and Archaeological Perspectives,* New York.

Hauser, G., and G. F. DeStefano. 1989. *Epigenetic Variants of the Human Skull,* Stuttgart.

Hawkey, D. 1998. "Disability, Compassion and the Skeletal Record: Using Musculoskeletal Stress Markers (MSM) to Construct an Osteobiography from Early New Mexico," *International Journal of Osteoarchaeology* 8, pp. 326–340.

Haydon, R. C. 1993. "Survey of Genetic Variation among the Chiribaya of the Osmore Drainage Basin, Southern Peru" (diss. Univ. of Chicago).

Hedges, R. E. M., and M. P. Richards. 1999. "How Chemical Analysis of Human Bones Can Tell Us the Diets of People Who Lived in the Past," in *Minoans and Mycenaeans: Flavors of Their Time,* ed. Y. Tzedakis and H. Martlew, Athens, pp. 213, 223, 227, 231.

Herrmann, B., and S. Hummel, eds. 1994. *Ancient DNA: Recovery and Analysis of Genetic Material from Paleontological, Archaeological, Museum, Medical, and Forensic Specimens,* New York.

Hill, M. C., and G. J. Armelagos. 1990. "Porotic Hyperostosis in Past and Present Perspective," in *A Life in Science: Papers in Honor of J. Lawrence Angel* (Center for American Archeology Scientific Papers 6), ed. J. E. Buikstra, Kampsville, Ill., pp. 52–63.

Hillson, S. 1996. *Dental Anthropology,* Cambridge.

———. 2000. "Dental Pathology," in *Biological Anthropology of the Human Skeleton,* ed. M. A. Katzenberg and S. R. Saunders, New York, pp. 249–286.

———. 2002. "Investigating Ancient Cemeteries on the Island of Astypalaia, Greece," *Archaeology International* 5, pp. 29–31.

Hodder, I. 1980. "Social Structure and Cemeteries: A Critical Appraisal," in *Anglo-Saxon Cemeteries* (*BAR-BS* 82), ed. P. Rahtz, T. Dickinson, and L. Watts, Oxford, pp. 161–169.

———. 1982. *Symbols in Action: Ethnoarchaeological Studies of Material Culture,* Cambridge.

Holden, K. 1996. "Ebola: Ancient History of 'New' Disease?" *Science* 272, p. 1591.

Ingvarsson-Sundström, A. Y. R. 2003. "Children Lost and Found: A Bioarchaeological Study of Middle Helladic Children in Asine with a Comparison to Lerna (Greece)" (diss. Univ. of Uppsala).

Jacobsen, T. W., and T. Cullen. 1990. "The Work of J. L. Angel in the Eastern Mediterranean," in *A Life in Science: Papers in Honor of J. Lawrence Angel* (Center for American Archeology Scientific Papers 6), ed. J. E. Buikstra, Kampsville, Ill., pp. 38–51.

Jameson M. H. 1977–1978. "Agriculture and Slavery in Classical Athens," *Classical Journal* 73, pp. 122–146.

Johnstone, S. 1998. "Cracking the Code of Silence: Athenian Legal Oratory and the Histories of Slaves and Women," in *Women and Slaves in Greco-Roman Culture: Differential Equations,* ed. S. R. Joshel and S. Murnaghan, New York, pp. 221–235.

Jurmain, R. 1999. *Stories from the Skeleton: Behavioral Reconstruction in Human Osteology,* Amsterdam.

Karali L., and A. Tsaliki. 2001. "Paleoanthropological Remarks on Skeletons from Neolithic Cyprus," in *Proceedings from the XIII European Meeting of the Paleopathology Association, Chieti, Italy, September 18–23, 2000,* ed. M. La Verghetta and L. Capasso, Teramo, pp. 135–138.

Kavadias, G., and A. Lagia. Forthcoming. "New Light on Old Finds: Two Classical Period Graves from Plateia Koumoundourou," in *Athenian Potters and Painters* II, ed. J. Oakley and O. Palagia, Oxford.

Kennedy, K. A. R. 1989. "Skeletal Markers of Occupational Stress," in *Reconstruction of Life from the Skeleton,* ed. M. Y. İşcan and K. A. R. Kennedy, New York, pp. 129–160.

King, R. J., S. Özcan, T. Carter, E. Kalfoğlu, S. Atasoy, C. Triantaphyllidis, A. Kouvatsi, A. A. Lin, C-E. T. Chow, L. A. Zhivotovsky, M. Michalodimitrakis, and P. A. Underhill. 2008. "Differential Y-Chromosome Anatolian Influences on the Greek and Cretan Neolithic," *Annals of Human Genetics* 72, pp. 205–214.

Konigsberg, L. W. 1988. "Migration Models of Prehistoric Residence," *American Journal of Physical Anthropology* 77, pp. 471–482.

Konigsberg, L. W., and J. E. Buikstra. 1995. "Regional Approaches to the Investigation of Past Human Biocultural Structure," in *Regional Approaches to Mortuary Analysis,* ed. L. Beck, New York, pp. 191–219.

Konigsberg, L. W., and S. R. Frankenberg. 1994. "Paleodemography: 'Not Quite Dead'," *Evolutionary Anthropology* 3, pp. 92–105.

Laffineur, R. 1989. *Transition: Le monde égéen du Bronze Moyen au Bronze Récent* (*Aegaeum* 3), Liège.

———. 1999. *POLEMOS: La contexte guerrier en Égée à l'âge du Bronze* (*Aegaeum* 19), Liège.

Laffineur R., and P. P. Betancourt, eds. 1997. *TEXNH: Craftswomen, Craftsmen, and Craftsmanship in the Aegean Bronze Age* (*Aegaeum* 16), Liège.

Lagia, A. 1993. "Differential Diagnosis of the Three Main Types of Anaemia (Thalassaemia, Sickle Cell Anaemia, and Iron Deficiency Anaemia) Based on Macroscopic and Radiographic Skeletal Characteristics" (M.Sc. diss. Univ. of Bradford).

———. 2002. "Ραμνούς, τάφος 8: Ανασύσταση της ταφικής συμπεριφοράς μέσα από το πρίσμα της ταφονομικής και ανθρωπολογικής ανάλυσης," *Ευλιμένη* 3, pp. 203–222.

———. 2007a. "Notions of Childhood in the Classical *Polis*: Evidence from the Bioarchaeological Record," in *Constructions of Childhood in Ancient Greece and Italy* (*Hesperia Suppl.* 41), ed. A. Cohen and J. B. Rutter, Princeton, pp. 293–306.

———. 2007b. "The Human Skeletal Remains," in *Die Submykenische Nekropole: Neufunde und Neubewertung* (*Kerameikos* XVIII), ed. F. Ruppenstein, Berlin, pp. 273–281.

———. Forthcoming. "A Bioarchaeological Survey of Social Structure in the Polis of Athens during the Classical, Hellenistic and Early Roman Periods" (diss. Univ. of Chicago).

Lagia, A., C. Eliopoulos, and S. Manolis. 2007. "Thalassemia: Macroscopic and Radiological Study of a Case," *International Journal of Osteoarchaeology* 17, pp. 269–285.

Lagia, A., I. Grigoropoulou, N. Kontoyiannis, I. Karali, and S. K. Manolis. 2003. "A Case of Syphilis from the Greek Mainland" (paper, Athens 2003).

Lagia, A., E. I. Petroutsa, and S. K. Manolis. 2002. "A Demographic and Paleopathological Assessment of a Paleochristian Population from the Island of Kos," in *Proceedings of the 24th Annual Conference of Hellenic Society for Biological Sciences, Eretria, Euboea Island, May 23–24, 2002*, pp. 163–164.

———. 2007. "Health and Diet during the Middle Bronze Age in the Peloponnese: The Site of Kouphovouno," in *Autour de la cuisine*, ed. C. Mee and J. Renard, Oxford, pp. 33–328.

Lagia, A., and M. Richards. In preparation. "Diet and the *Polis*: An Isotopic Study of Diet in Athens and Laurion during the Classical, Hellenistic, and Early Roman Periods," in *Stable Isotopic Studies in Greece*, ed. A. Papathanasiou and M. Richards.

Lambert, J. B., S. M. Vlasak, A. C. Thomet, and J. E. Buikstra. 1982. "A Comparative Study of the Chemical Analysis of Ribs and Femurs in Woodland Populations," *American Journal of Physical Anthropology* 59, pp. 289–294.

Lambrianides, K. 1995. "Present-Day Chora on Amorgos and Prehistoric Thermi on Lesbos: Alternative Views of Communities in Transition," in *Time, Tradition, and Society in Greek Archaeology: Bridging the "Great Divide,"* ed. N. Spencer, London, pp. 64–88.

Liston, M. A. 1993. "The Human Skeletal Remains from Kavousi, Crete: A Bioarchaeological Analysis (diss. Univ. of Tennessee).

———. 2007. "Secondary Cremation Burials at Kavousi Vronda, Crete: Symbolic Representation in Mortuary Practice," *Hesperia* 76, pp. 57–71.

Liston, M. A., and J. K. Papadopoulos. 2004. "The 'Rich Athenian Lady' Was Pregnant: The Anthropology of a Geometric Tomb Reconsidered," *Hesperia* 73, pp. 7–38.

Little, L. M. 1999. "Babies in Well G5:3: Preliminary Results and Future Analysis," *AJA* 103, p. 284 (abstract).

Little, L. M., and J. K. Papadopoulos. 1998. "A Social Outcast in Early Iron Age Athens," *Hesperia* 67, pp. 375–404.

Loth, S., and M. Y. İşcan. 1989. "Morphological Assessment of Age in the Adult: The Thoracic Region," in *Age Markers in the Human Skeleton*, ed. M. Y. İşcan, Springfield, Ill., pp. 105–136.

Macchinon, M. 2007. "Osteological Research in Classical Archaeology: State of the Discipline," *AJA* 111, pp. 473–504.

Manolis, S. K. 1990. "Ανθρωπολογική έρευνα της σύνθεσης του πληθυσμού της Νοτίου Ελλάδος στην εποχή του Χαλκού" (diss. Univ. of Athens).

Manolis, S. K., and A. A. Neroutsos. 1997. "The Middle Bronze Age Burial of Kolonna at Aegina Island, Greece: Study of the Human Skeletal Remains," in *Das Mittelbronzezeitliche Schachtgrab von Ägina*, ed. I. Kilian-Dirlmeier, Mainz, pp. 169–175.

Martinez, L., S. Mirabal, J. R. Luis, and R. J. Herrera. 2008. "Middle Eastern and European mtDNA Lineages Characterize Populations from Eastern Crete," *American Journal of Physical Anthropology* 137, pp. 213–223.

Martinez, L., P. A. Underhill, L. A. Zhivotovsky, T. Gayden, N. K. Moschonas, C. E. Chow, S. Conti, E. Mamolini, L. L. Cavalli-Sforza, and R. J. Herrera. 2007. "Paleolithic Y-Haplogroup Heritage Predominates in a Cretan Highland Plateau," *European Journal of Human Genetics* 15, pp. 485–493.

McHenry, H. 1968. "Transverse Lines in Long Bones of Prehistoric California Indians," *American Journal of Physical Anthropology* 29, pp. 1–17.

Merriwether, D. A., F. Rothhammer, and R. E. Ferrell. 1994. "Genetic Variation in the New World: Ancient Teeth, Bone, and Tissue as Sources of DNA," *Experientia* 50, pp. 592–601.

Milner, G. R., D. A. Humpf, and H. C. Harpending. 1989. "Pattern Matching of Age-at-Death Distributions in Paleodemographic Analysis," *American Journal of Physical Anthropology* 80, pp. 49–58.

Milner, G. R., J. W. Wood, and J. L. Boldsen. 2000. "Paleodemography," in *Biological Anthropology of the Human Skeleton*, ed. M. A. Katzenberg and S. R. Saunders, New York, pp. 467–497.

Moore, J., and E. Scott, eds. 1997. *Invisible People and Processes: Writing Gender and Childhood into European Archaeology*, London.

Morris, I. 1987. *Burial and Ancient Society: The Rise of the Greek City-State*, Cambridge.

———. 1991. "The Archaeology of Ancestors: The Saxe/Goldstein Hypothesis Revisited," *CAJ* 1, pp. 147–169.

———. 1992. *Death-Ritual and Social Structure in Classical Antiquity*, Cambridge.

———. 1994a. "Everyman's Grave," in *Athenian Identity and Civic Ideology*, ed. A. L. Boegehold and A. C. Scafuro, Baltimore, pp. 67–101.

———. 1994b. "Archaeologies of Greece," *Classical Greece: Ancient Histories and Modern Archaeologies*, ed. I. Morris, Cambridge, pp. 8–47.

———. 2000. *Archaeology as Culture History*, Cambridge.

Musgrave, J. H., and S. P. Evans. 1980. "By Strangers Honor'd: A Statistical Study of Ancient Crania from Crete, Mainland Greece, Cyprus, Israel, and Egypt," *Journal of Mediterranean Anthropology and Archaeology* 1, pp. 50–107.

Nafplioti, A. 2007. "Population Bio-Cultural History in the South Aegean during the Bronze Age" (diss. Univ. of Southampton).

———. 2008. "'Mycenaean' Political Domination of Knossos Following the Late Minoan IB Destructions on Crete: Negative Evidence from Strontium Isotope Ratio Analysis ($^{87}Sr/^{86}Sr$)," *JAS* 35, pp. 2307–2317.

Neils, J., and J.H. Oakley, eds. 2003. *Coming of Age in Ancient Greece: Images of Childhood from the Classical Past,* New Haven.

Nerlich, A. G., C. J. Haas, A. Zink, U. Szeimies, and H. G. Hagedorn. 1997. "Molecular Evidence for Tuberculosis in an Ancient Egyptian Mummy," *Lancet* (British Edition) 350, p. 1404.

Ober, J., W. Scheidel, B. D. Shaw, and D. Sanclemente. 2007. "Toward Open Access in Ancient Studies: The Princeton-Stanford Working Papers in Classics," *Hesperia* 76, pp. 229–242.

Ortner, D. J., and J. O. Kelley. 1990. "Contemporary Trends and Future Directions in Human Osteology and Paleopathology," in *A Life in Science: Papers in Honor of J. Lawrence Angel* (Center for American Archeology Scientific Papers 6), ed. J. E. Buikstra, Kampsville, Ill., pp. 17–23.

Pääbo, S., R. G. Higuchi, and A. C. Wilson. 1989. "Ancient DNA and the Polymerase Chain Reaction: The Emerging Field of Molecular Archaeology," *Journal of Biological Chemistry* 264, pp. 9709–9712.

Pääbo, S., H. Poinar, D. Serre, V. Jaenicke-Deprés, J. Hebler, N. Rohland, M. Kuch, J. Krause, L. Vigiland, and M. Hofreiter. 2004. "Genetic Analyses from Ancient DNA," *Annual Review of Genetics* 38, pp. 645–679.

Paine, R. R. 1989. "Life Table Fitting by Maximum Likelihood Estimation: A Procedure to Reconstruct Paleodemographic Characteristics from Skeletal Age Distributions," *American Journal of Physical Anthropology* 79, pp. 51–61.

———. 1997. "The Role of Uniformitarian Models in Osteological Paleodemography," in *Integrating Archaeological Demography: Multidisciplinary Approaches to Prehistoric Population* (Center for Archaeological Investigations Occasional Paper 24), ed. R. R. Paine, Carbondale, Ill., pp. 191–204.

Papathanasiou, A. 2000. "The Reconstruction of the Theopetra Cave Population Diet from Stable Isotope Analysis of Human Bone," in *Theopetra Cave: Twelve Years of Excavation and Research 1987–1998. Proceedings of the International Conference, Trikala 6–7 November 1998,* ed. N. Kyparissi-Apostolika, Athens, pp. 119–127.

———. 2001. *A Bioarchaeological Analysis of Neolithic Alepotrypa Cave, Greece* (*BAR-IS* 961), Oxford.

———. 2003. "Stable Isotope Analysis in Neolithic Greece and Possible Implications on Human Health," *International Journal of Osteoarchaeology* 13, pp. 314–324.

———. 2005. "Health Status of the Neolithic Population of Alepotrypa Cave, Greece," *American Journal of Physical Anthropology* 126, pp. 377–390.

———. In preparation. "Stable isotope analyses in Neolithic and Bronze Age Greece: an overview," in *Stable Isotopic Studies in Greece*, ed. Papathanasiou A. and M. Richards.

Papathanasiou, A., C. S. Larsen, and L. Norr. 2000. "Bioarchaeological Inferences from a Neolithic Ossuary from Alepotrypa Cave, Diros, Greece," *International Journal of Osteoarchaeology* 10, pp. 210–228.

Papathanasiou, A., and M. Richards. In preparation. *Stable Isotopic Studies in Greece.*

Parker Pearson, M. 1982. "Mortuary Practices, Society, and Ideology: An Ethnoarchaeological Study," in *Symbolic and Structural Archaeology,* ed. I. Hodder, Cambridge, pp. 99–113.

Petroutsa, E. I, 2007. "A Study of Diet in Greek Populations of the Bronze Age" (diss. Univ. of Athens).

Pickering, R. B. 1984. "Patterns of Degenerative Joint Disease in Middle Woodland, Late Woodland, and Mississippian Skeletal Series from the Lower Illinois Valley" (diss. Northwestern University).

Pike, S. 1997. "The Wiener Laboratory," *Paleopathology Newsletter* 100, pp. 8–9.

Powell, M. L. 1985. "The Analysis of Dental Wear and Caries for Dietary Reconstruction," in *The Analysis of Prehistoric Diets,* ed. R. I. Gilbert, and J. H. Mielke, Orlando, pp. 307–338.

Price, T. D., J. Blitz, J. H. Burton, and J. A. Ezzo. 1992. "Diagenesis in Prehistoric Bone: Problems and Solutions," *JAS* 19, pp. 513–529.

Price, T. D., G. Grupe, and P. Schrötter. 1994. "Reconstruction of Migration Patterns in the Bell Beaker Period by Stable Strontium Isotope Analysis," *Applied Geochemistry* 9, pp. 413–417.

Price, T. D., C. M. Johnson, J. A. Ezzo, J. E. Ericson, and J. H. Burton. 1994. "Residential Mobility in the Prehistoric Southwest United States: A Preliminary Study Using Strontium Isotope Analysis," *JAS* 21, pp. 315–330.

Ratip, S., and B. Modell. 1996. "Psychological and Sociological Aspects of the Thalassemias," *Seminars in Hematology* 33, pp. 53–65.

Renfrew, C., and J. F. Cherry. 1986. *Peer Polity Interaction and Sociopolitical Change,* Cambridge.

Richards, M. P., and R. E. M. Hedges. 1999. "A Neolithic Revolution? New Evidence of Diet in the British Neolithic," *Antiquity* 73, pp. 891–897.

———. 2008. "Evidence of Past Human Diet at the Sites of the Neolithic Cave of Gerani; the Late Minoan III Cemetery of Armenoi; Grave Circles A and B at the Palace Site of Mycenae; and Late Helladic Chamber Tombs," in *Archaeology Meets Science: Biomolecular and Site Investigations in Bronze Age Greece,* ed. Y. Tzedakis, H. Martlew, and M. K. Jones, Oxford, pp. 220–230.

Richards, M. P., J. A. Pearson, T. I. Molleson, N. Russell, and L. Martin. 2003. "Paleodietary Evidence from Neolithic Çatalhöyük, Turkey," *JAS* 30, pp. 67–76.

Richards, M. P., T. D. Price, and E. Koch. 2003. "The Mesolithic and Neolithic Transition in Denmark: New Stable Isotope Data," *CurrAnthr* 44, pp. 288–295.

Richards, M. P. Schulting, R. J., and R. E. M. Hedges. 2003. "Archaeology: Sharp Shift in Diet at Onset of Neolithic," *Nature* 425, p. 366.

Richards, M. P., and E. Vika. 2008. "Stable Isotope Results from New Sites in the Peloponnese: Cemeteries from Sykia, Kalamaki, and Spaliareika," in *Archaeology Meets Science: Biomolecular and Site Investigations in Bronze Age Greece,* ed. Y. Tzedakis, H. Martlew, and M. K. Jones, Oxford, pp. 231–235.

Rife, J. L., R. K. Dunn, M. M. Morrison, A. Barbet, D. H. Ubelaker, and F. Monier. 2007. "Life and Death at a Port in Roman Greece: The Kenchreai Cemetery Project, 2002–2006," *Hesperia* 76, pp. 143–181.

Roberts, C. A., C. Bourbou, A. Lagia, S. Triantaphyllou, and A. Tsaliki. 2005. "Health and Disease in Greece: Past, Present, and Future," in *Health in Antiquity,* ed. H. King, London, pp. 32–58.

Roberts, C. A., and J. E. Buikstra. 2003. *The Bioarchaeology of Tuberculosis: A Global View on a Reemerging Disease,* Gainesville.

Rose, J. C., K. Condon, and A. H. Goodman. 1985. "Diet and Dentition: Developmental Disturbances," in *The Analysis of Prehistoric Diets,* ed. R. I. Gilbert and J. H. Mielke, Orlando, pp. 281–305.

Rotroff, S. I., L. M. Little, and L. M. Snyder. 1999. "The Reanalysis of a Well Deposit from the Second Century B.C. in the Athenian Agora: Animal Sacrifice and Infanticide in Late Hellenistic Athens?" *AJA* 103, pp. 284–285 (abstract).

Ruff, C. B. 1992. "Biomechanical Analyses of Archaeological Human Skeletal Samples," in *The Skeletal Biology of Past Peoples: Advances in Research Methods,* ed. S. R. Saunders and M. A. Katzenberg, New York, pp. 37–58.

Salo, W. L., A. C. Aufderheide, J. E. Buikstra, and T. A. Holcomb. 1994. "Identification of *Mycobacterium tuberculosis* DNA in a Pre-Columbian Peruvian Mummy," *Proceedings of the National Academy of Sciences* 91, pp. 2091–2094.

Sattenspiel, L., and H. C. Harpending. 1983. "Stable Populations and Skeletal Age," *AmerAnt* 48, pp. 489–498.

Saunders, S. R., 2000. "Subadult Skeletons and Growth-Related Studies," in *Biological Anthropology of the Human Skeleton,* ed. M. A. Katzenberg and S. R. Saunders, New York, pp. 135–161.

Saxe, A. A. 1970. "Social Dimensions of Mortuary Practices" (diss. Univ. of Michigan).

Scheidel, W. 1995. "The Most Silent Women of Greece and Rome: Rural Labour and Women's Life in the Ancient World," *GaR* xlii, pp. 202–217.

Schoeninger, M. J. 1995. "Stable Isotope Studies in Human Evolution," *Evolutionary Anthropology* 4, pp. 83–98.

Sealy, J. C., R. Armstrong, and C. Schrire. 1995. "Beyond Lifetime Averages: Tracing Life Histories through Isotopic Analysis of Different Calcified Tissues from Archaeological Human Skeletons," *Antiquity* 69, pp. 290–300.

Sealy, J. C., N. J. van der Merwe, A. Sillen, F. J. Kruger, and H. Kruger. 1991. "$^{87}Sr/^{86}Sr$ as a Dietary Indicator in Modern and Archaeological Bone," *JAS* 18, pp. 399–416.

Smith, P., and G. Kahila. 1992. "Identification of Infanticide in Archaeological Sites: A Case Study of Late Roman–Early Byzantine Periods at Ashkelon, Israel," *JAS* 19, pp. 667–675.

Smith, P., G. Kahila, D. Filon, A. Oppenheim, E. Eisenburg, and M. Faerman. 1999. "The Application of Ancient DNA Analysis to Archaeological Problems: Its Role in Studies of Gender in Past Societies," in *The Practical Impact of Science on Near Eastern and Aegean Archaeology* (Wiener Laboratory Publication 3), ed. S. Pike and S. Gitin, London, pp. 71–74.

Spencer, N. 1995a. *Time, Tradition, and Society in Greek Archaeology: Bridging the "Great Divide,"* London.

———. 1995b. "Multi-Dimensional Group Definition in the Landscape of Rural Greece," in *Time, Tradition, and Society in Greek Archaeology: Bridging the "Great Divide,"* ed. N. Spencer, London, pp. 28–42.

Stears, K. 1995. "Dead Women's Society," in *Time, Tradition, and Society in Greek Archaeology: Bridging the "Great Divide,"* ed. N. Spencer, London, pp. 109–131.

Stewart, T. D. 1970. "Identification of the Scars of Parturition in the Skeletal Remains of Females," in *Personal Identification in Mass Disasters,* ed. T. D. Stewart, Washington, D.C., pp. 127–135.

Stone, A. C. 2000. "Ancient DNA from Skeletal Remains," in *Biological Anthropology of the Human Skeleton,* ed. M. A. Katzenberg and S. R. Saunders, New York, pp. 351–371.

Stone, A. C., G. R. Milner, S. Pääbo, and M. Stoneking. 1996. "Sex Determination of Ancient Human Skeletons Using DNA," *American Journal of Physical Anthropology* 99, pp. 231–238.

Stone, A. C., and M. Stoneking. 1993. "Ancient DNA from a Pre-Columbian Amerindian Population," *American Journal of Physical Anthropology* 92, pp. 463–471.

Stravopodi, E., and S. K. Manolis. 2000. "Το Βιοαρχαιολογικό προφίλ των Ανθρωπολογικών Ευρημάτων του Σπηλαίου Θεόπετρας: Ένα Πιλοτικό Πρόγραμμα στον Ελλαδικό Χώρο," in *Theopetra Cave: Twelve Years of Excavation and Research 1987–1998. Proceedings of the International Conference, Trikala 6–7 November 1998,* ed. N. Kyparissi-Apostolika, Athens, pp. 95–108.

Stravopodi, E., S. K. Manolis, and A. Neroutsos. 1997. "Μελέτη του Ανθρώπινου Σκελετικού Υλικού από το Σπήλαιο Λιμνών," in *Το Σπήλαιο των Λιμνών στα Καστριά Καλαβρύτων* (Εταιρεία Πελοποννησιακών Σπουδών 7), ed. A. Sampson, Athens, pp. 456–482.

Suchey, J. M., D. V. Wiseley, R. F. Green, and T. T. Noguchi. 1979. "Analysis of Dorsal Pitting in the *Os Pubis* in an Extensive Sample of Modern American Females," *American Journal of Physical Anthropology* 51, pp. 517–539.

Themelis, P. 2000. "Ταφικά Μνημεία Γυμνασίου," *Prakt* 2000, pp. 97–101.

Triantaphyllou, S. 1998. "Prehistoric Cemetery Populations from Northern Greece: A Breath of Life for the Skeletal Remains," in *Cemetery and Society in the Aegean Bronze Age,* ed. K. Branigan, Sheffield, 150–164.

———. 1999. "Prehistoric Makrigialos: A Story from the Fragments," in *Neolithic Society in Greece,* ed. P. Halstead, Sheffield, pp. 21–35.

———. 2001. *A Bioarchaeological Approach to Prehistoric Cemetery Populations from Central and Western Greek Macedonia* (*BAR-IS* 976), Oxford.

———. 2004. "Ανθρωπολογική ανάλυση του αρχαϊκού πληθυσμού Αγίας Παρασκευής," in *Ανθρωπολογικές Μελέτες από τη Βόρεια Ελλάδα* (Δημοσιεύματα του Αρχαιολογικού Ινστιτούτου Μακεδονικών και Θρακικών Σπουδών 5), ed. D. V. Grammenos and S. Triantaphyllou, Thessaloniki, pp. 89–269.

Triantaphyllou, S., M. P. Richards, C. Zerner, and S. Voutsaki. 2008. "Isotopic Dietary Reconstruction of Humans from Middle Bronze Age Lerna, Argolid, Greece" *JAS* 35, pp. 3028–3034.

Tritsaroli, P. 2006. "Pratiques funéraires en Grèce centrale à la période byzantine: Analyse à partir des données archéologiques et biologiques" (diss. Muséum national d'histoire naturèlle, Departément de prehistoire, Institut de paléontologie humaine).

Tsaliki, A. 2001. "Vampires beyond Legend: A Bioarchaeological Approach," in *Proceedings of the XIII European Meeting of the Paleopathology Association, Chieti, Italy, September 18–23, 2000,* ed. M. I. Verghetta and L. Capasso, Teramo, pp. 295–300.

———. 2005. "Ancient Human Skeletal Remains from Sifnos: An Overview," in *Proceedings of the 2nd International Sifnean Symposium, June 27–30, 2002,* Athens, pp. 139–154.

———. 2008. "An Investigation of Extraordinary Human Body Disposals, with Special Reference to Necrophobia" (diss. Univ. of Durham).

Tuross, N. 1994. "The Biochemistry of Ancient DNA in Bone," *Experientia* 50, pp. 530–535.

Ubelaker, D. H., and J. L. Rife. 2007. "The Practice of Cremation in the Roman-Era Cemetery at Kenchreai, Greece: The Perspective from Archeology and Forensic Science," *Bioarchaeology of the Near East* 1, pp. 35–57.

Vika, E. K. 2007. "A Diachronic Bioarchaeological Approach to the Society of Thebes, Greece. Paleopathological Investigations and Paleodietary Reconstruction Using $\delta^{13}C$, $\delta^{15}N$, and $\delta^{32}S$ Isotope Analysis" (diss. Univ. of Bradford).

Vlahogianni, E., A. Lagia, and V. Sabetai. Forthcoming. "Πολλαπλές ταφές από το νεκροταφείο της Αρχαίας Ακραιφίας: Αρχαιολογική και ανθρωπολογική διαπραγμάτευση," in *Proceedings of the Fourth International Conference of Boeotian Studies*.

Vlahogiannis, N. 2005. "'Curing' Disability," in *Health in Antiquity*, ed. H. King, London, pp. 180–191.

Wilkins, J., D. Harvey, and M. Dobson, eds. 1995. *Food in Antiquity*, Exeter.

Wittwer-Backofen, U., J. Wahl, V. Dresely, T. H. Schmidt-Schultz, and M. Schultz. 2000. "Das Spätbronzezeitliche Gräberfeld von Beşik-Tepe/Troas: Anthropologische Ansätze zur Sozial-struktur," in *Beşik-Tepe: Das Spätbronzezeitliche Gräberfeld*, ed. M. A. Basedow, Mainz, pp. 197–235.

Wood, J. W., D. J. Holman, K. M. Weiss, A. V. Buchanan, and B. LeFor. 1992a. "Hazard Models for Human Population Biology," *Yearbook of Physical Anthropology* 35, pp. 43–87.

Wood, J. W., G. R. Milner, H. C. Harpending, and K. M. Weiss. 1992b. "The Osteological Paradox: Problems of Inferring Prehistoric Health from Skeletal Samples," *CurrAnthr* 33, pp. 343–370.

Wright, L. E., and C. J. Yoder. 2003. "Recent Progress in Bioarchaeology: Approaches to the Osteological Paradox," *Journal of Archaeological Research* 11, pp. 43–70.

Zink, A., W. Grabner, and A. G. Nerlich. 2005. "Molecular Identification of Human Tuberculosis in Recent and Historic Bone Tissue Samples: The Role of Molecular Techniques for the Study of Historic Tuberculosis," *American Journal of Physical Anthropology* 126, pp. 32–47.

CHAPTER 2

Petralona: Link between Africa and Europe?

by Katerina Harvati

The Middle Pleistocene spans from 780,000 to 127,000 years before present (hereafter kya)[1] and marks the establishment of human presence in Europe. It is close to the early boundary of this time period that we first see evidence for a tentative human colonization in the Mediterranean zone,[2] and not until 500 kya that the human fossil record of northern Europe appears.[3] This time period witnessed several glaciation cycles, saw the last common ancestor between Neanderthals and modern humans, and gave root to the Neanderthal lineage in Europe.[4] With the exception of the unprecedented site of Sima de los Huesos, Sierra de Atapuerca, Spain, the human fossil record from this time period is relatively sparse in space and time. The classification and phylogenetic relationships of these specimens to each other and to later humans remain to this day unresolved.

Until the discovery of Sima de los Huesos,[5] the most anatomically complete and largely undistorted European specimen of probable Middle Pleistocene age was the Petralona specimen from Chalkidiki, northern Greece (Fig. 2.1). As such it has played a key role in discussions of human evolution during this epoch.[6] Discovered in 1960 by a group of local villagers,[7] the specimen has been surrounded by controversy regarding its classification and especially its age. Faunal material recovered from the Petralona cave has yielded contradictory and widely ranging ages,[8] with an estimate of about 700 kya put forth by Kurtén and Poulianos.[9] However, the association of these faunal remains with the human cranium, itself a surface find, is questionable. It is therefore not clear whether these faunal age estimates shed any light on the actual age of the specimen. The use of absolute dating has proven somewhat more successful, with a recent revision of Electron Spin Resonance and Uranium series dates on sediments associated with the cranium yielding an age of between 150 and 250 kya,[10] a date perhaps younger than its very primitive morphology would suggest.

The Petralona cranium was originally thought to represent a Neanderthal.[11] Later study led some authors to classify it within *Homo erectus*,[12] while others saw similarities with the African specimens from Bodo and Kabwe and with the European specimen Arago 21.[13] Stringer[14] compared the Petralona cranium with a large Neanderthal sample as well as with

1. See Delson et al. 2000, pp. 567–570.
2. Bermudez de Castro et al. 1997, pp. 1392–1395; Manzi, Mallegni, and Ascenzi 2001, pp. 10011–10016.
3. Roebroeks 2001, pp. 437–461.
4. Gamble 1999; Krings et al. 1997, pp. 19–30; Rightmire 1997, pp. 917–918.
5. See *Journal of Human Evolution* 33: 2–3, August 1997.
6. See Stringer, Howell, and Melentis 1979, pp. 235–253.
7. Kokkoros and Kanellis 1960, pp. 438–446.
8. See Tsoukala 1990.
9. Kurtén and Poulianos 1977, pp. 47–130; see also de Bonis and Melentis 1991, pp. 285–289.
10. Grün 1996, pp. 227–241. See also Schwarcz, Liritzis, and Dixon 1980, pp. 152–167; Hennig et al. 1981, pp. 533–536; Poulianos, Liritzis, and Ikeya 1982, pp. 280–282.
11. Kokkoros and Kanellis 1960, p. 442.
12. Poulianos 1981, p. 287; see also Wolpoff 1980, pp. 339–358.
13. Howells 1967; Murill 1975, pp. 176–187; 1981; Wolpoff 1980, pp. 339–358.
14. Stringer 1974, pp. 397–404.

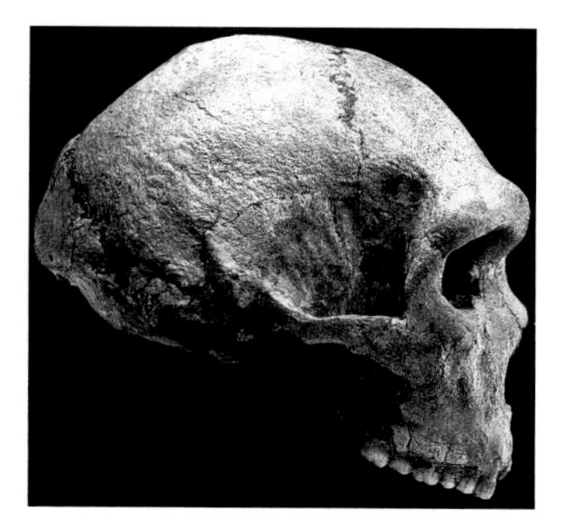

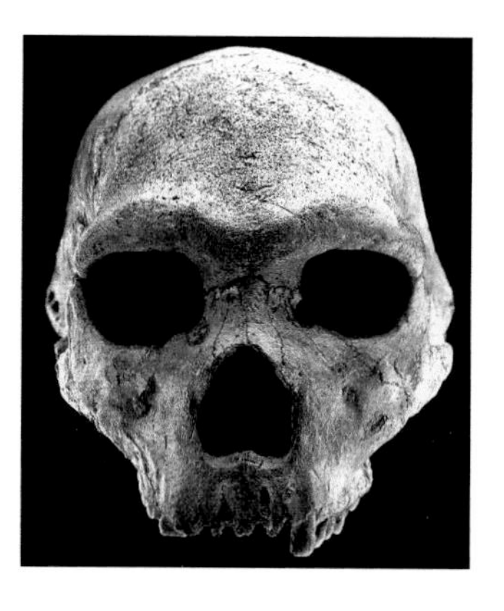

Figure 2.1. The Petralona specimen: lateral *(left)* and frontal *(right)* views.
Photo K. Havarti, courtesy E. Delson

other Middle and Late Pleistocene fossils from Europe and Africa. In his multivariate analysis Petralona was very distant from both Neanderthals and modern humans, its closest neighbors being Kabwe and the North African specimen Jebel Irhoud. Stringer concluded that, together with these fossil humans, it represented a very early form of *H. sapiens.* Stringer and colleagues[15] later revisited the issue of the classification of Petralona and other Middle Pleistocene fossils from Europe and Africa. They recommended the use of a grade system to accommodate what they termed the "archaic *H. sapiens*" sample. In this view Neanderthals were seen as a subspecies of *H. sapiens,* and the Middle Pleistocene European and African specimens were loosely considered as ancestral, "archaic" forms. They were thought to still belong to our own species (rather than to *H. erectus*) but to retain many plesiomorphic,[16] *erectus*-like features. They were grouped together based on what was perceived to be a similar level of adaptation or "grade." This grade classification of "archaic *Homo sapiens*" acknowledged affinities with later *Homo* but avoided confronting the thorny issues of the precise phylogenetic relationships and alpha taxonomy of these specimens.

In the last few decades the magnitude of morphological, genetic, and behavioral differences between Neanderthals and modern humans has become increasingly obvious.[17] The classification of this taxon as a subspecies of *H. sapiens* therefore appears increasingly untenable, with a new consensus favoring the assignment of Neanderthals to a distinct species, *H. neanderthalensis.* Evidence has also mounted for a recent, African origin for modern humans, excluding Neanderthals from the direct ancestry of our species. In this new configuration, the grade classification of the Middle Pleistocene human fossils as "archaic *H. sapiens*" also has become more difficult. Several questions remain unanswered: How are the European and African Middle Pleistocene hominids related to Neanderthals and to modern humans? Can they be considered as belonging to the same species, perhaps representing the last common ancestor of Neanderthals and modern humans, or have their evolutionary trajectories already diverged (the European specimens already on the Neanderthal lineage, and the African ones on the modern human one)? Furthermore, is there only one lineage in Pleistocene Europe, not to mention Africa? And finally, what is the relationship of the African and European "archaic *H. sapiens*" to the even less known Middle Pleistocene fossil human record from Asia?

15. Stringer, Howell, and Melentis 1979, pp. 235–253.

16. Plesiomorphic traits are shared by both the ancestral and descendant taxa. They are also termed *primitive* or *ancestral.* They contrast with autopomorphic traits, which are uniquely present in a taxon, and therefore can be used to define it.

17. See, for example, Krings et al. 1997, pp. 19–30; 2000, pp. 144–146; Pearson 2000, pp. 229–247; Ponce de León and Zollikofer 2001, pp. 534–538; Mellars 2002, pp. 31–47; Harvati 2002; 2003a, pp. 107–132; Harvati, Frost, and McNulty 2004, pp. 1147–1152; Serre et al. 2004, p. 57.

Continuity within the Middle–Late Pleistocene European record is often accepted and many researchers see a discrete, isolated lineage leading to the gradual evolution of Neanderthals in this continent, according to the "accretion" hypothesis of Neanderthal evolution.[18] Opinions differ over the placement of the earlier of these specimens within the taxon *H. heidelbergensis,* as advocated by some,[19] or within *H. neanderthalensis* itself. *H. heidelbergensis* can be viewed as a cross-continental taxon, including both the African and European (and possibly also Asian) Middle Pleistocene fossil humans, and representing the last common ancestor between Neanderthals (in Europe) and modern humans (in Africa).[20] Alternatively, one may consider *H. heidelbergensis* as a chronospecies, or an arbitrary segment of the Neanderthal lineage that includes only the European Middle Pleistocene record (with *H. rhodensiensis* or *H. helmei* applied to its African counterpart). In this latter view, the last common ancestor of Neanderthals and modern humans would have to be a different, older taxon, perhaps *H. antecessor* as suggested by Bermúdez de Castro and colleagues.[21] Finally, a few researchers see evidence for multiple taxa and not for strict continuity of a single evolving lineage even within the restricted European Middle Pleistocene record.[22]

The present study is a preliminary approach to the issues of Neanderthal and early modern human ancestry from the perspective of the Petralona specimen and with the use of the new methods of three-dimensional geometric morphometrics.[23] Its goal is to quantify proposed Neanderthal features in this specimen and to assess the degree of morphological similarity between Neanderthals and Petralona and among the African and European Middle Pleistocene fossil humans. This first analysis will elucidate some of the issues discussed in the literature and will point to the directions that future research must take. In so doing, it will lay the foundations for a more extensive and thorough study of the European and African Middle Pleistocene human fossil record.

MATERIALS AND METHODS

Ninety-seven modern human crania, representing five geographic populations,[24] as well as four Neanderthal and three Middle Pleistocene European and African specimens, were measured in this study (Table 2.1). Neanderthals were treated as a distinct species, *H. neanderthalensis.* Only adults were included, as determined by fully erupted permanent dentition. Sex was often unknown and was assigned based on cranial morphology and size. Since such sex assignment is imperfect for recent humans and even more problematic for fossil specimens,[25] sexes were pooled in the analysis. The presumed sex composition of both the recent and fossil samples is reported in Table 2.1. Specimens were chosen on the basis of the completeness of their face and braincases, the areas that are best preserved on the Petralona cranium. Landmarks of the basicranium are often considered to be very informative phylogenetically.[26] However, they were not measured here because this anatomical region is partly damaged and mostly covered

18. See Hublin 1998, pp. 295–310.
19. Arsuaga et al. 1997, pp. 219–281.
20. Rightmire 1998, pp. 218–227; 2001, pp. 77–84.
21. Bermúdez de Castro et al. 1997, pp. 1392–1395.
22. Schwartz and Tattersall 2005.
23. I thank the editors of this volume for inviting this contribution and for their helpful comments and suggestions. Special thanks to Georgia Tsartsidou for helping with the map illustration and Eric Delson for providing the photographs. This research was supported by the New York University and the Max Planck Institute for Evolutionary Anthropology.
24. It is important to note here that the present analysis does not aim to test hypotheses about continuity between Neanderthals and modern human populations (on this topic, see Harvati 2002, 2003a; Harvati, Frost, and McNulty 2004). A wider geographic representation in the modern human sample to include European populations and additional African specimens is desirable but not crucial to the results of this study.
25. See Howells 1973; Trinkaus 1983.
26. See Harvati and Weaver 2006.

TABLE 2.1. MODERN AND FOSSIL HUMAN SAMPLES

Modern Human Samples	*Specimens*	*Male*	*Female*
Andamanese	32	16	16
Chinese	20	10	9
East African	7	5	2
Near Eastern	19	12	7
Thai	19	9	10
Total	97	52	44

Fossil Human Samples	*Specimens*	*Male*	*Female*
Neanderthals (La Ferrasie 1, La Chapelle-aux-Saints, Gibraltar 1, La Quina 5*)	4	2	2
Kabwe	–	1	–
Sima de los Huesos 5	–	1	–
Petralona	–	1	–
Total	4	5	2

* La Quina 5 was used in calculating the mean Neanderthal configuration and in the analysis of neurocranial landmarks, but not in the analysis of facial landmarks.

by matrix on the Petralona specimen. These constraints on the landmarks employed resulted in severely limited fossil samples despite some reconstruction. The original Sima de los Huesos cranium 5 specimen was not available. Instead, measurements were taken on a stereolithograph housed in the Anthropology Department of the American Museum of Natural History, New York. In some cases, measurements on the original fossils were supplemented with measurements on high-quality casts from the collection of the Anthropology Department of the American Museum of Natural History.

The data were collected in the form of three-dimensional coordinates of osteometric landmarks on the cranium.[27] In geometric morphometrics landmarks are defined as homologous points that can be reliably and repeatedly located in all specimens under study.[28] Here, they mostly represent standard osteometric points. Thirty-nine landmarks were recorded for each specimen using the Microscribe 3D-X (Immersion Corp.) portable digitizer (Table 2.2, Fig. 2.2). They were chosen to reflect as much of the morphology of the cranium present in Petralona as possible. As a result, they were concentrated on the face and the braincase. As the facial morphology of Petralona has been claimed to show stronger Neanderthal affinities than its neurocranium, two analyses were performed: one using the subset of landmarks from the neurocranium, and another using the facial landmarks. Because the cranium is an integrated whole, such subdivisions are defined arbitrarily and are bound to overlap. In this case, however, the two data sets overlapped in only one point: glabella. Not all the Neanderthal specimens used preserve both data sets: La Quina 5 lacks a complete face. This specimen was therefore only used in the neurocranial landmarks analysis and in the computation of the Neanderthal mean configuration.[29]

27. For definitions of the standard osteometric landmarks used, see Howells 1973. Additional landmark definitions are listed in Table 2.2.

28. For the definition of landmarks and landmark types, see Bookstein 1990, pp. 216–225, and Valeri et al. 1998, pp. 113–124.

29. This means that for Neanderthals, the mean of each landmark was computed based on all available specimens for that particular landmark, whether they were four (including La Quina 5) or only three.

TABLE 2.2. LIST OF LANDMARKS AND THEIR DEFINITIONS

Landmark	Analysis / Definition
1. Inion	(2)
2–3. Asterion Right and Left (R and L)	(2)
4. Lambda	(2)
5, 8. Porion (R and L)	(2)
6, 9. Auriculare (R and L)	(2)
7, 10. Parietal Notch R	(2)
11–12. Distal M3 (R and L)	(2) Midpoint distal to M3
13. Bregma	(2)
14. Post-toral Sulcus	(2) Minima of concavity on midline post-toral frontal squama
15. Glabella	(1, 2)
16. Nasion	(1)
17. Rhinion	(1)
18. Nasospinale	(1)
19. Prosthion	(1)
20, 30. Mid-orbit Torus Superior (R and L)	(1) Point on superior aspect of supra-orbital torus, approximately at the middle of the orbit
21, 31. Mid-orbit Torus Inferior (R and L)	(1) Point on inferior margin of supra-obrital torus, approximately at the middle of the orbit
22, 32. Dacryon (R and L)	(1)
23, 33. Zygoorbitale (R and L)	(1)
24, 34. Frontomalare Orb (R and L)	(1)
25, 35. Infraorbital Foramen (R and L)	(1)
26, 36. Zygomaxillare (R and L)	(1)
27, 37. Alare (R and L)	(1)
28, 38. Jugale (R and L)	(1)
29, 39. Anterior Pterion (R and L)	(1)

Note: In parentheses are indicated the analyses in which each landmark was used: (1) facial landmarks analysis, (2) neurocranial landmarks analysis.

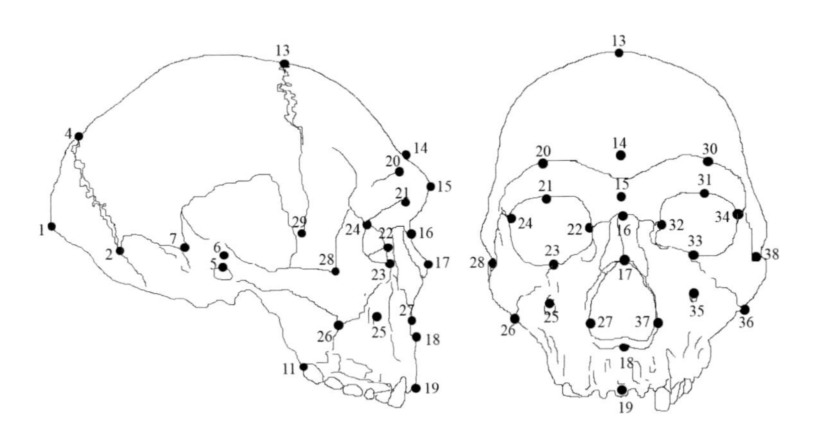

Figure 2.2. Landmarks collected shown in lateral and frontal views.
Drawing K. Havarti

The coordinate data for each data set were processed using the techniques of three-dimensional geometric morphometrics. These methods present several advantages over traditional morphometrics. They allow a much better preservation of the geometry of the object studied than traditional caliper measurements, so that far less information is lost about the original material. Furthermore, they allow one to identify the landmarks where shape differences occur, as well as the relative levels of difference at each landmark, after size has been corrected for as a factor. Geometric morphometrics also provide a way of quantifying shape differences of traits that are difficult to measure with linear or angle measurements and are therefore usually described qualitatively rather than quantitatively. For instance, this approach has been applied to the complex morphology of the basicranium that had not previously been evaluated quantitatively.[30] Finally, geometric morphometric methods can readily provide illustrations of the shape changes between specimens.[31]

Specimen landmark configurations were superimposed using Generalized Procrustes Analysis in the Morpheus software package.[32] This procedure translates the landmark configurations to common origin, scales them to unit centroid size (the square root of the sum of squared distances of all landmarks to the centroid of the object, the measure of size used here), and rotates them according to a best-fit criterion. It eliminates "size" as a factor (although size-related, allometric, shape differences may remain), so that "shape" can be analyzed separately.[33] Geometric morphometric methods have recently been successfully applied to primatology,[34] human variation,[35] and paleoanthropology.[36]

Morphometric analyses do not accommodate missing data. It was, therefore, necessary to perform some data reconstruction. Specimens with minimal damage were reconstructed during data collection through estimation of the position of the structure of interest using the morphology of the preserved surrounding areas. Missing landmarks were further reconstructed by reflecting the right and left sides of the specimen. Incomplete specimens were least-squares superimposed with their reflected equivalents using Morpheus, and missing data were reconstructed from their homologous counterparts on the other side. Further reconstruction involving the substitution of sample means was performed for the Neanderthal specimens and for midline landmarks such as rhinion (missing on all Neanderthals used here except Gibraltar 1). However, no landmark was reconstructed in this way in the three Middle Pleistocene specimens in this analysis.

The fitted landmark configuration of Petralona was visually compared with the Kabwe and Sima 5 configurations and with the modern human and Neanderthal mean configurations (Fig. 2.3). This allowed for an assessment of the described morphological similarities and differences among these samples after they had been scaled for size. The fitted coordinates were analyzed statistically using principal components, canonical variates, and Mahalanobis D^2 analyses with the use of the SAS statistical software program.[37] The principal components analysis achieved data reduction and the first few principal axes (accounting for approximately 85% of the total variance) were used as variables in the canonical variates analysis.[38] This latter analysis was conducted in order to find the canonical axes that

30. Harvati 2002, pp. 25–30; 2003b, pp. 323–328.

31. See Harvati 2003b, p. 114. Also see Rohlf and Marcus 1993, pp. 129–132; Dean 1993; O'Higgins 2000, pp. 103–120.

32. Slice 1992, 1994–1999.

33. See Rohlf 1990, pp. 227–236; Rohlf and Marcus 1993, pp. 129–132; Slice 1996, pp. 179–199; O'Higgins and Jones 1998, pp. 251–272.

34. See, e.g., O'Higgins and Jones 1998, pp. 251–272; Collard and O'Higgins 2000, pp. 322–331; Singleton 2002, pp. 547–578; Frost et al. 2003, pp. 1048–1072.

35. Ahlström 1996, pp. 415–421; Wood and Lynch 1996, pp. 407–414; Rosas and Bastir 2002, pp. 236–245; Hennessy and Stringer 2002, pp. 37–48; Nicholson and Harvati 2006, p. 158.

36. Dean 1993; Yaroch 1996, pp. 43–89; Dean et al. 1998, pp. 485–508; Bookstein et al. 1999, pp. 217–224; Ponce de León and Zollikofer 2001, pp. 534–538; Delson et al. 2001, pp. 380–297; Harvati 2002, pp. 25–30; 2003a, pp. 107–132; 2003b, pp. 323–328; Bruner, Manzi, and Arsuaga 2004, pp. 15335–15340; Harvati, Frost, and McNulty 2004, pp. 1147–1152; Nicholson and Harvati 2006, p. 158.

37. SAS Institute, release 8.02.

38. Bennett and Bowers 1976; Lestrel 2000.

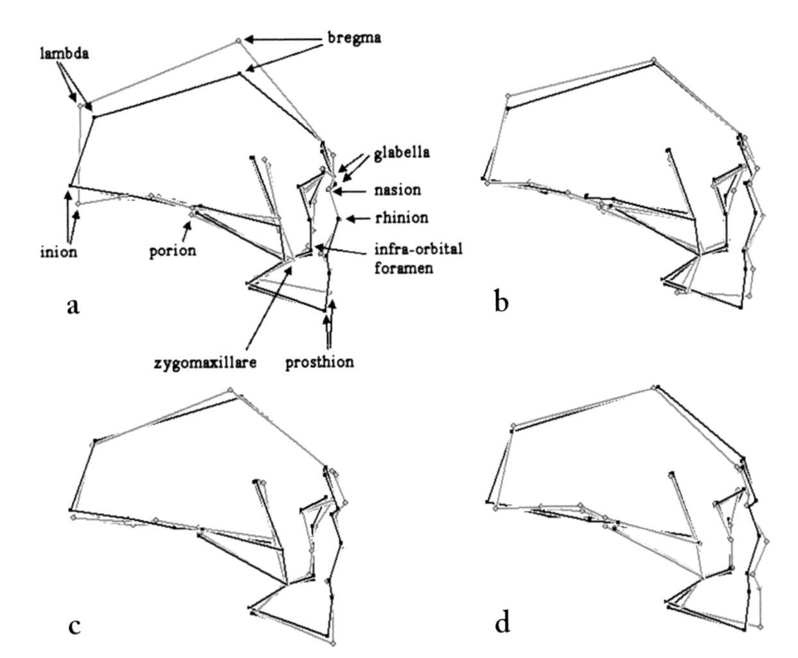

Figure 2.3. The Petralona landmark configuration (black) compared to (a) the modern human mean landmark configuration; (b) the Neanderthal mean landmark configuration; (c) the Kabwe landmark configuration; and (d) the Sima 5 landmark configuration (all in gray). Drawing K. Havarti

maximize the differences between the samples included.[39] A Bonferroni t-test was performed on all pairwise comparisons of the sample means of these axes, so as to detect significant differences between groups. Shape differences along the canonical axes were explored with the software GRF-ND.[40] The scores of each canonical variate were regressed against the fitted coordinates and the regression coefficients imported as eigenvector coefficients in GRF-ND.[41] This allowed visualization of hypothetical specimens along each axis and was used to illustrate shape differences.

The similarity among groups was evaluated by the Mahalanobis D^2 statistic, corrected for unequal sample sizes.[42] The larger the values of the D^2 distance, the farther the group centroids are from each other.[43] When performing a Mahalanobis distance analysis, the position of singletons and groups comprising very few specimens relative to other groups can be misleading due to misrepresentation of the variation in their groups. For single specimens, such as the Middle Pleistocene fossils used here, this can occur despite the adjustment for unequal sample sizes. False singleton situations, produced by randomly taking specimens out of the modern human sample and treating them as unknowns, showed that these specimens exhibited a very large Mahalanobis squared distance to their own population in some, but not all, cases. Therefore, it must be born in mind that the Mahalanobis D^2 involving the single fossil specimens (Petralona, Kabwe, and Sima 5) may be somewhat exaggerated. A discriminant analysis was conducted in order to test the predictive power of neurocranial and facial traits in classifying these groups. In order to better visualize the distance results, a sequential agglomerative hierarchical nested cluster analysis was performed using the program NT-Sys.[44] Finally, the combined landmark configurations of the three Middle Pleistocene specimens were compared to each other and to the mean configurations of Neanderthals and modern humans.

39. Bennett and Bowers 1976.
40. Generalized rotational fitting of *n*-dimensional landmark data. Slice 1992, 1994–1999.
41. Slice 1996, pp. 179–199.
42. Marcus 1993, pp. 99–130.
43. Lestrel 2000.
44. 1986–2000 Applied Biostatistics Inc., version 2.10.

RESULTS

Landmark Configuration Comparison

Figure 2.3 shows Petralona (gray) compared to the mean configurations for modern humans (A) and for Neanderthals (B), as well as to Kabwe (C) and Sima 5 (D). Many similarities have been noted previously between Petralona and Kabwe, but also some differences. The latter include a more projecting nasal spine and puffy maxilla, indicating increased midfacial prognathism; reduced lower face prognathism; and an oblique orientation of the maxilla lateral to the nasal aperture and the zygomatic process in Petralona relative to the African specimen.[45] All of these features are considered to be incipient manifestations of Neanderthal autapomorphic traits. Several similarities between Petralona and Sima 5, as well as between Sima 5 and the Neanderthals, have also been noted, also regarding incipient Neanderthal traits.[46]

Examination of the fitted configurations of these specimens allows for a visual evaluation of these traits after the specimens have been scaled to size. When Kabwe is compared to Petralona, the greatest difference is the reduced alveolar prognathism in the latter. Petralona also shows a slightly higher rhinion and lower nasion, resulting in a slightly higher nasal bridge, as well as infra-orbital foramina that are positioned more anteriorly, suggesting an inflated maxilla. The infra-orbital foramina of Kabwe are more anteriorly positioned than those of modern humans but less than those of Petralona, Sima 5, and Neanderthals. Even though Petralona has a less projecting glabella and a shorter face, it has a more projecting midface, in keeping with the above descriptions. When Petralona is compared to the mean Neanderthal configuration, it can be seen that its midface is nevertheless much less projecting, with rhinion and nasospinale much more posteriorly positioned. The nasal bridge, although shorter, is more or less parallel to the Neanderthal one.

In the morphology of the cheekbone, however, both Petralona and Kabwe contrast sharply with the condition observed in modern humans. The latter show a coronally oriented lateral part of the maxilla and medial part of the zygomatic bone. This results in a relatively sharply angled zygomatic bone. Neanderthals, on the other hand, are described as having a lateral maxilla and a zygomatic that sweep posteriorly (more sagitally rather than coronally oriented face),[47] without a sharply angling zygomatic bone. Petralona has been described as showing Neanderthal-like morphology in this respect.[48] When compared to modern humans, it can clearly be seen that the line linking the infra-orbital foramen to zygomaxillare is coronally oriented in modern humans and more sagittaly oriented in Petralona and in Neanderthals, reflecting an oblique zygomatic process in the latter, a proposed Neanderthal-derived trait. However, this condition is also found in Kabwe, even though the maxilla is more posteriorly positioned (not inflated). In fact, in this analysis, the *orientation* of the lateral face and the angle between the face and the zygomatic arch, insofar as they can be captured by the landmarks employed, are almost indistinguishable between Neanderthals and Kabwe. This is inconsistent with previous views that see this feature as a Neanderthal autapomorphic trait, whose incipient develop-

45. Wolpoff 1980, p. 347; Hublin 1998, p. 299.
46. Arsuaga et al. 1997, pp. 219–281.
47. Hublin 1998, p. 299.
48. Hublin 1998, p. 299.

ment in Petralona would constitute evidence for a European evolutionary lineage leading to Neanderthals from the Middle Pleistocene onward. If this trait, incipient or otherwise, is present in Kabwe, an African Middle Pleistocene fossil, then it cannot be regarded as a Neanderthal-derived feature, but is most likely an ancestral condition to both Neanderthals and modern humans.

When compared to Petralona and Kabwe, Sima 5 shows an even greater degree of midfacial projection, with both rhinion and nasospinale placed more anteriorly and the palate rotated upward, as in the Neanderthals. This is consistent with previous descriptions.[49] Although Sima 5 shows a level of midfacial projection almost equivalent to the Neanderthals, this specimen is also much more prognathic in the lower face. In this respect it is similar to Petralona and to a lesser extent Kabwe. Sima 5 also shares with Neanderthals and the other two fossil specimens the oblique orientation of the face and the wide angle between the face and the zygomatic arch. Although it has been described as having a horizontally concave infra-orbital plate,[50] differing in this respect from Neanderthals, the position of its infra-orbital foramen as recorded here was very forward and similar to Petralona, Neanderthals, and to a lesser extent Kabwe. The nasal apertures of both Sima 5 and Petralona were very broad and similar to those of the Neanderthals, but wider than Kabwe. The nasal aperture of the latter specimen was most similar in shape to the modern human sample.

In terms of the vault, Petralona, along with Kabwe, Sima 5, and the Neanderthals, displays the plesiomorphic configuration of a low, elongated braincase. All of the fossil specimens differ from modern humans, who show a very high position of bregma, lambda, and the post-toral sulcus point, a low and anteriorly placed inion, and more lateral anterior pteria. These differences describe the high, round, and not receding vault in modern humans, a derived condition for our species. Neanderthals differ from all three Middle Pleistocene specimens in having a higher lambda, indicating a higher posterior part of the vault. Petralona, and to a lesser extent Kabwe, additionally differ in their more posteriorly projecting inion. Sima 5 resembles Neanderthals most in its posterior vault morphology as reflected in the landmarks used, having a higher lambda than Petralona and Kabwe. The degree of projection of inion relative to the occipital squama, as well as the presence of an occipital "bun," a proposed Neanderthal-derived trait, cannot be evaluated from the landmarks used in this study. In the frontal bone, Sima 5 differs from Neanderthals and even the other two Middle Pleistocene specimens in having a lower position of the supra-torus point, suggesting a more receding frontal squama.

Inspection of the specimen landmark configurations and comparisons among samples confirmed the somewhat greater degree of midfacial prognathism in Petralona, suggesting some degree of Neanderthal affinities. Midfacial projection is even greater in Sima 5. The oblique orientation of the maxilla and the zygomatic in Neanderthals and the Middle Pleistocene European specimens, although shared, is also present in Kabwe, suggesting that this condition may in fact be plesiomorphic rather than derived for Neanderthals. All three Middle Pleistocene specimens share the plesiomorphic condition of the shape of the vault and differ from modern humans, but also from Neanderthals.

49. Arsuaga et al. 1997, pp. 219–281.
50. Arsuaga et al. 1997, pp. 219–281.

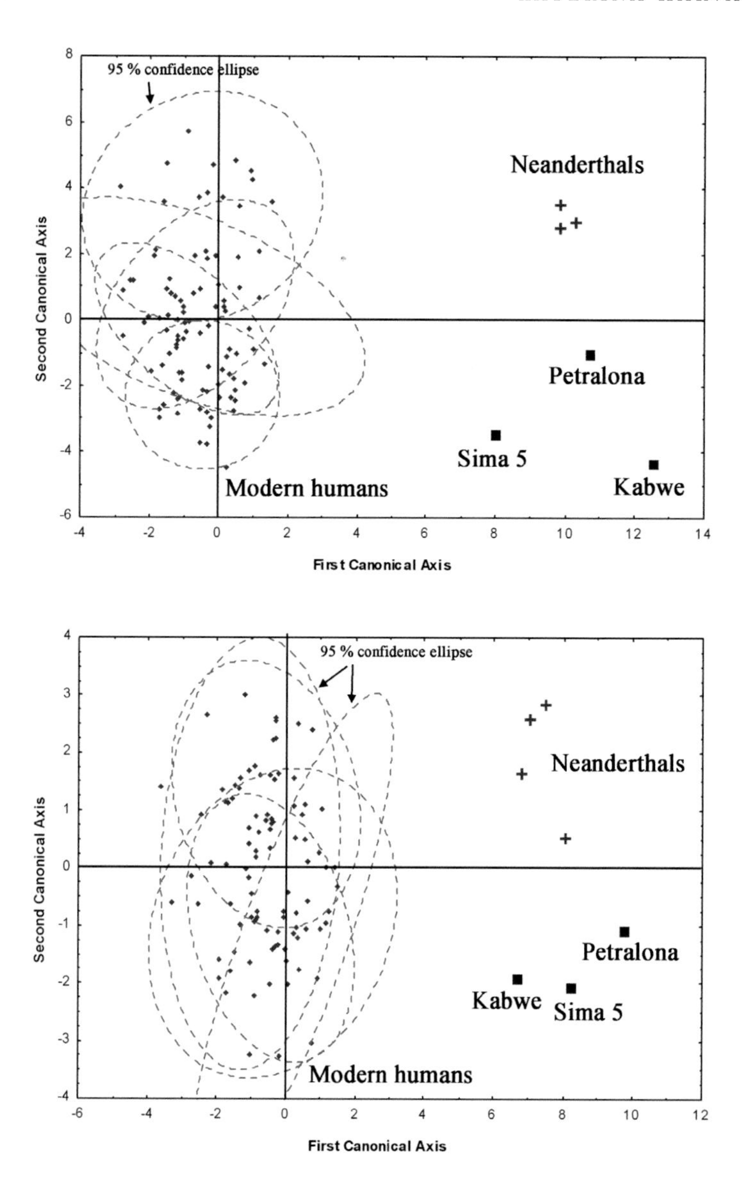

Figure 2.4. Canonical variates analysis, axes 1 and 2: facial landmarks analysis *(top)*; neurocranial landmarks analysis *(bottom)*

Analysis of Facial Landmarks

The analysis of facial landmarks used as variables the first 18 principal components calculated on the basis of the facial landmark coordinates. They accounted for 85.94% of the total variance. The first canonical axis explained 42.79% of the total variance and separated modern humans from all fossil specimens (Fig. 2.4). The mean score of the modern human sample was significantly different from the mean Neanderthal score and from the scores of the three Middle Pleistocene specimens. Furthermore, Kabwe was significantly different from Sima 5 along this axis, but not from Neanderthals or Petralona. The second canonical axis explained 29.11% of the total variance. It separated Neanderthals from all the Middle Pleistocene fossil humans. The mean Neanderthal score on this axis was significantly different from that of all the other fossils. Petralona also differed from Kabwe, but not from Sima 5.

TABLE 2.3. MAHALANOBIS SQUARED DISTANCES AMONG MODERN AND FOSSIL HUMAN SAMPLES

	Kabwe	*Petralona*	*Sima 5*	*Neanderthals*	*Andamanese*	*Chinese*	*East African*	*Near Eastern*	*Thai*
Facial Landmarks									
Kabwe	0.00								
Petralona	36.41	0.00							
Sima 5	51.47	25.64	0.00						
Neanderthals	109.08	102.14	104.25	0.00					
Andamanese	171.95	149.92	87.34	138.72	0.00				
Chinese	193.71	149.18	103.46	124.91	11.58	0.00			
E. African	191.42	154.86	103.91	147.51	17.38	12.46	0.00		
Nr. Eastern	217.98	157.71	123.36	120.83	29.99	13.98	18.54	0.00	
Thai	202.26	154.98	108.86	131.32	10.79	4.09	23.68	21.28	0.00
Neurocranial Landmarks									
Kabwe	0.00								
Petralona	15.43	0.00							
Sima 5	11.61	7.89	0.00						
Neanderthals	17.74	40.18	23.73	0.00					
Andamanese	58.01	114.49	83.81	59.60	0.00				
Chinese	49.10	108.03	71.24	61.67	6.59	0.00			
E. African	45.90	116.45	80.17	65.52	12.44	12.66	0.00		
Nr. Eastern	47.19	106.14	72.67	62.37	5.56	3.18	10.37	0.00	
Thai	58.82	116.72	85.09	69.14	8.47	5.33	27.89	7.45	0.00

The shape differences along the first canonical axis included the following: All fossil specimens showed a longer and more forwardly projecting lower part of the face, a more projecting and thicker supra-orbital torus, narrow and posteriorly sloping cheekbones (as also described above), and a more forward position of the landmarks around the nose. Modern humans, in contrast, showed a much shorter lower face, a flat and thin brow ridge, wide and coronally oriented cheekbones, and a more posteriorly positioned nasal area.

The shape differences along the second canonical axis characterized Neanderthals relative to Middle Pleistocene fossils. Neanderthals showed a very strong projection of the top of the nose and a flat lower face. The Middle Pleistocene specimens were characterized by a more subdued nose and a forwardly projecting lower face. These differences were less pronounced in Petralona, which fell nearest to the Neanderthals along this axis.

The Mahalanobis squared distances among samples are reported in Table 2.3. All three Middle Pleistocene specimens are very distant from both Neanderthals and modern humans, although Kabwe and Petralona are closer to Neanderthals than they are to modern humans. Neanderthals are also very distant from both modern humans and all Middle Pleistocene

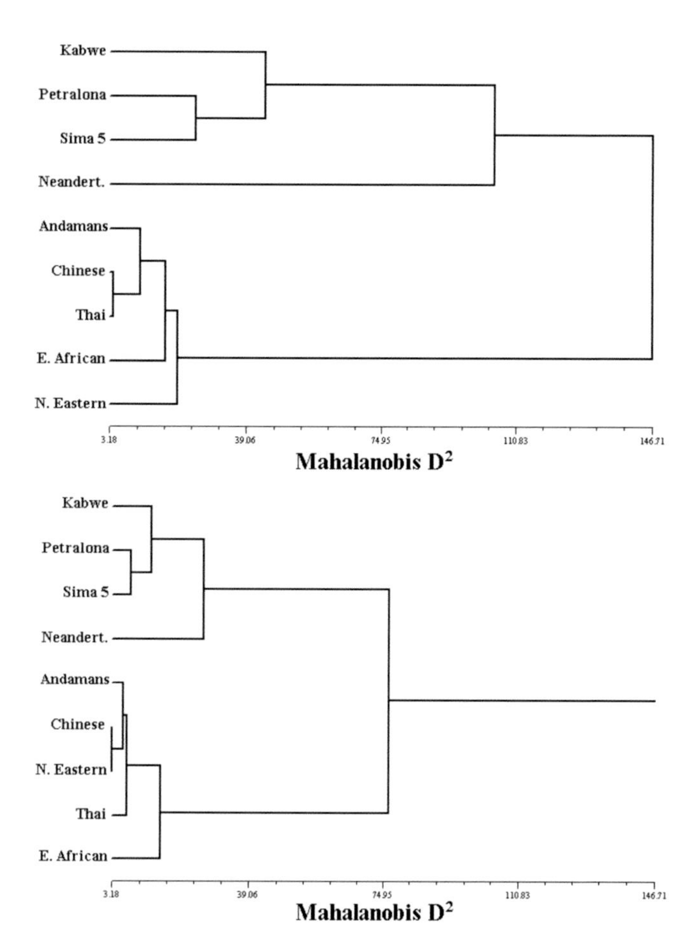

Figure 2.5. Cluster analysis of the Mahalanobis squared distances: facial landmarks analysis *(top)*; neurocranial landmarks analysis *(bottom)*

specimens, but closer to the latter. Finally, Petralona, Kabwe, and Sima 5 are relatively close to each other, the smallest distance being between Petralona and Sima 5. The cluster tree calculated from the facial landmark distances is shown in Figure 2.5 (top). The three Middle Pleistocene specimens cluster together, with Petralona closer to Sima 5. Neanderthals cluster with the other fossils, but are quite distant from them. Modern humans are on a completely different branch. In the discriminant analysis, all fossils were correctly classified and no fossil human was classified as modern or vice versa. Among the modern human groups, classification success was somewhat higher than in the neurocranial landmarks analysis, ranging from 100% to 80%.

Analysis of Neurocranial Landmarks

This analysis used as variables the first 13 principal components calculated on the basis of the neurocranial landmark coordinates. Together they accounted for 85.37% of the total variance. The first canonical axis accounted for 52.09% of the total variance and again separated all modern humans from all fossil samples (Fig. 2.4). The mean score of the modern human sample was significantly different along this axis from the Neanderthal mean score, as well as from the scores of all three Middle Pleistocene specimens. Petralona and Sima 5 differed significantly from Kabwe, but not from each other or from Neanderthals along this axis. Kabwe also did not differ significantly from Neanderthals. The third canonical axis separated Neanderthals from the other fossil specimens. Here Neanderthals were significantly different from Kabwe and Sima 5, but not from Petralona. The latter was also not significantly different from the other two Middle Pleistocene specimens.

The shape differences (Fig. 2.4) separating modern humans from all fossils included the greater height and width of the braincase in modern humans and its more rounded shape. All fossils were characterized by a lower, narrower, and more elongated braincase with a posteriorly projecting inion and anteriorly projecting glabella. The shape differences that separated Neanderthals from all three Middle Pleistocene specimens included a somewhat higher braincase and less posteriorly projecting inion in the former.

Mahalanobis squared distances for the neurocranial analysis are reported in Table 2.3. Results differ from those of the facial analysis in that the Middle Pleistocene specimens are very distant from modern humans but relatively close to Neanderthals, with Petralona being the most distant of the three to the latter group and Kabwe the closest. Neanderthals are very distant from all modern human groups, but closer to the Middle Pleistocene sample. Finally, the three Middle Pleistocene samples are very close to each other; the shortest distance is between Petralona and Sima 5. The UPGMA (Unweighted Pair Group Method with Arithmetic mean) and neighbor joining cluster trees calculated from these distances are shown in Figure 2.5 (bottom). It is remarkably similar to that obtained using the facial measurements. The greatest difference is that Neanderthals and Middle Pleistocene fossils cluster together more tightly. This is probably due to the plesiomorphic nature of the aspects of their neurocranial morphology captured in this analysis. As reflected by the landmarks used, the Neanderthal vault retains the major primitive traits found in the earlier specimens. The Middle Pleistocene specimens are also more tightly clustered, with Petralona again clustering with Sima 5 rather than Kabwe. In the discriminant analysis, all the fossil specimens were classified correctly when using posterior probabilities resubstitution classification. No fossil human was misclassified as modern and vice versa. Classification success within modern humans ranged from 87.5% to 65%.

DISCUSSION AND CONCLUSIONS

The major finding of this analysis is the close morphological similarity between the two European (Petralona and Sima 5) and the one African (Kabwe) Middle Pleistocene humans examined here. This similarity was obvious both in the neurocranial anatomy and in the morphology of the face, despite the presence of certain similarities to the Neanderthals in the facial region of Petralona and especially Sima 5. Neither of the European specimens was closer to Neanderthals than Kabwe was, as would be expected from the predictions of the accretion hypothesis. Instead, the three specimens emerge almost as a unit in both analyses. This preliminary finding supports the view of the African and European Middle Pleistocene fossil human sample as a single taxon, ancestral to both Neanderthals and modern humans. It does not support the placement of the European specimens within *H. neanderthalensis*.

Human evolution in Europe is often thought to have been shaped by a series of alternating episodes of isolation from and contact with the Near East and Africa.[51] The close morphological affinities between the African and European Middle Pleistocene fossil human specimens points to such contact between African and European populations. These populations later diverged in their evolutionary paths, the European one leading to Neanderthals and the African ones to modern humans. Greece and the Balkans are situated at the crossroads of Africa, Europe, and the Near East. This area has historically been the gateway into the continent through which migrant populations have repeatedly passed. Furthermore, its place among the three major southern European peninsulas has led some authors to consider it a refugium of comparatively mild climatic conditions during glacial episodes.[52] This geographic position, therefore, makes the fossil human record from this area ideal to test hypotheses about the timing and nature of population contact and about the course of Neanderthal and modern human evolution in Europe.

Nonetheless, Greek paleolithic archaeology and paleoanthropology remain largely unexplored.[53] There are no fossil human finds from the country dating to the Early Pleistocene or to the early Middle Pleistocene, and the Lower Paleolithic stone tool industry is virtually unknown.[54] The middle and late parts of the Middle Pleistocene are relatively well represented, with Petralona, as well as the two well-preserved, albeit little-known, crania from Apidima.[55] Despite the possible role of the country as a refugium during glacial times, the presence of Neanderthals in Greece was only recently confirmed with the findings from Lakonis, and possibly also Kalamakia, both situated in the Mani area.[56] Finally, despite the position of Greece on the most commonly hypothesized course of migration of early modern humans from Africa into Europe around 40,000 years ago,[57] no early modern human specimens are known. The Theopetra Upper Paleolithic skeleton is the earliest modern human in Greece, dating as late as 16,620–16,380 before present.[58] It has long been suspected that the rarity of such findings reflects the lack of research in this region rather than the absence of such material.[59] This suspicion has been confirmed by the important strides achieved in the field in the last few years.[60]

This preliminary analysis of the Petralona specimen is a first example of the importance of the Greek paleoanthropological record in elucidating the course of events of human evolution in Europe. It points to several directions for future research, including the further quantitative assessment of the shape of proposed unique Neanderthal features, the evaluation of the polarity of these traits, and the analysis of an extended and more finely divided sample of fossil human specimens. Finally, further paleoanthropological and Paleolithic field research in southeastern Europe is essential for a deeper understanding of both Greek and European prehistory.

51. See Finlayson 2004.

52. See Gibbons 2001, pp. 1725–1729.

53. Darlas 1994, pp. 305–328; 1995, pp. 51–57; Runnels 1995, 2001; Otte et al. 1998, pp. 413–431; Bar-Yosef 2000, pp. 9–18; Bailey et al. 1999; Panagopoulou et al. 2001, pp. 121–151.

54. Runnels 1995, pp. 699–728; 2001, pp. 225–258; Runnels and van Andel 1999, pp. 215–220; Runnels et al. 1999, pp. 120–129.

55. Harvati and Delson 1999, pp. 343–348.

56. Harvati et al. 2003, pp. 465–473; Darlas and DeLumley 1998, pp. 655–661.

57. Runnels 1995, pp. 699–728; 2001, pp. 225–258; Bar-Yosef 1998, pp. 141–163.

58. Stravopodi et al. 1999, pp. 271–281.

59. E.g., Runnels 1995, pp. 699–728; 2001, pp. 225–258.

60. See, e.g., Runnels 1995, pp. 699–728; 2001, pp. 225–258; Darlas and DeLumley 1998, pp. 655–661; Bailey et al. 1999; Panagopoulou et al. 2001, pp. 121–151; Harvati et al. 2003, pp. 465–473; Panagopoulou et al. 2004.

REFERENCES

Ahlström, T. 1996. "Sexual Dimorphism in Medieval Human Crania Studied by Three-Dimensional Thin-Plate Spline Analysis," in *Advances in Morphometrics,* ed. L. F. Marcus, M. Corti, A. Loy, G. J. P. Naylor, and D. E. Slice, New York, pp. 415–421.

Arsuaga, J. L., I. Martinez, A. Gracia, and C. Lorenzo. 1997. "The Sima de los Huesos Crania (Sierra de Atapuerca, Spain). A Comparative Study," *Journal of Human Evolution* 33, pp. 219–281.

Bailey G., E. Adam, E. Panagopoulou, C. Perlès, and K. Zachos, eds. 1999. *The Palaeolithic Archaeology of Greece and Adjacent Areas: Proceedings of the ICOPAG Conference, Ioannina, September 1994,* London.

Bar-Yosef, O. 1998. "The Nature of Transitions: The Middle to Upper Paleolithic and the Neolithic Revolution," *CAJ* 8, pp. 141–163.

———. 2000. "A Mediterranean Perspective on the Middle/Upper Paleolithic Revolution," in *Neanderthals on the Edge,* ed. C. B. Stringer, R. N. E. Barton, and J. C. Finlayson, Oxford, pp. 9–18.

Bennett, S., and D. Bowers. 1976. *An Introduction to Multivariate Techniques for Social and Behavioural Sciences,* New York.

Bermúdez de Castro J. M., J. L. Arsuaga, E. Carbonell, A. Rosas, I. Martinez, and M. Mosquera. 1997. "A Hominid from the Lower Pleistocene of Atapuerca, Spain: Possible Ancestor to Neandertals and Modern Humans," *Science* 276, pp. 1392–1395.

Bookstein, F. L. 1990. "Introduction to Methods for Landmark Data," in *Proceedings of the Michigan Morphometrics Workshop,* ed. F. J Rohlf and F. L. Bookstein, Ann Arbor, pp. 216–225.

Bookstein, F., K. Schäfer, H. Prossinger, H. Seidler, M. Fieder, C. Stringer, G. Weber, J.-L. Arsuaga, D. Slice, J. Rohlf, W. Recheis, A. J. Mariam, and L. F. Marcus. 1999. "Comparing Frontal Cranial Profiles in Archaic and Modern *Homo* by Morphometric Analysis," *Anatomical Record* 257, pp. 217–224.

Bruner, E., G. Manzi, and J. L. Arsuaga. 2004. "Encephalization and Allometric Trajectories in the Genus *Homo:* Evidence from the Neandertal and Modern Lineages," *Proceedings of the National Academy of Sciences* 100, pp. 15335–15340.

Collard, M., and P. O'Higgins. 2000. "Ontogeny and Homoplasy in the Papionin Monkey Face," *Evolution and Development* 3, pp. 322–331.

Darlas, A. 1994. "Le paléolithique inférieur et moyen de Grèce." *L'anthropologie* (Paris) 98, pp. 305–328.

———. 1995. "The Earliest Occupation of Europe: The Balkans," in *The Earliest Occupation of Europe,* ed. W. Roebroeks and T. van Kolfschoten, Leiden, pp. 51–57.

Darlas, A., and H. de Lumley. 1998. "Fouilles franco-helléniques de la grotte de Kalamakia (Aréopolis; Péloponnèse)," *BCH* 122, pp. 655–661.

Dean, D. 1993. "The Middle Pleistocene *Homo erectus/Homo sapiens* Transition: New Evidence from Space Curve Statistics" (diss. City Univ. of New York).

Dean D., J.-J. Hublin, R. Holloway, and R. Ziegler. 1998. "On the Phylogenetic Position of the Pre-Neanderthal Specimen from Reilingen, Germany," *Journal of Human Evolution* 34, pp. 485–508.

de Bonis, L., and J. Melentis. 1991. "Age et position phylétique du Crâne de Petralona (Grèce)," in *Les premiers Européens: Actes du 114e Congrès national des Sociétés Savantes,* ed. E. Bonifay and B. Vandermeersch, Paris, pp. 285–289.

Delson, E., I. Tattersall, J. A. Van Couvering, and A. S. Brooks. 2000. *Encyclopedia of Human Evolution and Prehistory,* New York.

Delson, E., K. Harvati, D. Reddy, L. F. Marcus, K. Mowbray, G. J. Sawyer, T. Jacob, and S. Marquez. 2001. "Sambungmachan 3 *Homo erectus* calvaria: A Comparative Morphometric and Morphological Analysis," *Anatomical Record* 262, pp. 380–297.

Finlayson, C. 2004. *Neanderthals and Modern Humans.* Cambridge.

Frost, S. R., L. F. Marcus, F. Bookstein, D. P. Reddy, and E. Delson. 2003. "Cranial Allometry, Phylogeography and Systematics of Large Bodied Papionins (Primates: Cercopithecinae) Inferred from Geometric Morphometric Analysis of Landmark Data," *Anatomical Record* 275A, pp. 1048–1072.

Gamble, C. 1999. *The Palaeolithic Societies of Europe,* Cambridge.

Gibbons, A. 2001. "The Riddle of Coexistence," *Science* 291, pp. 1725–1729.

Grün, R. 1996. "A Re-Analysis of Electron Spin Resonance Dating Results Associated with the Petralona Hominid," *Journal of Human Evolution* 30, pp. 227–241.

Harvati, K. 2002. "Models of Shape Variation between and within Species and the Neanderthal Taxonomic Position: A 3D Geometric Morphometrics Approach," in *Three Dimensional Imaging in Paleoanthropology and Prehistoric Archaeology* (*BAR-IS* 1049), ed. B. Mafart and H. Delingette, Oxford, pp. 25–30.

———. 2003a. "The Neanderthal Taxonomic Position: Models of Intra- and Inter-Specific Craniofacial Variation," *Journal of Human Evolution* 44, pp. 107–132.

———. 2003b. "Quantitative Analysis of Neanderthal Temporal Bone Morphology Using Three-Dimensional Geometric Morphometrics," *American Journal of Physical Anthropology* 120, pp. 323–328.

Harvati, K., and E. Delson. 1999. "Conference Report: Paleoanthropology of the Mani Peninsula (Greece)," *Journal of Human Evolution* 36, pp. 343–48.

Harvati, K., S. R. Frost, and K. P. McNulty. 2004. "Neanderthal Taxonomy Reconsidered: Implications of 3D Primate Models of Intra- and Interspecific Differences," *Proceedings of the National Academy of Sciences* 101, pp. 1147–1152.

Harvati, K., E. Panagopoulou, and P. Karkanos. 2003. "First Neanderthal Remains from Greece: The Evidence from Lakonis," *Journal of Human Evolution* 45, pp. 465–473.

Harvati, K., and T. D. Weaver. 2006. "Reliability of Cranial Morphology in Reconstructing Neanderthal Phylogeny," in *Neanderthals Revisited: New Approaches and Perspectives,* ed. K. Harvati and T. Harrison, New York, pp. 239–254.

Hennessy, R. J., and C. B. Stringer. 2002. "Geometric Morphometric Study of the Regional Variation of Modern Human Craniofacial Form," *American Journal of Physical Anthropology* 117, pp. 37–48.

Hennig, G. J., W. Herr, E. Weber, and N. I. Xirotiris. 1981. "ESR-Dating of the Fossil Hominid Cranium from Petralona Cave, Greece," *Nature* 292, pp. 533–536.

Howells, W. W. 1967. *Mankind in the Making,* Garden City.

———. 1973. *Cranial Variation in Man: A Study by Multivariate Analysis of Patterns of Difference among Recent Human Populations* (Papers of the Peabody Museum of Archaeology and Ethnology 67), Boston.

Hublin J.-J. 1998. "Climatic Changes, Paleogeography and the Evolution of Neanderthals," in *Neanderthals and Modern Humans in Western Asia,* ed. T. Akazawa, K. Aoki, and O. Bar-Yosef, New York, pp. 295–310.

Kokkoros P., and A. Kanellis. 1960. "Découverte d'un crâne d'homme paléolithique dans la péninsule Chalchidique," *L'anthropologie* 64, pp. 438–446.

Krings, M., C. Capelli, F. Tschentscher, H. Geisert, S. Meyer, A. von Haeseler, K. Grossschmidt, G. Possnert, M. Paunovic, and S. Pääbo 2000. "A View of Neandertal Genetic Diversity," *Nature Genetics* 26, pp. 144–146.

Krings, M., A. Stone, R. W. Schmitz, H. Krainitzki, M. Stoneking, and S. Pääbo. 1997. "Neandertal DNA Sequences and the Origin of Modern Humans," *Cell* 90, pp. 19–30.

Kurten, B., and A. N. Poulianos. 1977. "New Stratigraphic and Faunal Material from Petralona Cave, with Special Reference to the Carnivora," *Anthropos* 4, pp. 47–130.

Lestrel, P. E. 2000. *Morphometrics for the Life Sciences,* River Edge, N.J.

Manzi G., F. Mallegni, and A. Ascenzi. 2001. "A Cranium for the Earliest Europeans: Phylogenetic Position of the Hominid from Ceprano, Italy," *Proceedings of the National Academy of Sciences* 98, pp. 10011–10016.

Marcus, L. F. 1993. "Some Aspects of Multivariate Statistics for Morphometrics," in *Contributions to Morphometrics,* ed. L. F. Marcus, E. Bello, and A. García-Valdecasas, Madrid, pp. 99–130.

Mellars, P. 2002. "Archaeology and the Origins of Modern Humans: European and African Perspectives," in *The Speciation of Modern Homo sapiens,* ed. T. J. Crow, Oxford, pp. 31–47.

Murrill, R. I. 1975. "A Comparison of the Rhodesian and Petralona Upper Jaws in Relation to Other Pleistocene Hominids," *Zeitschrift für Morphologie und Anthropologie* 66, pp. 176–187.

———. 1981. *Petralona Man: A Descriptive and Comparative Study, with New Important Information on Rhodesian Man,* Springfield, Ill.

Nicholson, E., and K. Harvati. 2006. "Quantitative Analysis of Modern Human and Fossil Mandibles Using 3-D Geometric Morphometrics," *American Journal of Physical Anthropology* 126, pp. 368–383.

O'Higgins, P. 2000. "The Study of Morphological Variation in the Hominid Fossil Record: Biology, Landmarks and Geometry," *Journal of Anatomy* 197, pp. 103–120.

O'Higgins, P., and N. Jones. 1998. "Facial Growth in *Cercocebus torquatus:* An Application of Three-Dimensional Geometric Morphometric Techniques to the Study of Morphological Variation," *Journal of Anatomy* 193, pp. 251–272.

Otte, M., I. Yalçinkaya, J. K. Kozlowski, O. Bar-Yosef, I. López Bayón, and H. Taskiran. 1998. "Long-Term Technical Evolution and Human Remains in the Anatolian Paleolithic," *Journal of Human Evolution* 34, pp. 413–431.

Panagopoulou, E., E. Kotjabopoulou, and P. Karkanas. 2001. "Geoarchaeological Research in Alonnissos: New Evidence on the Palaeolithic and the Mesolithic in the Aegean," in *Archaeological Research in the Northern Sporades,* ed. A. Sampson, Alonnissos, pp. 121–151.

Pearson, O. M. 2000. "Postcranial Remains and the Origin of Modern Humans," *Evolutionary Anthropology* 9, pp. 229–247.

Ponce de León, M., and C. Zollikofer. 2001. "Neanderthal Cranial Ontogeny and Its Implications for Late Hominid Diversity," *Nature* 412, pp. 534–538.

Poulianos, A. N. 1981. "Pre-Sapiens Man in Greece," *CurrAnthr* 22, pp. 187–288.

Poulianos, A. N., Y. Liritzis, and M. Ikeya. 1982. "Petralona Cave Dating Controversy," *Nature* 299, pp. 280–282.

Rightmire, G. P. 1997. "Deep Roots for the Neanderthals," *Nature* 389, pp. 917–918.

———. 1998. "Human Evolution in the Middle Pleistocene: The Role of *Homo heidelbergensis*," *Evolutionary Anthropology* 6, pp. 218–227.

———. 2001. "Patterns of Hominid Evolution and Dispersal in the Middle Pleistocene," *Quaternary International* 75, pp. 77–84.

Roebroeks, W. 2001. "Hominid Behaviour and the Earliest Occupation of Europe: An Exploration," *Journal of Human Evolution* 41, pp. 437–461.

Rohlf, J. F. 1990. "Rotational Fit (Procrustes) Methods," in *Proceedings of the Michigan Morphometrics Workshop,* ed. F. J. Rohlf and F. L. Bookstein, Ann Arbor, pp. 227–236.

Rohlf, J. F., and L. F. Marcus. 1993. "A Revolution in Morphometrics," *Trends in Ecology and Evolution* 8, pp. 129–132.

Rosas, A., and M. Bastir. 2002. "Thin-Plate Spline Analysis of Allometry and Sexual Dimorphism in the Human Craniofacial Complex," *American Journal of Physical Anthropology* 117, pp. 236–245.

Runnels, C. 1995. "Review of Aegean Prehistory IV: The Stone Age of Greece from the Palaeolithic to the Advent of the Neolithic," *AJA* 99, pp. 699–728.

———. 2001. "Review of Aegean Prehistory IV: The Stone Age of Greece from the Palaeolithic to the Advent of the Neolithic with an Addendum," in *Aegean Prehistory: A Review,* ed. T. Cullen, Boston, pp. 225–258.

Runnels, C., and T. H. van Andel. 1999. "The Paleolithic in Larisa, Thessaly," in *The Palaeolithic Archaeology of Greece and Adjacent Areas: Proceedings of the ICOPAG Conference, Ioannina, September 1994,* ed. G. Bailey, E. Adam, E. Panagopoulou, C. Perlès, and K. Zachos, London, pp. 215–220.

Runnels C., T. H. van Andel, K. Zachos, and P. Paschos. 1999. "Human Settlement and Landscape in the Preveza Region, Epirus, in the Pleistocene and Early Holocene," in *The Palaeolithic Archaeology of Greece and Adjacent Areas: Proceedings of the ICOPAG Conference, Ioannina, September 1994,* ed. G. Bailey, E. Adam, E. Panagopoulou, C. Perlès, and K. Zachos, London, pp. 120–129.

Schwarcz, H. P., Y. Liritzis, and A. Dixon. 1980. "Absolute Dating of Travertines from the Petralona Cave, Khalkidiki, Greece," *Anthropos* 7, pp. 152–167.

Schwartz, J. H., and I. Tattersall. 2005. *The Human Fossil Record,* 4: *Craniodental Morphology of Early Hominids (Genera Australopithecus, Paranthropus, Orrorin) and Overview,* New York.

Serre, D., A. Langaney, M. Chech, M. Teschler-Nicola, M. Paunovic, P. Mennecier, M. Hofreiter, G. Possnert, and S. Pääbo. 2004. "No Evidence of Neandertal mtDNA Contribution to Early Modern Humans," *PLoS Biology* 2, p. 57

Singleton, M. 2002. "Patterns of Cranial Shape Variation in the Papionini (Primates: Cercopithecinae)," *Journal of Human Evolution* 42, pp. 547–578.

Slice, D. E. 1992. *GRF-ND: Generalized Rotational Fitting of n-Dimensional Landmark Data,* Department of Ecology and Evolution, State University of New York.

———. 1994–1999 Copyright. *Morheus et al.: Software for Morphometric Research,* Department of Ecology and Evolution, State University of New York.

———. 1996. "Three-Dimensional Generalized Resistance Fitting and the Comparison of Least-Squares and Resistant Fit Residuals," in *Advances in Morphometrics,* ed. L. F. Marcus, M. Corti, A. Loy, G. J. P. Naylor, and D. E. Slice, New York, pp. 179–199.

Stravopodi, E., S. Manolis, and N. Kyparissi-Apostolika. 1999. "Paleoanthropological Findings from Theopetra Cave in Thessaly: A Preliminary Report," in *The Palaeolithic Archaeology of Greece and Adjacent Areas: Proceedings of the ICOPAG Conference, Ioannina, September 1994,* ed. G. Bailey, E. Adam, E. Panagopoulou, C. Perlès, and K. Zachos, London, pp. 271–281.

Stringer, C. B. 1974. "A Multivariate Study of the Petralona Skull," *Journal of Human Evolution* 3, pp. 397–404.

Stringer, C. B., F. C. Howell, and J. Melentis. 1979. "The Significance of the Fossil Hominid Skull from Petralona, Greece," *JAS* 6, pp. 235–253.

Trinkaus, E. 1983. *The Shanidar Neanderthals,* New York.

Tsoukala, E. 1990. "Contribution to the Study of the Pleistocene Fauna of Large Mammals (Carnivora, Perissodactyla, Artiodactyla) from Petralona Cave, Chalkidiki (N. Greece)" (diss. Aristotle Univ. of Thessaloniki).

Valeri, C. J., T. H. Cole III, S. Lele, and J. T. Richtsmeier. 1998. "Capturing Data from Three-Dimensional Surfaces Using Fuzzy Landmarks," *American Journal of Physical Anthropology* 107, pp. 113–124.

Wolpoff, M. H. 1980. "Cranial Remains of Middle Pleistocene European Hominids," *Journal of Human Evolution* 9, pp. 339–358.

Wood, C. G., and J. M. Lynch. 1996. "Sexual Dimorphism in the Craniofacial Skeleton of Modern Humans," in *Advances in Morphometrics,* ed. L. F. Marcus, M. Corti, A. Loy, G. J. P. Naylor, and D. E. Slice, New York, pp. 407–414.

Yaroch, L. A. 1996. "Shape Analysis Using the Thin-Plate Spline: Neanderthal Cranial Shape as an Example," *American Journal of Physical Anthropology* (Yearbook) 39, pp. 43–89.

CHAPTER 3

"In This Way They Held Funeral for Horse-Taming Hector": A Greek Cremation Reflects Homeric Ritual

by Philippe Charlier, Joël Poupon, Murielle Goubard, and Sophie Descamps

Figure 3.1. The Attic *lebes* BR 2590, second quarter of the 5th century B.C. (Musée du Louvre, Paris). Diam. 31 cm, H. 20 cm. Photo courtesy Réunion des Musées Nationaux/Art Resource, N.Y.

A vase in the Louvre Museum collections[1] (Fig. 3.1) found to contain cremated human remains is an Attic *lebes* (BR 2590) dating from the 5th century B.C. The vessel was discovered in the Athens neighborhood of Ambelokipi, and comes from the Rayet collection in France. On the surface of the vase is a text written in Greek:

Ἀθεναῖοι ἆθλ' ἐπὶ τοῖς ἐν τôι πο{π}λέμοι

This inscription (loosely translated as "the Athenians gave prizes for those fallen in war") is an official sentence written by magistrates during celebrations for funeral games. Some festivals took place after the Persian wars, such as the Epitaphia, the Herakleia, or the Eleutheria, and included athletic races. A bronze vase was then given to the winners that could be used, some years afterwards, as a funeral urn. It is therefore very probable that the bones in the vase were those of an athlete, similar to some other skeletons known archaeologically: one from Tarentum in south Italy (studied by Gaspare Baggieri and Marina Di Giacomo)[2] and another from Ayios Nikolaos in east Crete (yet unpublished but on display in the municipal archaeological museum).

MATERIALS AND METHODS

Our project, following previous works by Gejvall, Duday, and Grévin,[3] consisted of the extraction and analysis of all the remains contained in the vase, particularly the human bones.

The bones were carefully washed, then classified by anatomical region: skull, tooth, mandible, maxilla, foot, phalanges (foot and hand), ribs, pelvic bones, scapulae, vertebrae, long bones (superior limb, inferior limb, indeterminate limb), and indeterminate fragments. A weighing of all groups was then performed. The comparison of these data with tables of theoretical bone representation by Krogman[4] gave us important information about the funeral practice.

Concerning the anthropological data, the sex determination was made according to Bruzek's method.[5] The evaluation of the age at death was

1. This research was conducted under the authority of Alain Pasquier (Chief Conservator of the Department of Greek, Roman, and Etruscan Antiquities, Louvre Museum). The inscription is published as *IG* I² 524. With the exception of Fig. 3.1, all photographs are by P. Charlier.
2. Gaspare Baggieri and Marina Di Giacomo (pers. comm.).
3. Gejvall 1971; Duday 1990; Grévin 1990. See also Charlier 2008.
4. Krogman 1978.
5. Bruzek 2002.

achieved following Ferembach, Schwidetzky, and Stloukal[6] integrating age maturation, cranial synostosis, tooth abrasion, and degenerative disease. Last, the determination of the cremation temperature was possible using the method published by Shipman, Foster, and Schoeninger.[7]

RESULTS

Weight Data

The *lebes* contained at least 1.813 kg of human cremated bones. It has been shown that the mean total weight of a human cremated corpse is around 1.600 kg.[8] As the total weight of the human bones was somewhat greater than this theoretical weight and no skeletal elements were duplicated, we may conclude that the recovery of the anatomical pieces was essentially complete.

Numerous anatomical elements were identifiable, representing a total of 1.373 kg from the 1.813 kg of cremated bones, i.e., 75% of the material. The results of the weighing by anatomical regions are given in Table 3.1.

The total weight of the head fragments (skull, mandible, maxilla, and teeth) is 178 g. According to Krogman's tables,[9] this weight should be 20% of the total cremated body weight, i.e., a theoretical weight for our case should be 362 g (out of a total of 1,813 g). This means that a significant portion of the skull was not identifiable and that cranial fragments are very probably in the indeterminate bone pieces.

The total weight of the limbs is 985 g. According to Krogman's tables,[10] this weight should be 63% of the total cremated body, or a theoretical weight for our case of 1.142 kg. As our recovered material is similar in weight to the theoretical predicted amount, this suggests that very few bones from this anatomical region are missing. Once again, they are probably among the indeterminate bone pieces.

TABLE 3.1. WEIGHT OF MATERIAL BY ANATOMICAL REGION

Anatomical Region	*Weight (g)*
Skull	148
Dentition, mandible, and maxilla	30
Foot	30
Phalanges (hand and foot)	20
Ribs	105
Pelvic bones	75
Scapulae	15
Vertebrae	105
Long bones	695
Indeterminate long bones	150
Indeterminate fragments	440

Anthropological Data

All of the human remains are most likely part of only one individual, as indicated by the absence of redundant bones. No faunal pieces were found during the examination.

The sex determination was difficult due to the significant fragmentation of the skeleton during cremation and the poor state of preservation. The pelvic bones were particularly destroyed and useless for such a determination.[11] However, the great stature, robust bone structure, and marked muscular insertions (the existence of an enthesopathy on the lesser trochanter of the left femur is a sign of significant physical activity or stress on the hip joint) are important factors indicating a male individual—but we cannot be sure.[12]

The determination of age at death was also difficult due to the presence of conflicting evidence: The complete synostosis of epiphyseal surfaces indicates that the individual is an adult. The left pubic symphysis is entirely smooth with a posterior and anterior crest, corresponding to a stage 4 according to Acsàdi and Nemeskéri.[13] This stage indicates an age between

6. Ferembach, Schwidetzky, and Stloukal 1979.
7. Shipman, Foster, and Schoeninger 1984.
8. McKinley 1993.
9. Krogman 1978.
10. Krogman 1978.
11. Bruzek 2002.
12. Ferembach, Schwidetzky, and Stloukal 1979.
13. Acsàdi and Nemeskéri 1970.

Figure 3.2. Two pieces of cranial vault with patent and complex sutures. The bone is a grayish color due to the incineration of the body at a very early stage of putrefaction (hours or days after death). Scale 3:2

50 and 70 years old. No degenerative disease was present on the bones that are often affected by degenerative diseases after age 45 (particularly the vertebrae and the hip).

The patent and complex cranial sutures (Fig. 3.2) are a good argument in favor of a young age (young adult), even if this method is best applied to populations rather than individuals.

No antemortem tooth loss was observed on the six mandibular and six maxillary alveolar sockets still preserved. The occlusal abrasion was not visible because of enamel destruction during the cremation process. No temporo-mandibular degenerative disease was observed. All these observations are arguments for a young age of the subject.

To conclude, it seems logical to accept the evidence supporting a younger adult age, as only the pubic symphysis suggests a substantially older age. Two explanations exist for the appearance of the pubic bone that indicates an older age estimate: The high temperature of cremation may have changed the structural and morphological aspects of this bone (leveling the surface of the pubic symphysis). It may also be possible that this bone belongs to another subject cremated at the same place days, weeks, or months before, whose bone was recovered by accident. However, the absence of any other redundant bones could be an argument against this explanation. For us, these bones appear to belong to an adult, broadly aged between 25 and 45 years old, of undeterminable sex (but probably a male).

Cremation Mode

The subject was cremated as a fresh body, i.e., in the immediate hours or days after death, as shown by transversal fractures on the bone pieces.[14] Fragments were recovered without any secondary fragmentation.

Many fragments originating from the same anatomical region present different coloration: black as charcoal, white, and brown with a ligneous

14. Guillon 1986.

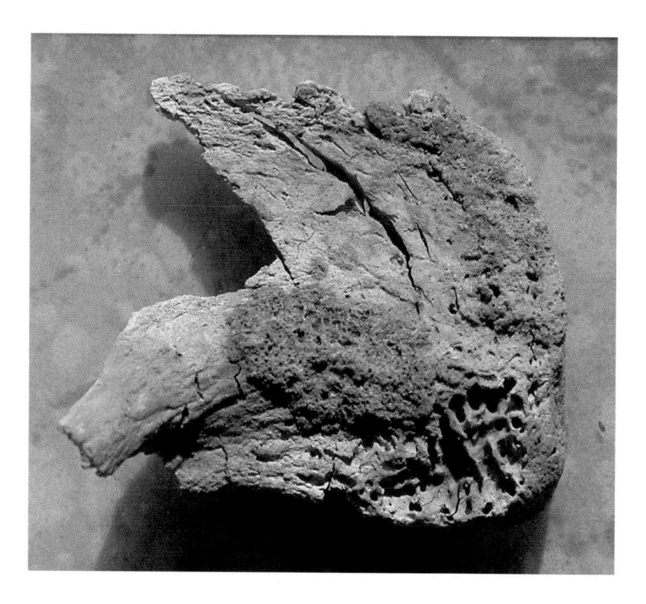

Figure 3.3. Traces of corrosion and fusion on the surface of a scapula. The metal item had a high composition of copper (maybe bronze?). Scale 3:1

Figure 3.4. Traces of corrosion and fusion of iron on the external surface of a rib. Scale 3:1

appearance. This diversity indicates the bursting of bone pieces with high temperature and differential combustion according to their place in the pyre. Cremation temperatures varied from 285° to 940°, according to the Shipman, Foster, and Schoeninger scale.[15]

Green and purple traces were seen on many bone fragments stemming from the fusion and corrosion of metals during the cremation. Moreover, iron items were present in the pyre during the cremation as indicated by brownish traces on the surface of bone fragments (Figs. 3.3, 3.4). A piece of sword 15 cm long was identified inside the vase and directly associated with the bones.

Many black centimeter-sized concretions were also recovered (Fig. 3.5), upon which elemental analysis was performed according to a methodology already applied to other archaeological substances (see below).[16]

A piece of charcoal 8 mm long was also identified, coming very probably from the wood used during the cremation (Fig. 3.6). It is much less probable that it belongs to an artifact made of wood placed in the pyre. The microscopic study by Stéphanie Thiébaut[17] indicated the species origin to be an olive tree.

15. Shipman, Foster, and Schoeninger 1984.

16. Charlier and Poupon 2003.

17. CNRS, Université de Paris 10-Nanterre.

Figure 3.5. Two black centimeter-sized concretions associated with the cremation, very friable and with a spongy aspect. Scale 4:1

Figure 3.6. The only fragment of charcoal recovered in the cremation. Scale 4:1

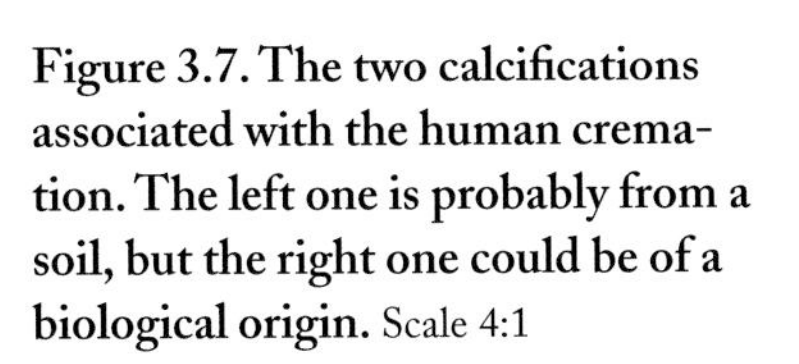

Figure 3.7. The two calcifications associated with the human cremation. The left one is probably from a soil, but the right one could be of a biological origin. Scale 4:1

Two mineral formations were associated with the human remains (Fig. 3.7). One is a white, regular, and smooth oval piece, 6 mm long, corresponding very probably to a limestone piece taken by accident during the recovery of the cremated bones. The other is a yellow polyhedric crystalline form, 5 mm long, which may be a calculus of undeterminable origin.[18]

Finally, very small fragments of bronze were found, originating from fragmentation of the *lebes*.

Elemental Analysis of Black Concretions Recovered in the Cremation

Thirty fragments, with lengths varying from 2–20 mm, were studied by two different laboratories in order to determine their elemental composition.[19]

18. Baud and Kramar 1990.

19. Joël Poupon from the Laboratoire de Biochimie et Toxicologie, Hôpital Lariboisière, Paris; and Murielle Goubard, Groupe Rhodia, France.

Indeed, a single macroscopic examination was not able to provide information on the origin of these formations. Were they organic—anatomical (pseudomorphic) carbonized fragments—or mineral (slag made of an alloy of different metallic fragments during cremation)? The association of significant concentrations of carbon, nitrogen, and oxygen with high levels of copper, silver, traces of gold and tin, and magnesium in these fragments confirms a mixed origin: mineral matter fused with organic material during the cremation, and then solidified. In such a case, the mineral objects could be jewelry (such as a fibula, ring, or belt buckle), weapons (sword, lance), or parts of the funeral offerings (a strigil, spit, or standard).

DISCUSSION

The *Iliad* offers us information on two cremations of famous heroes (Patroclus and Hector). Both descriptions partially repeat themselves, illustrating commonalties in the rituals.

> . . . but they who were nearest and dearest to the dead remained there, and heaped up the wood, and made a pyre of one hundred feet this way and that, and on the topmost part they set the dead man, inwardly grieving. And many noble sheep and many sleek cattle of shambling gait they flayed and dressed before the pyre; and from them all great-hearted Achilles gathered the fat, and enfolded the dead in it from head to foot, and about him heaped the flayed bodies. And on it he set two-handled jars of honey and oil, leaning them against the bier; and four horses with high-arched necks he cast swiftly upon the pyre, groaning aloud. . . . Twelve noble sons of the great-hearted Trojans, all these together with you the flame devours; but Hector, son of Priam, I will not give to the fire to feed on, but to dogs. . . . First they quenched the pyre with ruddy wine, so far as the flame had spread, and the ash had settled deep; and weeping they gathered up the white bones of their gentle comrade into a golden urn, and wrapped them in a double layer of fat, and placing the urn in the hut they covered it with a soft linen cloth.[20]
>
> . . . then they carried bold Hector out, shedding tears, and on the topmost pyre they laid out the dead man and cast fire on it. . . . The bones they took and placed in a golden urn, covering them over with soft purple robes, and quickly laid the urn in a hollow grave, and covered it over with great close-set stones. . . . In this way they held funeral for horse-taming Hector.[21]

We may summarize some important features of this methodology for cremation:

1. The pyre is made of wood; the dead individual is placed on the top of it, either naked or dressed.
2. Animals and other organic substances (wine, oil, and animal fat) are placed on the pyre.
3. The day after the cremation, the steaming pyre is quenched using wine.

20. *Il.* 23.161–261, trans. A. T. Murray, Cambridge, Mass., 1999.
21. *Il.* 24.782–805, trans. A. T. Murray, Cambridge, Mass., 1999.

4. Family and friends recover the human remains and place them in a container, sometimes mixing them with fat to temporarily delay putrefaction.
5. Metallic or organic offerings are placed close to the bones in a funerary urn.

Each aspect of these behaviors has been directly or indirectly observed in the Louvre cremation. This means that all the observations performed on this cremation permit us to relate it to a Homeric funeral ritual as described in the funerals of Hector and Patroclus.

CONCLUSION

The anthropological and paleopathological study of the cremated bones preserved in the *lebes* BR 2590 provides information enabling us to integrate an archaeological object into a human and sociological context: The cremated individual was very probably a robust male, 25 to 45 years old. While the circumstances of his last hours are not known, the analysis of human remains and the mineral and organic formations associated with the cremation allow us to answer some questions about a Greek burial from the 5th century B.C. The funeral pyre was probably made of olive trees. A very precise selection was made of the pieces placed in the container, as only one fragment of wood was recovered in the remains. Metallic artifacts containing significant parts of iron and copper were placed close to the body. These melted in the high temperature, blended with the organic matter coming from the body or its offerings (animals, fat, skins, tissues, perfumes, or clothes), and then hardened when the temperature cooled. After the quite complete recovery of the bones, they were placed into a bronze vase that corroded and partially stained the bones green. All these observations show that this cremation from the 5th century B.C. reflects a ritual process matching Homeric descriptions of the funerals of Hector and Patroclus.

REFERENCES

Acsàdi, G. Y., and J. Nemeskéri. 1970. *History of Human Life Span and Mortality,* Budapest.

Baud, C. A., and C. Kramar. 1990. "Les calcifications biologiques en archéologie," *Bulletins et mémoires de la Société d'Anthropologie de Paris* 3–4 (2), pp. 163–170.

Bruzek, J. 2002. "A Method for Visual Determination of Sex, Using the Human Hip Bone," *American Journal of Physical Anthropology* 117, pp. 157–168.

Charlier, P., ed. 2008. *Ostéo-archéologie et techniques médico-légales: Tendances et perspectives,* Paris.

Charlier, P., and J. Poupon. 2003. "Étude toxicologique du contenu du vase de Malluro," in *Maternité et petite enfance en Gaule romaine,* ed. D. Gourevitch, Bourges, pp. 190–192.

Duday, H. 1990. "L'étude anthropologique des sépultures à incinération," *Nouvelles de l'archéologie* 40, pp. 28–40.

Ferembach, D., I. Schwidetzky, and M. Stloukal. 1979. "Recommandations pour déterminer l'âge et le sexe sur le squelette," *Bulletins et mémoires de la Société d'Anthropologie de Paris* 6 (13), pp. 7–45.

Gejvall, N. G. 1971. "Cremations," in *Science in Archaeology,* ed. D. Brothwell and E. Higgs, London, pp. 468–479.

Grévin, G. 1990. "La fouille en laboratoire des sépultures à incinération. Son apport à l'archéologie," *Bulletins et mémoires de la Société d'Anthropologie de Paris* 3–4 (2), pp. 67–74.

Guillon, F. 1986. "Brûlés secs ou brûlés frais?," in *Anthropologie physique et archéologie,* ed. H. Duday and C. Masset, Paris, pp. 191–194.

Krogman, W. M. 1978. *The Human Skeleton in Forensic Medicine,* Springfield, Ill.

McKinley, J. I. 1993. "Bone Fragment Size and Weights of Bone from Modern British Cremations and the Implications for the Interpretations of Archaeological Cremations," *International Journal of Osteoarchaeology* 3, pp. 283–287.

Shipman, P., G. Foster, and M. Schoeninger. 1984. "Burnt Bones and Teeth: An Experimental Study of Color, Morphology, Crystal Structure and Shrinkage," *JAS* 11, pp. 307–325.

CHAPTER 4

It Does Take a Brain Surgeon: A Successful Trepanation from Kavousi, Crete

by Maria A. Liston and Leslie Preston Day

Trepanation (or trephination)[1] is known throughout antiquity in many parts of the world. The practice involves the surgical removal of a portion of the cranial vault in order to create an opening in the skull. There are many examples of trepanations identified from the resulting cranial wounds, but there is little evidence for specialized surgical instruments other than round bores, or trephines, used in drilling techniques. The Vronda excavations at Kavousi, Crete, recovered an Early Iron Age cranium with a healed lesion on the right side of the cranium. [2] The location, shape, cross sectional morphology, and other characteristics of the lesion suggest that it is the result of deliberate surgical intervention, not some other biological or traumatic process. The body had been cremated, but the calvarium was reconstructed from the well-preserved fragments. This appears to be the first case of trepanation identified in cremated remains from an archaeological site.

Also at Vronda, two graves produced groups of iron tools similar to later assemblages identified as medical kits at other sites. From these two graves there were three iron objects, unknown elsewhere, that appear to have been designed as scrapers. These iron tools would be quite effective in scraping human bone, and the size and curvature of the functional surfaces of these scrapers is consistent with the lesion identified as a trepanation. Therefore, we suggest that the cranial lesion may aid us in identifying these scrapers as elements of medical kits intended for surgery.

TREPANATION IN ANTIQUITY

The preserved medical literature provides a particularly rich record of the uses and techniques of trepanation as a significant part of medical practice.[3] The Hippocratic corpus considers the procedure repeatedly (e.g., *Cap. Vul.* 14; *Loc. Hom.* 32; *Morb.* 2.23, 25). The Roman medical writers also discuss cranial surgery and trepanation, generally following the guidelines set out by Hippocrates, although the surgical instruments had evolved by the Roman period (Celsus, *Med.* 8.3.1–2, 7–9; Galen, *Meth. Med.* 10, 445 K.).[4]

1. The terms *trephination* and *trepanation* are often used interchangeably, but *trepanation* appears to be the older and more general term. For a brief discussion of the etymology and usage of these terms, see Cook 2000; Ortner 2003, p. 171.

2. The excavations at Kavousi were conducted from 1987 to 1992 by the Universities of Tennessee and Minnesota and Wabash College under the auspices of the American School of Classical Studies at Athens, and directed by Geraldine C. Gesell, Leslie Preston Day, and the late William D. E. Coulson. Financial support for this analysis was provided by the Institute for Aegean Prehistory. Facilities for analysis were provided by the INSTAP Study Center for East Crete and the Wiener Laboratory. Maria Liston also received support as the Malcolm H. Wiener Laboratory Visiting Research Professor at the American School of Classical Studies.

3. Martin-Araguz et al. 2002, pp. 1183–1185; Nerlich et al. 2003, pp. 191–192; Rose 2003, pp. 347–351.

4. Rocca 2003, pp. 254–258.

The earliest archaeological evidence of trepanation may date to the Mesolithic;[5] it is well documented in the Neolithic and Bronze Age,[6] and continues throughout antiquity. Indeed, the practice continues into the present, as a part of both traditional and modern Western medical practices. Trepanation has been employed for the treatment of headache, severe head and scalp injuries,[7] as well as for ritual or spiritual purposes.[8]

The widespread practice of this surgery is attested by ancient trepanations found all over the world, including North America,[9] Anatolia,[10] Africa,[11] Australia,[12] Europe, and the circum-Mediterranean region.[13] However, the most frequent ancient applications of trepanation are found in skulls from Incan archaeological sites of the Andes, and from sites in Mesoamerica.[14] For example, one early archaeological expedition to Peru found that 46 out of 298 skulls (15%) had received at least one trepanation.[15]

While the procedure was unquestionably dangerous to the patient, survival rates for this surgery, as evidenced by healed trepanations, sometimes exceeded 50% in the Americas.[16] In a series of ancient trepanations documented in Great Britain, 9/24 (38%) survived in the Neolithic to Roman periods, while 19/24 (79%) show some evidence of healing in the early medieval period.[17]

The purpose of trepanation in antiquity can be difficult to determine. Explanations from literature or ethnographic accounts may address both therapy and prevention of disease, involving physical or psychological conditions, magic, or religious practice.[18] Ritual and magic, as well as psychological aspects of behavior, are normally difficult or impossible to discern in the archaeological record, leaving trauma, infection, or other diseases as the conditions identified on bones in association with this surgery.[19] Other potentially identifiable causes may be hydrocephalus and other cranial diseases.[20] However, in a recent review of 62 accounts of trepanations from archaeological sites in Great Britain, in 43 cases (69%) there was no physical cause for the surgery noted in the reports.[21]

When cranial trauma was present, the therapeutic intention of the trepanation may have been to relieve pressure caused by swelling of the brain or accumulating fluids. The injuries to the bone associated with the precipitating trauma may be recognized even after significant healing has

5. Capasso and Di Tota 1996, p. 316; Crubézy et al. 2001, pp. 419–420.

6. Mogle and Zias 1995, pp. 77–79; Arnott 1997; Arnott 1999; Ortner 2003, pp. 169–170. Ortner provides a summary of the bibliography on the practice and history of trepanation.

7. E.g., Oakley et al. 1959, pp. 93–96; Marino and Gonzales-Portillo 2000, pp. 940–942; Terzioğlu, Aslan, and Saydam 1999, pp. 204–206.

8. E.g., www.trepanationguide.com. A Google search for "trepanation OR trephination" results in about 74,000 citations, many of which are sites advocating the practice for a variety of purposes. Even self-trepanation is attested, purportedly for self-fulfillment; the consequences of this practice are also reported in medical literature (e.g., Wadley, Smith, and Shieff 1997, pp. 156–158).

9. Stone and Miles 1990, pp. 1015–1020; Stone and Urcid 2003, pp. 237–249.

10. Angel 1968, p. 262; 1976, p. 385; Bazarsad 2003, pp. 203–208.

11. El Khamlichi 1998, pp. 222–227.

12. Dan 1994, pp. 280–284.

13. Loughborough 1946, pp. 416–422; Crubézy et al. 2001, pp. 417–423.

14. Christensen and Winter 1997, pp. 467–476; Janssens 1970, pp. 125–139; Marino and Gonzales-Portillo 2000, pp. 940–950; Stewart 1958, pp. 461–491; Verano 2003, pp. 224–227.

15. Loughborough 1946, p. 416.

16. Buikstra 1993, p. 307; Verano 1999.

17. Roberts and Cox 2003, pp. 59, 88, 103, 161, 217.

18. Janssens 1970, pp. 133–138; Lisowski 1967, pp. 657–659; Aufderheide and Rodríguez-Martín 1998, p. 32.

19. Martin 2003, pp. 323–326; Roberts and McKinley 2003, pp. 61–62.

20. Jackson 2005, p. 101.

21. Roberts and McKinley 2003, pp. 61–62.

taken place,[22] and many Greek trepanned skulls have associated cranial fractures.[23]

Headaches, including migraines, have long plagued humans, and are well attested in Hippocrates (*Loc. Hom.* 47; *Prorrh.* 2.30) and presumably interrupted the activities of ancient life, whether the headache was real or feigned (e.g., Ov. *Am.* 1.8.73, 2.19.11). Trepanation to relieve severe headaches is attested ethnographically,[24] and has also been suggested to be the cause of trepanations in antiquity. However, neither severe headaches nor most of their underlying causes leave any evidence on the cranial bones.

A few archaeological skulls do provide evidence for conditions, other than trauma, potentially leading to trepanation. One of these is the trepanned skull of a young woman from central Italy, radiocarbon dated to A.D. 1030–1470. In addition to a healed 4.5 mm trepanation on the occipital bone, there is a large depressed area on the internal surface of the occipital caused by an abnormal mass pressing against the bone during life. The authors speculate that endocranial pressure associated with this mass would have caused a variety of symptoms, including headache. The location of the trepanation would have relieved this pressure, alleviating the symptoms.[25] Likewise, a medieval archaeological specimen from the Czech Republic, an adult male over 50 years old at the time of his death, preserves a large trepanation on the frontal bone above the right eye. The opening is approximately 29 × 24 mm, and is well healed. The internal table of this skull preserves evidence for a benign meningioma tumor in the right middle meningeal artery, which could have caused elevated endocranial pressure and associated headache. Meningiomas are also strongly correlated with epilepsy, Parkinson's syndrome, speech disorders, and behavioral disruptions. Any of these could have prompted an attempt to alleviate the symptoms by cranial surgery.[26]

Other trepanations for which there is a possible connection to identifiable medical conditions include a rare trepanation on a child's skull from the Middle Bronze Age in Israel that may have been intended to relieve subperiosteal and intracranial hemorrhage associated with scurvy.[27] Endocranial infections also may be associated with trepanation to relieve pressure. An undated Peruvian cranium from an archaeological collection at Cambridge University has three healed and healing trepanations and surgery on the right side of the head. All of these were apparently associated with an ear infection, which may have caused foul-smelling discharge from the ear, partial deafness, impaired sense of balance, and facial paralysis.[28] Finally, a Chalcolithic skull from Jericho exhibits a series of three trepanations on the left side of the frontal bone. The third surgery was apparently unsuccessful, as the individual died soon after the procedure. In this individual, the frontal squama and frontal sinus show signs of chronic sinusitis that perforated the wall of the sinus, spreading the infection into the epidural space above the frontal lobe of the brain. The first trepanation drained this area but also spread the infectious osteomyelitis to the outer surface of the skull. The trepanations relieved the pressure of the infection but did not cure it. The three surgeries probably took place over a period of several months, during which time the patient may have been significantly incapacitated and possibly bedridden.[29]

22. Janssens 1970, p. 136.

23. Agelarakis 2006, pp. 7–8; *Lerna* II, pp. 43–44, 93, pl. XXII; Angel 1976, p. 385; 1982, p. 109; Grmek 1989, pp. 64–65.

24. Kemp 1935, p. 279; Bennike 1987, p. 101.

25. Capasso and Di Tota 1996, pp. 316–319.

26. Smrčka, Kuželka, and Melková 2003, pp. 325–320.

27. Mogle and Zias 1995, pp. 77–79.

28. Mann 1991, pp. 165–166.

29. Zias and Pomeranz 1992, pp. 183–185.

There are a number of different techniques of trepanation documented by the resulting wounds on the skull and in the ancient written sources. Galen indicates that the preferred technique changed through time (*Introduction seu Medicus* 394–395 [14, 783 K.]). Cutting, drilling, grooving, and scraping are known to have been used, but scraping, using a variety of sharp implements, is found worldwide and is the most common method employed by European practitioners in antiquity. Scraping is known from archaeological specimens as early as the Neolithic.[30] In this method, the bone surface is gradually shaved away until a thin layer is left or until the bone is removed and the membranes surrounding the brain are exposed. Scraping probably offered the greatest chance of survival because it carries the lowest risk of damaging the underlying membranes and the brain itself.[31] Hippocrates (*Cap. Vul.* 14.13–24, 19.33–39, 21.1–7), Galen (*Meth. Med.* 10, 444–455 K.), and Celsus (*Med.* 8.4.6) indicate that scraping is the best method of trepanning when infection is present in the bone and for exploring the nature of a previous injury.

TREPANATION IN THE EASTERN MEDITERRANEAN

Despite its worldwide use, there is relatively little evidence for cranial surgery in the eastern Mediterranean region, but a small number of possible or definite trepanations have been recorded.[32] J. L. Angel identified a number of trepanations from the Bronze Age. At Karataş, two skulls (81 Ka and 522 Ka) have trepanations. The trepanation of 522 Ka is associated with a cranial fracture on which there are no signs of healing.[33] At Lerna, a young adult (31 Ler) from a Middle Bronze Age tomb has a large possible trepanation, but the presence in the grave of the bone fragments that had been cut away suggests that this may be a traumatic injury, not deliberate surgery. Angel compares this to a skeleton from the shaft grave at Mycenae (51 Myc) which also retains the circular portion of bone that was removed. The wound is also associated with a perimortem fracture on the frontal bone.[34] There is a clearer ancient case of trepanation at Asine; this is a young adult male who survived for some time after the trepanation was performed. This case also appears to have been associated with a cranial fracture.[35] From 7th century B.C. Abdera, the skull of a woman preserves a scraped trepanation on the lambdoidal suture associated with a depressed fracture.[36] Much later, a possible trepanation has been identified from medieval Corinth. The surgery appears to have been in response to lesions associated with eosinophilic granuloma caused by histiocytosis,[37] an autoimmune disease that has been present in Greece since at least the Early Iron Age.[38]

ARCHAEOLOGICAL CONTEXT OF THE KAVOUSI TREPANATION

The excavations of the Early Iron Age site of Vronda at Kavousi in eastern Crete revealed a number of intrusive Geometric (8th century B.C.) graves in and around the remains of abandoned houses from a Late Bronze Age

30. Lorkiewicz et al. 2005; Roberts and Manchester 2005, pp. 125–126.
31. Ortner 2003, pp. 171–173; Lisowski 1967, pp. 660–664.
32. Grmek 1989, pp. 64–65.
33. Angel 1968, p. 262; 1976, p. 385.
34. *Lerna* II, pp. 43–44, 93, pl. XXII.
35. Angel 1982, p. 109.
36. Agelarakis 2006, pp. 6–7.
37. Barnes and Ortner 1997, pp. 542–547.
38. Gesell, Day, and Coulson 1995, p. 84; Liston 1993, pp. 149–151; 1991.

village (Fig. 4.1).[39] Most were stone-lined cist graves containing multiple cremation burials, occasional inhumations, and quantities of pottery and iron and bronze objects. The presence of anatomically ordered fragments of cremated bone makes it clear that the cremations took place in the cist graves, and the burials were simply covered with stones and soil after the pyre had burned out.[40] The graves were reused repeatedly, and the earlier material was pushed to the sides of the cist. In some cases, bone was redeposited outside the cist to make way for the new burials, as is indicated by the presence of bone and pottery fragments from earlier burials found in discrete deposits outside the cist but which join with pieces found inside the grave. Two of these intrusive Geometric graves, 5 and 9, contained iron scrapers in the deposit, along with apparent medical or surgical instruments, and grave 5 also contained the individual with a healed cranial lesion identified as a trepanation.

Grave 5

Grave 5 is located just west of the Vronda ridge in building C (Fig. 4.1).[41] The grave was a large (2.03 × 1.08 m) rectangular pit that had been dug into the rock tumble and roofing debris in the center of room 4 of the earlier LM IIIC building C and then lined with large stones. It had been used on several occasions, and some joins were made with material from the earlier burials in the grave that had been dumped outside to the southwest. This deposit was originally designated grave 11. Pottery and metal objects comprised the offerings associated with grave 5. The pottery included eight cups, a two-handled cup or kantharos, an aryballos, two amphoras, two large decorated vessels, a jug, and two pyxides, one of which accompanied the last inhumation. Metal artifacts were plentiful, but most had been pushed to the northwest and southwest sides of the cist and could not be associated with specific burials. In addition to a single bronze pin, many iron objects were found, including at least five spearheads, three knives, a sickle, a scraper, and a set of tweezers.

The grave contains seven burials[42] that date to the Geometric period (8th century B.C.). Some interments were made very late in that period. The first burials in the grave were later pushed to the sides of the cist when it was reused for later cremations. Burial 7, an adult male 60+ years old at death, is the only undisturbed primary adult inhumation found at Kavousi. The six cremations in grave 5 include two adult males, one adult female, one juvenile (11–12 years), and two infants. Interestingly, all three of the adult male skeletons show signs of healed trauma on the skull or arms typical of interpersonal conflict,[43] a condition that is otherwise unusual among the Kavousi skeletons. The first burial in grave 5 was a cremated adult male, whose reconstructed cranium showed clear evidence of a large, healed, beveled lesion on the right side of the head.

Grave 9

Grave 9 is located on the northeast side of the summit of the Vronda ridge in building J (Fig. 4.1). The grave was a large (1.1 × 1.8 m), rectangular, stone-lined cist that utilized as its southern boundary a wall from the

39. Gesell, Day, and Coulson 1988, pp. 297–298; Gesell, Coulson, and Day 1991; Gesell, Day, and Coulson 1995, p. 117; Day 1995.

40. Liston 1993, pp. 44–46.

41. Gesell, Day, and Coulson 1988, pp. 285–286.

42. At the time of the preliminary publication (Gesell, Day, and Coulson 1988, pp. 285–286), three cremations and one inhumation had been identified during excavation and analysis of the grave. Further reconstruction and analysis of the water-sieved residue of the grave soils revealed the three additional infant and juvenile skeletons in this grave.

43. Galloway 1999b, p. 145.

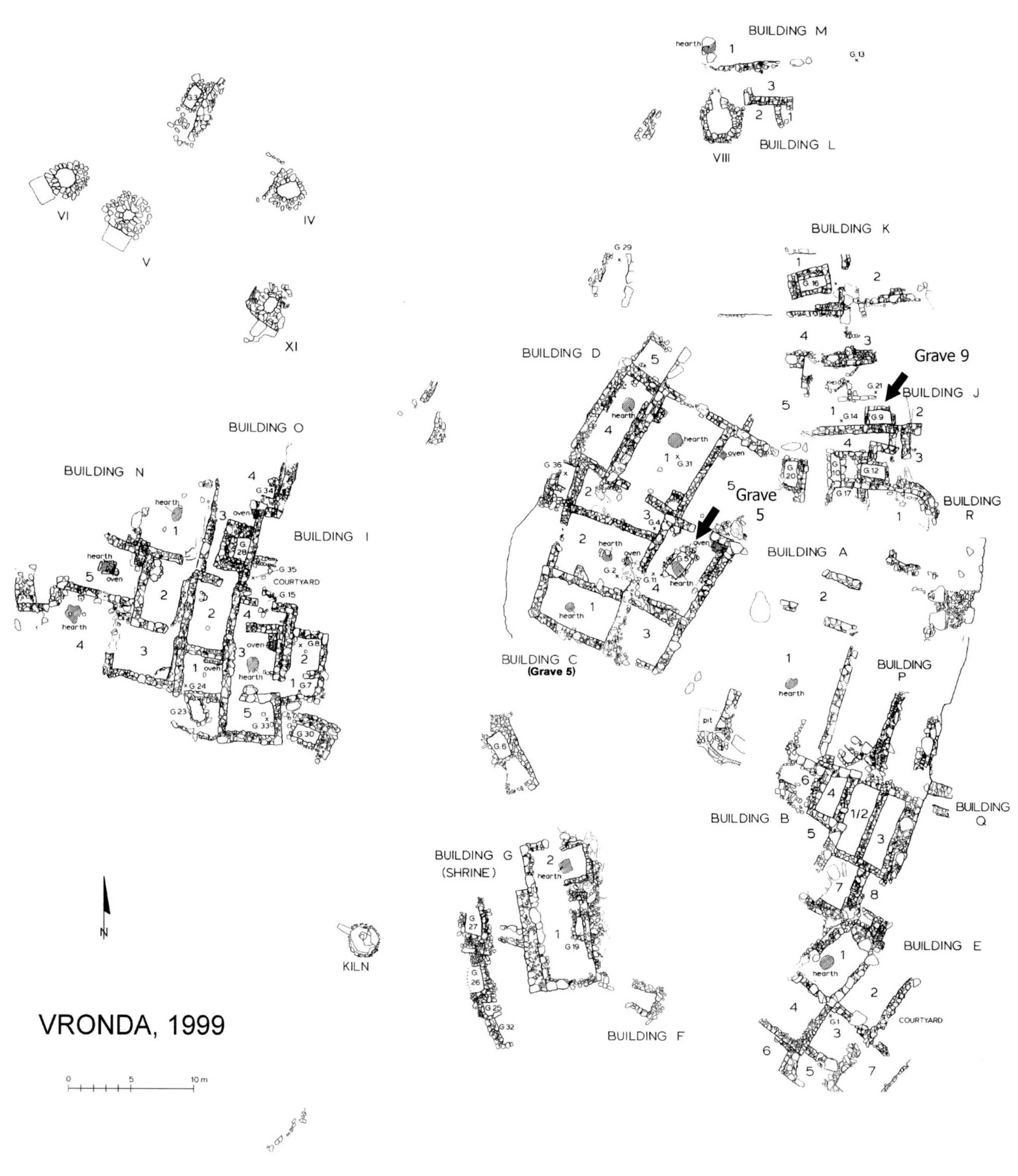

Figure 4.1. Plan of the Vronda settlement, showing the intrusive Geometric graves. Arrows indicate the locations of grave 5 and grave 9. Drawing R. Docsan

earlier LM IIIC building J, which must have been visible at the time the area was used for burial.[44] The large quantity of objects and pottery from grave 9 was associated with seven cremation burials. These included three adult males, two adult females, a child less than two years of age, and a five- to six-month fetus.[45]

Grave 9 was one of the richest graves at the site, as assessed in terms of the numbers of objects found with the burials. It contained at least thirty-four cups, two skyphoi, two kalathoi, two bowls, a krater, seven jugs, an aryballos, a pyxis, a basket vase, and four large decorated amphoras, some of which may have stood as markers over the grave.[46] Metal artifacts were also plentiful, and included a large number of bronze objects: three fibulas, and numerous fragments of thin bronze plate with rivets used for attachment to some perishable object. The more common iron items included sixteen spearheads, five daggers, three axe heads, three knives, two sickles, two scrapers, a chisel, and five awls. There were other tools whose precise function has not been previously identified, but which we argue are medical instruments.

Grave 5, Burial 1 Skeletal Evidence for Trepanation

The trepanation from Vronda was identified on the skeleton of burial 1, grave 5, the first individual buried in the cist grave. The burial was disturbed by the later cremations in the cist. However, 1,264 g of bone were recovered from this burial, and most portions of the skeleton are represented in the inventoried bone.[47]

The skeleton was an adult male, an unusually large and robust individual compared to the rest of the Vronda population. The cranium has a typical male morphology, including a pronounced supraorbital ridge, a prominent nuchal line, and large mastoid processes.[48] The condition of the preserved portions of the skeleton suggests an age of 25–40 years at death, probably at the upper end of the range. All of the preserved epiphyses are completely fused, including rib heads and ring epiphyses on the centra of the vertebrae. The vertebral bodies exhibit signs of osteoarthritic lipping and the articular facets of the lumbar and first sacral vertebrae had developed extensive subchondral pitting. The cranial sutures are partially obliterated on the endocranial surface, particularly along the preserved sections of the sagittal, coronal, and lambdoid sutures.

The cremation of this individual was typical of the Early Iron Age burials at Kavousi.[49] When the cist was reused, the earlier remains were pushed aside and reburned under the subsequent pyres. The bone from burial 1 is thoroughly cremated and very fragmentary, but the cranial bone is somewhat less burned than the postcranium. It is not completely calcined, as indicated by the mottled dark gray color. Very little of the face was preserved, but large portions of the cranial vault were collected and reconstructed.

The postcranial skeleton exhibits a number of traumatic and degenerative pathologies. In addition to the osteoarthritic changes on the vertebrae noted above, there was a healed fracture of the humerus shaft and a healed

44. Gesell, Coulson, and Day 1991, pp. 152–154.

45. The preliminary report (Gesell, Coulson, and Day 1991, p. 152) lists four cremated adults and an infant. As in grave 5, later analysis and examination of the soil residue from the water sieve revealed additional burials, in this case one adult and one fetus.

46. Gesell, Coulson, and Day 1991, pp. 152–153.

47. Liston 1993, p. 129.

48. Bass 1987, pp. 81–82

49. Liston 1993, pp. 43–45.

lesion on the preserved fragment of a patella. The side of the body on which these injuries occurred could not be identified due to the fragmentary nature of the bone and the warping associated with cremation. The fact that both the postcranial injuries and the trepanation are old, healed injuries indicates that they could be associated with a single episode of trauma in this man's life, but this cannot be determined with any certainty.

The skull has a number of pathological lesions. The most obvious is the large lesion that appears to be the result of a surgical procedure. This is located high on the right side of the skull behind the coronal suture on the parietal above the ear (Figs. 4.2, 4.3). The lesion is located above the temporal line, avoiding damage to the temporal muscle.[50] The reconstructed fragments of the cremated skull preserve an elongated oval depression, approximately 6.5 × 2.0 cm in size. A portion of the center and lower border of the lesion was not recovered in excavation or in the sorting of the heavy fraction from the water-sieved soil. The preserved portions of the lesion are symmetrically beveled inward, creating a slightly curved, scooped-out area, sloping steeply on the long sides, and at a less steep angle on the ends. The result is a surface that slopes downward from the outer table of the skull at a 30- to 40-degree angle from all sides. The width of the slope is approximately 7.5 mm wide on the long sides and 10 mm wide on the ends.

The lesion appears most similar to other archaeological and modern trepanations, created by the scraping or circular grooving methods, probably using a curved blade.[51] The preserved portion of the lesion clearly penetrated the outer table of the skull and continued through the layer of diploë to the inner table. The morphology of the wound, sloping inward from all directions, is inconsistent with a slicing wound from a blade. Since all the preserved edges in the center of the wound appear to have been broken during the cremation and subsequent taphonomic history of this burial, it is unclear whether the trepanation completely penetrated the entire parietal bone at any point, or if it was terminated, leaving a thin layer of bone from the inner cortex of the parietal.

The wound is well healed, with smooth, remodeled bone on the beveled edges. There is no evidence of active infection or bone activity directly associated with most of the lesion. However, there is a depressed area approximately 1 cm in diameter on the posterior inferior border of the lesion (Fig. 4.3). This area has a rough, irregular surface that appears to be less completely healed, although there is no evidence of active periosteal bone formation. This may be an area where healing was delayed due to a lingering infection associated with the surgical wound. Alternatively, it may be evidence of an injury that led to surgical intervention by trepanation. Hippocrates recommends scraping the bone in the area of both a ἕδρα (literally "seat"—a depression or bone bruise) and a σφάκελος (caries or infection of the skull) (*Cap. Vul.* 14.30; *Morb.* 2.23). This area of the lesion may be the injury that led to the trepanation.

There is also an area of irregular cortical bone on the interior aspect of the frontal bone, extending across most of the frontal squama. The abnormal bone is most severe to the right side of the frontal crest. Here, too, there is no evidence of active bone formation or loss, but a rough surface that

50. These muscles extend onto the mandible and are necessary for chewing. Celsus (*Med.* 8.4.9) notes the necessity to avoid these in cranial surgery.

51. Roberts and McKinley 2003, pp. 59–60, figs. 2–4; Verano 2003, p. 229, fig. 2c.

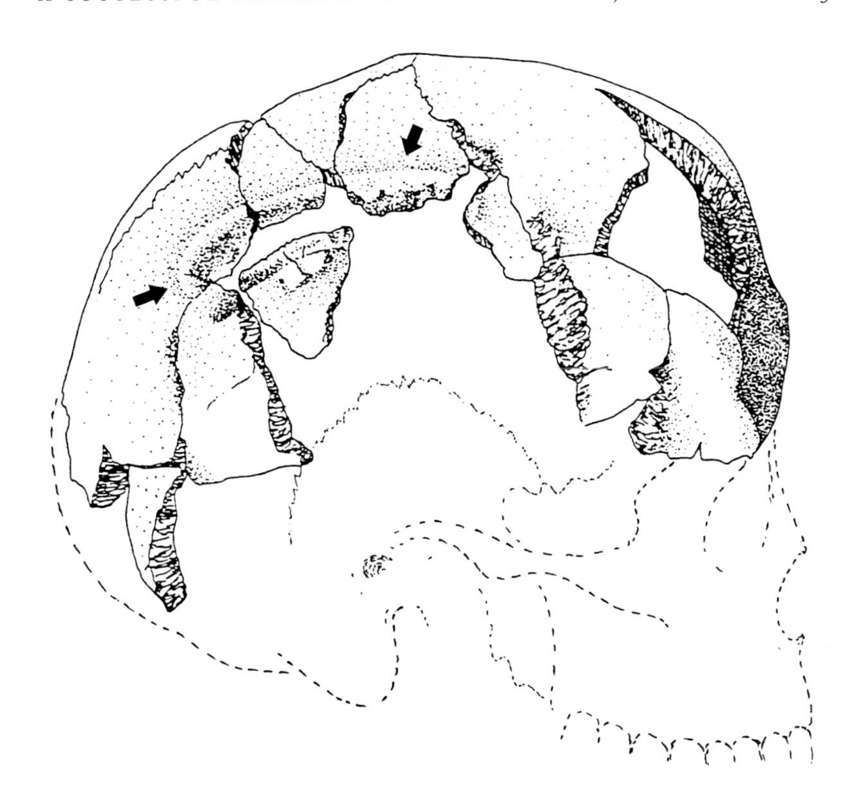

Figure 4.2. Reconstructed cranial vault showing the trepanation on the right side of the cranium (Vronda grave 5, burial 1). Drawing G. Houston

Figure 4.3. Trepanation detail (Vronda grave 5, burial 1). Photo M. Liston

appears to be the remodeled remnant of an earlier pathological process. Such lesions are occasionally associated with subdural hematomas or extradural empyaema or abscess, all of which may be associated with traumatic injury to the skull.[52] If this area of pathological bone indicates the presence of a hematoma or cranial infection in the area of the right frontal lobe, this could well account for the decision to conduct cranial surgery, or be associated with a subsequent infection. While the lesion does not appear to be actively remodeling, it may be associated with the surgery on the right parietal bone, because both appear to be largely, or completely, healed.

52. Anslow 2004, p. E147.

The absence of associated fracture in the area of the trepanation may indicate some continuity of tradition with the surgical protocol explained in later Hippocratic writings. These state that trepanation should be used only if there is not a visible fracture of the cranial bone. Apparently, the presence of a fracture was believed sufficient to release pressure or drain pus from the skull without further surgical intervention. The location of the trepanation, avoiding impingement of the coronal suture, also conforms to the protocol of the Hippocratic sources which stipulate that cranial sutures should not be trepanned (*Cap. Vul.* 14. 33–66).

DIFFERENTIAL DIAGNOSIS

There are a number of conditions that may produce openings into the cranium or depressed areas penetrating one or both tables of the bone. In addition to surgical procedures, these include developmental defects, pathological processes such as infection and tumors, traumatic injury, and postmortem damage. The clear evidence of healing obscuring the diploë eliminates postmortem causes in the Vronda specimen. The identification of the lesion as a trepanation is based on a number of factors, including the location, morphology, and the symmetrical shape of the lesion.[53]

Holes in the calvarium resulting from developmental defects or congenital anomalies normally occur in consistent locations. These include the fetal fontanelles, where cranial sutures intersect, and adjacent to the sagittal suture in the area of the parietal foramina.[54] The mid-parietal location of the lesion on the skull from Vronda grave 5 does not correspond to these. Bilateral holes in the mid-parietal also occasionally occur associated with severe biparietal thinning due to either bone loss in the elderly or congenital failure of the ossification process.[55] While the location of the grave 5 lesion on the right parietal is similar, the preservation of the left parietal makes it clear that this lesion is unilateral, not bilateral. In addition, the clearly delimited boundaries of the lesion do not resemble parietal thinning, which normally is more diffuse and involves a larger area of the bone.

The smoothly beveled edges of the lesion do not resemble a depressed fracture in which the diploë has compressed and the outer table has failed, and there is no evidence of crushing or of radiating, stellate, or hinge fractures often associated with cranial fractures.[56] The lesion also differs from penetrating wounds caused by a projectile or other object, which normally are wider at the inner surface of the bone than at the entrance on the outer table. In some cases, a penetrating wound exhibits associated crushing around the entrance, producing depressed fractures around the entrance wound and radiating fractures, but these are absent in the Vronda specimen.[57] A glancing blow or slice from a sharp instrument normally will pass through the bone in a single plane, although it may terminate in a hinge fracture.[58] The lesion found at Vronda clearly is cut into the bone surface at a steep angle from all sides, creating a scooped-out cross section. It is not possible to create an injury of this shape with a single cut or slice from an edged weapon.

53. Roberts and Manchester, 2005, pp. 125–126.

54. Barnes 1994, pp. 140–146; Kaufman, Whitaker, and McTavish 1997, pp. 193, 195–198.

55. Aufderheide and Rodríguez-Martín 1998, p. 316; Barnes 1994, pp. 146–148; Kaufman, Whitaker, and McTavish 1997, pp. 195–197; Mann and Hunt 2005, pp. 49–50.

56. Bennike 2003, pp. 100–101; Mann and Hunt 2005, pp. 27–29; Galloway 1999a, pp. 65–69.

57. Novak 2000a, pp. 97–99; Novak 2000b, pp. 242, 246–247, 261, 263; Stirland 1996, pp. 94–96.

58. Bennike 2003, pp. 102–103; Brothwell 1981, p. 119. Liston's recent (2006) examination of cranial injuries on skulls excavated at the Lion Monument associated with the Battle of Chaironeia also shows that oval or rounded lesions produced by sharp force trauma striking the calvarium at an acute angle will create a defect lying in a single plane, removing an arc shaped segment from the curved surface of the cranium. This differs markedly from the inwardly beveled shape of the lesion on the Vronda cranium.

The shape and size of the healed wound on the Vronda specimen probably reflects the original shape of the affected area. Documentation of modern surgical trepanations clearly indicates that the remodeling expected in major wounds to the postcranial skeleton does not occur in cranial injuries. The shape and angle of a surgical incision are preserved in the wound, and while the cut surface is remodeled to the extent of sealing off the exposed diploë, the defect normally does not fill with new bone, even after postsurgical survival of many years.[59]

SURGICAL INSTRUMENTS

The identification of this trepanation also leads to a possible interpretation of three iron tools and the accompanying assemblages found in the cremation graves at Vronda. The identification of these objects is complicated by the very small number of medical tools known from ancient Greece. While a number of large collections of medical instruments are known from Roman period sites,[60] primarily from graves, burial with a tool assemblage was much rarer in Greek practice.[61] In addition, while some instruments are described in the ancient medical literature, many objects could have been used in a variety of tasks, making it difficult to identify specifically medical tool kits. Finally, instruments from the prehistoric period may have been replaced by different tools by the time medical knowledge was collected and written down.

Greek and Roman medical writers provide many descriptions and uses for medical instruments. These are found in Hippocrates, Galen, Soranus, and many later writers such as Paulus Aegineta (late 6th–early 7th centuries A.D.) who compiled information from earlier sources and produced a medical encyclopedia in seven books.[62] The known assemblages of medical instruments from the Roman world contain a number of common objects. Approximately 60% of the tools are probes and/or small spatulas, 13% are tweezers, and 11% are scalpels.[63] Ancient scalpels often are double ended, with two blades, or a blade and a probe or scoop. The blades are often curved, with the cutting edge set in line with the handle. In addition, there are rasps and scrapers with the working edge set perpendicular to the handle. Finally, there were specialized objects such as catheters and crown trephines, which were used in drilled trepanations.[64]

Interpretation of the Vronda grave materials is complicated by their preservation and by the mixing of funerary assemblages that occurred when the graves were reused. In addition, with time, the metal objects have shifted downward through the ash deposits, and most were found in accumulations on the bottom of the grave or pushed to the sides by later burials. It is impossible in most cases to associate specific objects or grave assemblages with the individual burials.[65]

Although badly corroded, most of the metal objects found in the Vronda graves could be restored and their original shape and dimensions recovered. Many of the iron objects were weapons, including spear points, arrowheads, and axes, although some burials may have included craftsmen's

59. Nerlich et al. 2003, pp. 49–50.
60. Jackson 2005, pp. 101–103; Künzl 1983; Künzl 2002; Krämer 2000.
61. Salazar 2000, pp. 237–239.
62. Milne 1907, pp. 5–7. Milne's work remains one of the most comprehensive detailing ancient surgical instruments.
63. Salazar 2000, pp. 238–239.
64. Milne 1907, pp. 26–27, 121–123; Tabenelli 1958, pls. XVII–XXI; Crummy 2002, p. 51.
65. Gesell, Day, and Coulson 1988, pp. 285–286.

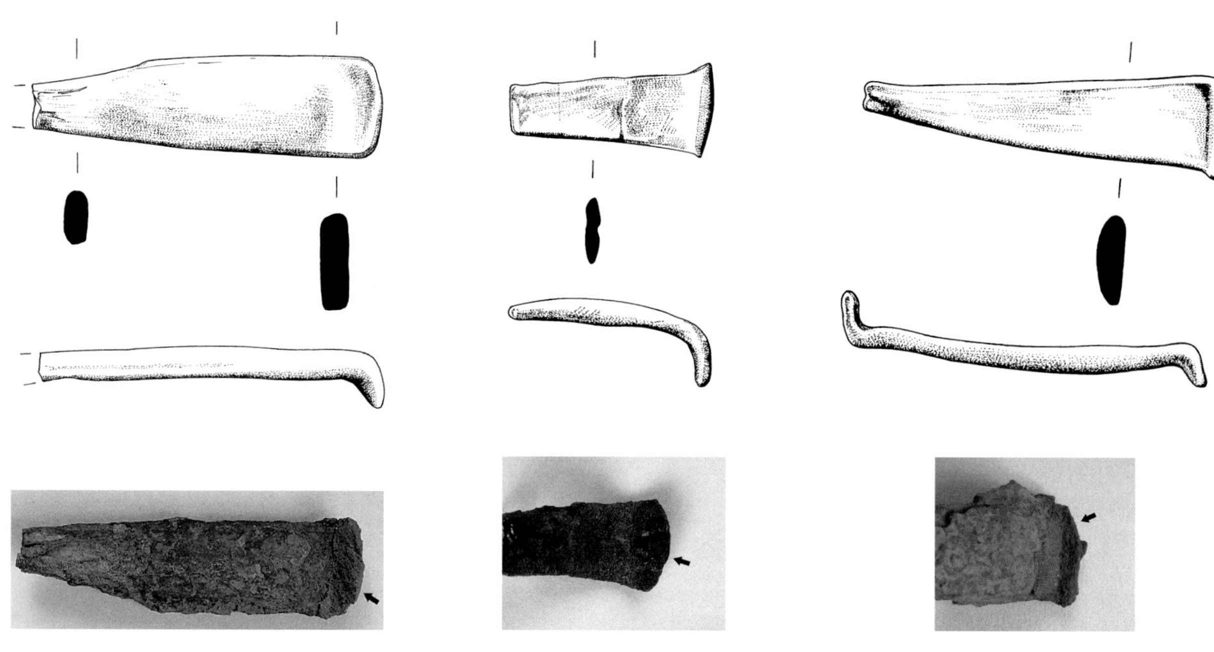

Figure 4.4 ***(left).*** **Iron scraper V87.97 from Vronda grave 5, back** ***(top),*** **side** ***(middle),*** **and front** ***(bottom)*** **views. Arrow indicates sharpened edge.** Scale 1:2. Drawing L. P. Brock, photo C. Papanikolopoulos

Figure 4.5 ***(center).*** **Iron scraper V88.181 from Vronda grave 9, front** ***(top and bottom)*** **and side** ***(middle)*** **views. Arrow indicates sharpened edge.** Scale 1:2. Drawing L. P. Brock, photo C. Papanikolopoulos

Figure 4.6 ***(right).*** **Iron scraper V88.192 from Vronda grave 9, front** ***(top and bottom)*** **and side** ***(middle)*** **views. Arrow indicates sharpened edge.** Scale 1:2. Drawing L. P. Brock, photo C. Papanikolopoulos

tools as well. The precise functions of some of these objects are not readily identifiable, but in graves 5 and 9, there are groups of objects that resemble assemblages identified as medical kits for surgery or wound treatment.

Among the artifacts in graves 5 and 9 were three iron tools of similar type (V87.97, V88.181, V88.192) that were originally labeled "scrapers." The example from grave 5 (V87.97)[66] (Fig. 4.4) was shaped rather like a curved chisel, with the broad end bent at a right angle; the narrow end, which was broken, looks as though it had been folded around a flat handle, perhaps of wood. The preserved length was 9.3 cm. Object V88.192,[67] from grave 9 (Fig. 4.6), was similar in shape and size, measuring 10.1 cm in length. The broad end was similarly bent at a right angle, and the narrower end was bent in the opposite direction, also at a right angle. Another iron tool, V88.181, also from grave 9, was smaller, measuring only 5.7 cm in length (Fig. 4.5). Like the others, its broad end was bent, but the other end was straight, and lacking either the bend found on V88.192 or the folding on V87.97. Although badly corroded, it appears to have had a slot through the narrow end for insertion of a flat object, such as a handle. On each of these iron tools, the broad curved ends were sharpened, as was the narrow pointed end of V88.192. Clearly, these were the functional surfaces of the tools, and the shape indicates that they were intended for cutting or scraping a surface.

Although we have been unable to locate parallels for these tools in other sites in Greece, the curvature of the sharpened edge resembles that of chipped stone scrapers. Therefore, possible suggestions for their functions could have included wood or leather working. The size and shape of the broad curved edge of these iron scrapers, however, would also have made them appropriate instruments for producing a scraped trepanation of the type found on grave 5, burial 1. The width and slope of the scraped wound is consistent with the sharpened surface of the tools. The smaller pointed

66. Gesell, Day, and Coulson 1988, p. 286.

67. Gesell, Coulson, and Day 1991, p. 152, pl. 58.

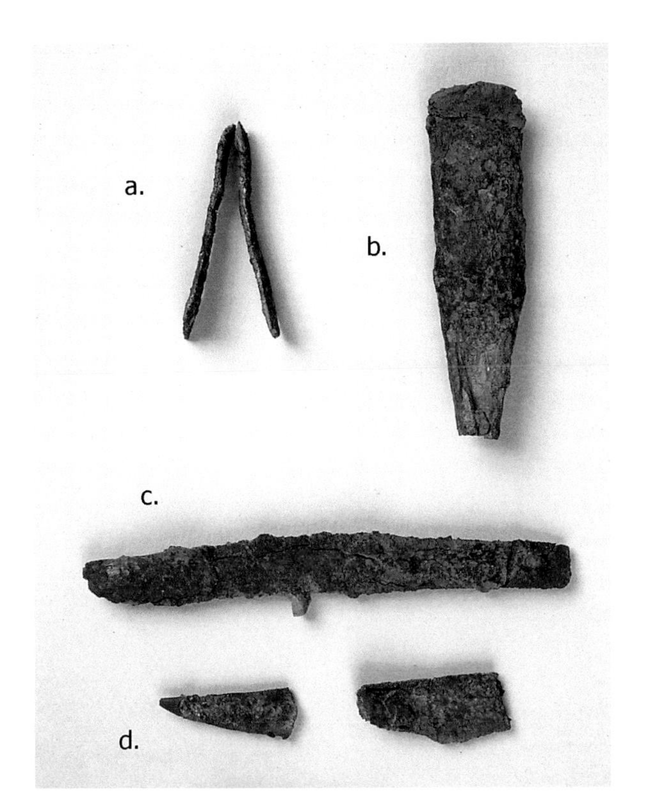

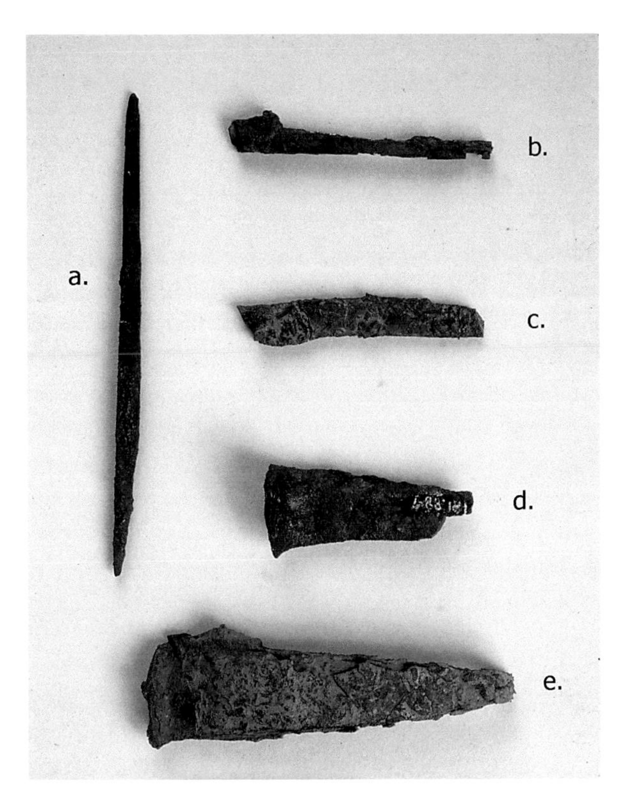

Figure 4.7. Medical tool assemblage from Vronda grave 5: (a) iron tweezers V87.100; (b) iron scraper V87.97; (c) iron knife V87.104; (d) iron knife fragments V87.103. Scale 1:2. Photo C. Papanikolopoulos

Figure 4.8. Medical tool assemblage from Vronda grave 9: (a), (b) iron probes; (c) iron knife V88.213; (d), (e) iron scrapers V88.181, V88.192. Scale 1:2. Photo C. Papanikolopoulos

end of V88.192 presumably could have been used for finer work, including cleaning the edges of the wound or lifting chips and fragments of bone.

Similar instruments with the functional ends bent at right angles to the handle and with working surfaces at either end of the handle are well known from Roman medical kits and sculptural reliefs from gravestones.[68] However, the ends of these objects do not appear to be sharpened. Rather, they appear to be a variety of hooks, lifters, and retractors. On the other hand, descriptions of medical instruments that are perhaps similar to the Vronda scrapers are found in a number of sources. Hippocrates (*Cap. Vul.* 3.366) and Galen (*Meth. Med.* 10, 445 K.) describe the use of chisels, scrapers, or raspers (ξυστήρ, *scalper excisorius*) on cranial wounds, and the employment of different-sized implements for specific purposes.

The interpretation of these iron scrapers as medical instruments is supported by the presence in graves 5 and 9 of other instruments typically found in ancient surgical kits.[69] In grave 5 (Fig. 4.7) there were two small knives similar to scalpels, V87.103 and V87.104, and a pair of tweezers, V87.100, in addition to one of the iron scrapers, V87.97. Grave 9 (Fig. 4.8) contained, along with the two iron scrapers, V88.181 and V88.192, tools that may have been used as probes, V88.226 and possibly V88.215 and V88.217. There is also a curved probe, V88.227, and one small knife, V88.213. The latter is possibly a scalpel, although its blade is less broad and curved than many common later Greek and Roman scalpels. It does, however, differ from many other knives found at Vronda, which may reflect a specialized use. Similar blades are occasionally included in collections of Roman surgical instruments.[70] All of these items were rare or absent

68. Jackson 1988, pp. 114–115, figs. 27, 28; Jackson 2005, p. 102, fig. 5.2; Jackson and La Niece 1986, p. 169, pl. XII, fig. B; Geroulanos and Bridler 1994, figs. 97–100.

69. Arnott 1999; Arnott 1997; Jackson 1988, pp. 113–118.

70. E.g., Künzl 1983, p. 82, pl. 56, nos. 6, 7; Milne 1907, pl. V, nos. 1, 2.

in other graves at Vronda, and together constitute functional groups of instruments appropriate for wound care and simple surgery.

CONCLUSIONS

The analysis of the cremation graves at Kavousi has provided a wealth of information about life and health in Early Iron Age Crete. In particular, the cranium of one individual from grave 5 has yielded evidence of a trepanation performed on the right side of the cranial vault, possibly associated with an area of endocranial infection. The surgery was performed by scraping the cranial bone, rather than by cutting or drilling. The surgeon thinned or removed a large area of the parietal bone. The patient survived the surgery and lived long enough afterward for the bone to have largely healed and remodeled along the edges of the wound. In addition, the analysis of the techniques used to produce this incision provides an interpretation of a type of iron scraper found in two of the Vronda Early Iron Age graves. We propose that these scrapers may be identified as the instruments used in the surgical trepanation, attested in the remains of the skeleton from grave 5, burial 1. Together with the knives, probes, and tweezers found in graves 5 and 9, there appear to have been two assemblages of tools that may have been medical/surgical instruments among the materials deposited in the Geometric period graves at Kavousi Vronda.

REFERENCES

Agelarakis, A. P. 2006. "Early Evidence of Cranial Surgical Intervention in Abdera, Greece: A Nexus to *On Head Wounds* of the Hippocratic Corpus," *Mediterranean Archaeology and Archaeometry* 6, pp. 5–18.

Angel, J. L. 1968. "Appendix: Human Remains at Karataş," in M. J. Mellink, "Excavations at Karataş-Semayük in Lycia, 1967," *AJA* 72, pp. 243–263.

———. 1976. "Excavations in the Elmali Area, Lycia, 1975. Appendix: Early Bronze Age Karataş, People and Their Cemeteries," *AJA* 80, pp. 385–391.

———. 1982. "Ancient Skeletons from Asine," in *Asine* II: *Excavations East of the Acropolis 1970–1974,* ed. S. Dietz, Copenhagen, pp. 105–138.

Anslow, P. 2004. "Cranial Bacterial Infection," *European Radiology 14,* pp. E145–E154.

Arnott, R. 1997. "Surgical Practice in the Prehistoric Aegean," *Medizinhistorisches Journal* 32, pp. 249–278.

———. 1999. "Healing Cult in Minoan Crete," in *Meletemata: Studies in Aegean Archaeology Presented to Malcolm H. Wiener as He Enters His 65th Year* (*Aegaeum* 20), ed. P. P. Betancourt, V. Karageorghis, R. Laffineur, and W.-D. Niemeier, Liège, pp. 1–6.

Arnott, R., S. Finger, and C. U. M. Smith, 2003. *Trepanation: History, Discovery, Theory,* Lisse.

Aufderheide, A. C., and C. Rodríguez-Martín. 1998. *The Cambridge Encyclopedia of Human Paleopathology,* Cambridge.

Barnes, E. 1994. *Developmental Defects of the Axial Skeleton in Paleopathology,* Niwot, Colo.

Barnes, E., and D. J. Ortner. 1997. "Multifocal Eosinophilic Granuloma with a Possible Trepanation in a Fourteenth Century Greek Young Skeleton," *International Journal of Osteoarchaeology* 7, pp. 542–547.

Bass, W. M. 1987. *Human Osteology: A Laboratory and Field Manual,* 3rd ed., Columbia, Mou.

Bazarsad, N. 2003. "Four Cases of Trepanation from Mongolia, Showing Surgical Variation," in Arnott, Finger, and Smith 2003, pp. 203–208.

Bennike, P. 1987. Rev. of K. Jappe and L. Jappe, *The Peasant Physician,* in *Journal of Paleopathology* 1, pp. 101–102.

———. 2003. "Ancient Trepanations and Differential Diagnoses: A Reevaluation of Skeletal Remains from Denmark," in Arnott, Finger, and Smith 2003, pp. 95–115.

Brothwell, D. R. 1981. *Digging Up Bones,* 3rd ed., Ithaca, N.Y.

Buikstra, J. 1993. "Diseases of the Pre-Columbian Americas," in *Cambridge World History of Human Disease,* ed. K. Kiple, Cambridge, pp. 305–317.

Capasso, L., and G. Di Tota. 1996. "Possible Therapy for Headaches in Ancient Times," *International Journal of Osteoarchaeology* 6, pp. 316–319.

Christensen, A. F., and M. Winter. 1997. "Culturally Modified Skeletal Remains from the Site of Huamelulpan, Oaxaca, Mexico," *International Journal of Osteoarchaeology* 7, pp. 467–480.

Cook, D. C. 2000. "Trepanation or Trephination? The Orthography of Paleopathology" *Paleopathology Newsletter* 109, pp. 9–12.

Crubézy, É., J. Bruzek, J. Guilaine, E. Cunha, D. Rougé, and J. Jelinek. 2001. "The Antiquity of Cranial Surgery in Europe and in the Mediterranean Basin," *Comptes rendus de l'Academie des Sciences Series IIA: Earth and Planetary Science* 332, pp. 417–423.

Crummy, P. 2002. "A Preliminary Account of the Doctor's Grave at Stanway, Colchester, England," in *Practitioner, Practices and Patients: New Approaches to Medical Archaeology and Anthropology,* ed. P. A. Baker and G. Carr, Oxford, pp. 47–57.

Dan, N. G. 1994. "Neurosurgery in the Pre-European Era in Australia," *Journal of Clinical Neuroscience* 1, pp. 280–284.

Day, L. P. 1995. "The Geometric Cemetery at Vronda, Kavousi," in *Πεπραγμένα του Ζ΄ Διεθνούς Κρητολογικού Συνεδρίου Α2: Τμήμα αρχαιολογικό,* ed. N. E. Papadogiannakis, Rethymnon, pp. 789–796.

El Khamlichi, A. 1998. "African Neurosurgery Part I: Historical Outline," *Surgical Neurology* 49, pp. 222–227.

Galloway, A. 1999a. "Fracture Patterns and Skeletal Morphology: Introduction and the Skull," in *Broken Bones: Anthropological Analysis of Blunt Force Trauma,* ed. A. Galloway, Springfield, Ill., pp. 63–80.

———. 1999b. "Fracture Patterns and Skeletal Morphology: The Upper Extremity," in *Broken Bones: Anthropological Analysis of Blunt Force Trauma,* ed. A. Galloway, Springfield, Ill., pp. 113–159.

Geroulanos, S., and R. Bridler. 1994. *Trauma: Wound-Entstehung und Wund-Pflege in antiken Griechenland,* Mainz.

Gesell, G. C., W. D. E. Coulson, and L. P. Day. 1991. "Excavations at Kavousi, Crete, 1988," *Hesperia* 60, pp. 145–177.

Gesell, G. C., L. P. Day, and W. D. E. Coulson. 1988. "Excavations at Kavousi, Crete, 1987," *Hesperia* 57, pp. 279–301.

———. 1995. "Excavations at Kavousi, Crete, 1989 and 1990," *Hesperia* 64, pp. 67–120.

Grmek, M. D. 1989. *Diseases in the Ancient Greek World,* trans. M. Muellner and L. Muellner, Baltimore.

Jackson, R. 1988. *Doctors and Diseases in the Roman Empire,* London.

———. 2005. "Holding on to Health? Bone Surgery and Instrumentation in the Roman Empire," in *Health in Antiquity,* ed. H. King, London, pp. 97–119.

Jackson, R., and S. La Niece. 1986. "A Set of Roman Medical Instruments from Italy," *Britannia* 17, pp. 119–167.

Janssens, P. A. 1970. *Paleopathology: Diseases and Injuries of Prehistoric Man,* London.

Kaufman, M. H., D. Whitaker, and J. McTavish. 1997. "Differential Diagnosis of Holes in the

Calvarium: Application of Modern Clinical Data to Palaeopathology," *JAS* 24, pp. 193–218.

Kemp, P. 1935. *Healing Ritual: Studies in the Technique and Tradition of the Southern Slavs,* London.

Krämer, K. 2000. *Die antiken Instrumente des Deutschen Medizinhistorischen Museums Ingolstadt: Realienkundliche Untersuchung und Katalog,* Ingolstadt.

Künzl, E. 1983. *Medizinische Instrumente aus Sepulkralfunden der Römischen Kaiserzeit,* Cologne.

———. 2002. *Medizinische Instrumente der Römischen Kaiserzeit im Römisch-Germanischen Zentralmuseum,* Mainz.

Lerna II = J. L. Angel, *The People,* Princeton 1971.

Lisowski, F. P. 1967. "Prehistoric and Early Historic Trepanation," in *Diseases in Antiquity,* ed. D. Brothwell and A. T. Sandison, London, pp. 651–672.

Liston, M. A. 1991. "Histiocytosis in Crete? A Possible Case from the Early Iron Age Site of Kavousi" (paper, Milwaukee 1991).

———. 1993. "Human Skeletal Remains from Kavousi, Crete: A Bioarchaeological Analysis" (diss. Univ. of Tennessee, Knoxville).

Lorkiewicz, W., H. Stolarczyk, A. Śmiszkiewicz-Skwarska, and E. Żądzińska. 2005. "An Interesting Case of Prehistoric Trepanation from Poland: Re-evaluation of the Skull from the Franki Suchodolskie Site," *International Journal of Osteoarchaeology* 15, pp. 115–123.

Loughborough, J. L. 1946. "Notes on the Trepanation of Prehistoric Crania," *American Anthropologist* 48, pp. 416–422.

Mann, G. 1991. "Chronic Ear Disease as a Possible Reason for Trepanation," *International Journal of Osteoarchaeology* 1, pp. 165–168.

Mann, R. W., and D. R. Hunt. 2005. *Photographic Regional Atlas of Bone Disease,* 2nd ed., Springfield, Ill.

Marino, R., and M. Gonzales-Portillo. 2000. "Preconquest Trepanation and the Art of Medicine in ancient Peru," *Neurosurgery* 47, pp. 940–950.

Martin, G. 2003. "Why Trepan? Contributions from Medical History and the South Pacific," in Arnott, Finger, and Smith 2003, pp. 323–346.

Martin-Araguz, A., C. Bustamante-Martinez, M. T. Emam-Mansour, and J. M. Moreno-Martinez. 2002. "Neuroscience in Ancient Egypt and in the School of Alexandria," *Revista de Neurologica* 34, pp. 1183–1194.

Milne, J. S. 1907. *Surgical Instruments in Greek and Roman Times,* Oxford.

Mogle, P., and J. Zias. 1995. "Trephination as a Possible Treatment for Scurvy in a Middle Bronze Age (ca. 2200 B.C.) Skeleton," *International Journal of Osteoarchaeology* 5, pp. 77–81.

Nerlich, A. G., O. Peschel, A. Zink, and F. W. Rösing. 2003. "The Pathology of Trepanation: Differential Diagnosis, Healing and Dry Bone Appearance in Modern Cases," in Arnott, Finger, and Smith 2003, pp. 43–51.

Nerlich, A. G., A. Zink, U. Szeimies, H. G. Hagedorn, and F. W. Rösing. 2003. "Perforating Skull Trauma in Ancient Egypt and Evidence for Early Neurosurgical Therapy," in Arnott, Finger, and Smith 2003, pp. 191–202.

Novak, S. A. 2000a. "Battle-Related Trauma," in *Blood Red Roses: The Archaeology of a Mass Grave from the Battle of Towton A.D. 1461,* ed. V. Fiorato, A. Boylston, and C. Knusel, Oxford, pp. 90–102.

———. 2000b. "Case Studies," in *Blood Red Roses: The Archaeology of a Mass Grave from the Battle of Towton A.D. 1461,* ed. V. Fiorato, A. Boylston, and C. Knusel, Oxford, pp. 240–268.

Oakley, K. P., W. M. A. Brooke, A. R. Akester, and D. R. Brothwell. 1959. "Contributions on Trepanning or Trepanation in Ancient and Modern Times," *Man* 59, pp. 93–96.

Ortner, D. J. 2003. *Identification of Pathological Conditions in Human Skeletal Remains,* 2nd ed., San Diego.

Roberts, C. A., and M. Cox. 2003. *Health and Disease in Britain: From Prehistory to the Present Day,* Stroud.

Roberts, C. A., and K. Manchester. 2005. *The Archaeology of Disease,* 3rd ed., Ithaca, N.Y.

Roberts, C. A., and J. McKinley. 2003. "Review of Trepanations in British Antiquity Focusing on Funerary Contexts to Explain Their Occurrence," in Arnott, Finger, and Smith 2003, pp. 55–78.

Rocca, J. 2003. "Galen and the Uses of Trepanation," in Arnott, Finger, and Smith 2003, pp. 253–271.

Rose, F. C. 2003. "An Overview from Neolithic Times to Broca," in Arnott, Finger, and Smith 2003, pp. 247–363.

Salazar, C. 2000. *The Treatment of War Wounds in Graeco-Roman Antiquity,* Leiden.

Smrčka, V., V. Kuželka, and J. Melková 2003. "Meningioma Probably Reason for Trepanation," *International Journal of Osteoarchaeology* 13, pp. 325–330.

Stewart, T. D. 1958. "Stone Age Skull Surgery: A General Review with Emphasis on the New World," *Smithsonian Institution Annual Report 1957,* pp. 461–491.

Stirland, A. 1996. "Patterns of Trauma in a Unique Medieval Parish Cemetery," *International Journal of Osteoarchaeology* 6, pp. 92–100.

Stone, J. L., and M. L. Miles. 1990. "Skull Trepanation among the Early Indians of Canada and the United States," *Neurosurgery* 26, pp. 1015–20.

Stone, J. L., and J. Urcid. 2003. "Pre-Columbian Skull Trepanation in North America," in Arnott, Finger, and Smith 2003, pp. 237–249.

Tabenelli, M. 1958. *Lo strumento chirurgico e la sua storia, dalle epoche greca e romana al secolo decimosesto,* Milan.

Terzioğlu, A., G. Aslan, and M. Saydam. 1999. "Trepanation in the Treatment of Scalp Avulsion: Successful Application of a Historical Method," *Journal of Oral and Maxillofacial Surgery* 57, pp. 204–206.

Verano, J. W. 1999. "Health and Medical Practices in the Pre-Columbian Americas," *Perspectives in Health* 4 (1), pp. 9–12.

———. 2003. "Trepanation in Prehistoric South America: Geographic and Temporal Trends over 2,000 Years," in Arnott, Finger, and Smith 2003, pp. 223–236.

Wadley, J. P., G. T. Smith, and C. Shieff 1997. "Self-Trepanation of the Skull with an Electric Power Drill," *British Journal of Neurosurgery* 11, pp. 156–158.

Zias, J., and S. Pomeranz. 1992. "Serial Craniectomies for Intracranial Infection 5.5 Millennia Ago," *International Journal of Osteoarchaeology* 2, pp. 183–186.

CHAPTER 5

The Malleable Body: Headshaping in Greece and the Surrounding Regions

by Kirsi O. Lorentz

Headshaping[1] is a form of body modification involving the intentional altering of human cranial form in infancy (Fig. 5.1). This modification is based on the rapid growth and plasticity of the human cranium during the first few years of life. Headshaping is permanent and irreversible once the rapid growth of the cranium ceases. Various devices including cradleboards, bindings, and headdresses have been used to intentionally modify infant head shape, and different types of headshaping arise from the differential restriction of the growth vectors of the cranial plates. The employment of headshaping indicates a profound understanding of the human body. It requires a detailed knowledge of the timing of cranial growth and development. Headshaping is an ancient form of body modification far more sophisticated than plastic surgery today in that it does not cause pain or damage hard tissues. It simply redirects the growth vectors of the cranium. A current pediatric practice, the so-called dynamic orthotic cranioplasty[2] employed for "correcting" plagiocephaly (cranial asymmetry) is a modern form of headshaping.

Headshaping has been practiced on all continents over periods from prehistory to the present.[3] It has been used to denote gender, ethnicity, high social status, or other group affiliations.[4] Aesthetic reasons for headshaping have also been cited.[5] Headshaping in its different forms has thus been used as a medium for ascribing various kinds of identities on the body, with no possibility of reversal or acquisition in later life.[6]

Physical anthropological studies of headshaping have concentrated on its effects on the craniofacial complex, the basicranium, the intracranial vessel form, sutural complexity in the form of increased interdigitation, premature sutural fusion, cranial asymmetry, and the expression of extra-sutural bones.[7] Headshaping practices and their physical anthropological effects have been investigated intensely in the Americas, but little work beyond mere identification or brief descriptions has been done in the Mediterranean and the Near East.[8]

Until recently, no substantial evidence for headshaping had been found in prehistoric Greece. Brief early mentions of recent practices of headbinding by Dingwall[9] cannot be verified at present. Given the evidence for headshaping, both archaeological and recent, in the Balkans,

1. The term headshaping is preferred here over such terms as "artificial cranial deformation" in that the latter implies that this cultural practice of body modification was performed to "deform," rather than to shape according to cultural aesthetics and social norms. For more extensive discussion on terminology, see Lorentz 2003b.

2. Littlefield, Pomatto, and Kelly 2000.

3. White and Folkens 1991; Ubelaker 1989; Ortner and Putschar 1981; Brown 1981; Kiszely 1978; Dingwall 1931.

4. Lorentz 2003a, 2003b.

5. Dingwall 1931.

6. Lorentz 2003b.

7. Cheverud and Midkiff 1992; Cheverud et al. 1992, 1993; Holliday 1993; Kohn et al. 1993; Ossenberg 1970; Schendel, Walker, and Kamisugi 1980.

8. But see, e.g., Lorentz 2003b, 2004, 2005, 2007, forthcoming.

9. Dingwall 1931.

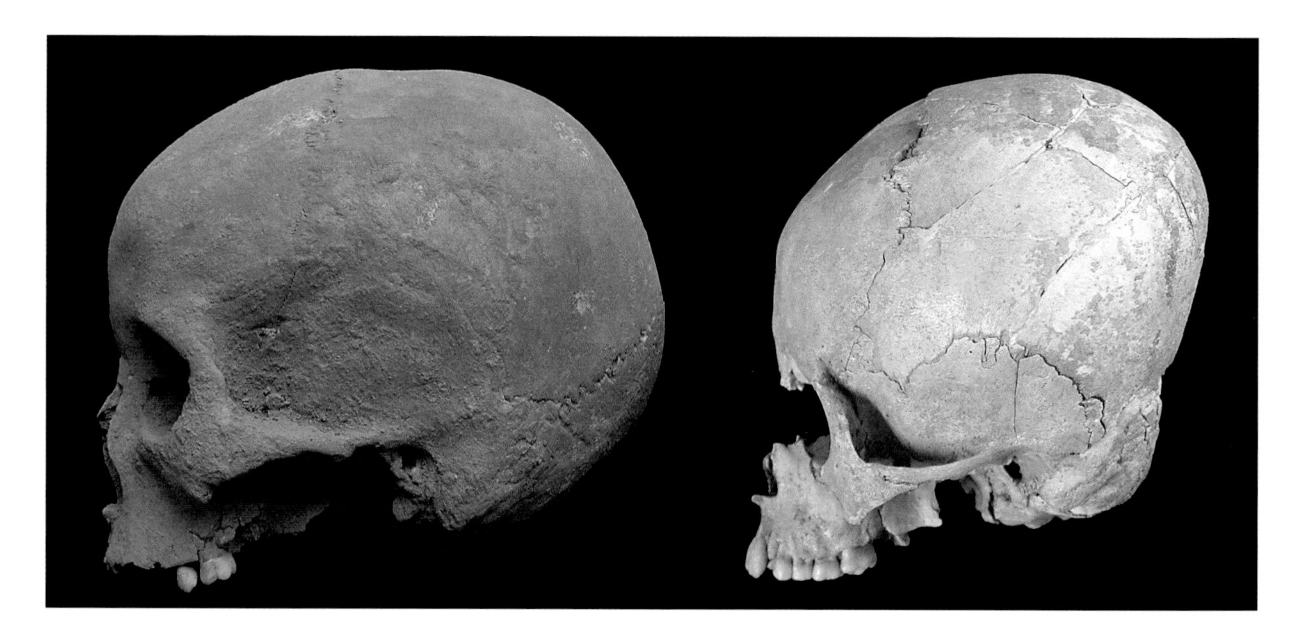

Figure 5.1. Unmodified cranium *(left)* and headshaped cranium of the antero-posterior type *(right)*.
Scale 1:2. Photos K. O. Lorentz, courtesy Department of Antiquities, Cyprus

Italy, Turkey, and Cyprus,[10] the lack of evidence from Greece has been a curious lacuna.[11]

MATERIALS AND METHODS

The discovery of a series of crania of clearly abnormal shape from the Neolithic cave cemetery in Tharrounia, Euboia,[12] provided the opportunity to investigate the potential presence of intentional cultural modification of head form in Greece.[13]

Skotini Cave, in the locality of Tharrounia, is one of the most important Neolithic sites in Euboia. The excavations in the cave and the surrounding area took place over a period of five years (1986–1991), and were conducted under the direction of A. Sampson. During the 1989 field season, a cemetery located near the Skotini Cave was excavated, dating to the early phases of the Neolithic era.[14]

The majority of the graves at this cemetery at Tharrounia were affected by various destructive taphonomic processes, including erosion caused by long-term agricultural activities. The shape of the burial facilities was pentagonal or roughly trapezoidal, constructed by placing flagstones upright. According to the excavators, the structure and shape are atypical, differing from the more or less rectangular cist graves of subsequent periods. Although the burials at Tharrounia are contemporaneous with the burials at Kephala, on the island of Kea, they differ in structure, if not shape (Kephala burial facilities have similar trapezoidal and pentagonal, as well as circular, shapes). The large quantities of pottery found at Tharrounia date from the earlier Neolithic period (phase II).

All excavated burial facilities contained skeletal remains from multiple individuals. The minimum number of individuals within the recovered skeletal series is 25.[15] The number of crania recovered and available for study is 13, and these are of varying degrees of completeness (see Table 5.1).

10. Lorentz 2003a, 2003b.

11. Kiszely 1978; Dingwall 1931; Lorentz 2003b; Özbek 2001.

12. Sampson 1993; Stravopodi 1993; Lorentz and Manolis, forthcoming.

13. I am grateful to Sotiris Manolis of the University of Athens for his kind invitation to study this material. Sherry Fox, the director of the Wiener Laboratory (ASCSA), conducted the radiography of the material. The research on which this paper is based has benefited at its various stages from funding from the following sources: The Leverhulme Trust, the Arts and Humanities Research Board, the British Institute of Archaeology in Ankara (the British Academy), the Rouse Ball Foundation (Trinity College, Cambridge), the Smuts Memorial Fund (University of Cambridge), the Oskar Huttunen Foundation (Finland), and the Dorothy Garrod Fund (University of Cambridge).

14. Sampson 1993.

15. Stravopodi 1993.

TABLE 5.1. COMPLETENESS SCORES FOR THE THARROUNIA CRANIA CALVARIAL REGIONS AND INDIVIDUAL CRANIAL ELEMENTS OF THE CALVARIUM

Cranium	*Calvarial Score (%)**	*Completeness Scores (%) for Separate Elements***
1	50	frontal and left parietal 25, right parietal 75
2	75	occipital and parietals 75, frontal 50, temporals 25
3	75	parietals 75, frontal and occipital 50, right temporal 25
4	25	frontal 25, right parietal 50
5	50	left parietal and occipital 50, right parietal 75
6	25	left parietal 75, right parietal 25
7	25	small frontal, parietal, and occipital fragments; very few joins, overall shape cannot be inferred
8	75	calvarium complete apart from a large part of left parietal, and small fragments all over calvarium
9	100	frontal, parietals, occipital, and left temporal 75; right temporal 50 (facial area very fragmentary and not restored)
10	100	frontal and parietals 100, occipital almost 100, left temporal 50, right temporal 75 (right zygomatic present, but most facial elements missing)
11	100	frontal, parietals, occipital, and temporals 75 (some facial bones present and joining the restored calvarium)
12	50	frontal 75, parietals and left temporal 50
13	<25	fragments of all calvarial elements

* The calvarial completeness scores for each cranium relate to the calvarial portion (braincase) of the cranium only, without taking into account the facial region, which in all crania was either fragmentary, or missing (where present, details of the facial region are given in brackets). Providing completeness scores for both individual elements and the calvarium as a whole is useful in explaining why certain observations could only be made on one side of the cranium (e.g., sutural complexity), or for particular crania only (e.g., presence of headshaping).

** The completeness scores are derived by a procedure similar to that described by Buikstra and Ubelaker 1994, p. 7, allocating an approximate percentage for each bone element (or the calvarium as a whole) based on visual assessment of completeness. <25% = less than ¼ of the bone element (or calvarium) present; 25% = approximately ¼ of the bone element (or calvarium) present; 50%: approximately ¼–½ of the bone element (or calvarium) present; 75%: approximately ½–¾ of the bone element (or calvarium) present; 100%: complete or almost complete (more than ¾ of the bone element, or calvarium present).

HEADSHAPING IDENTIFICATION, CLASSIFICATION, AND ANALYTICAL METHODS

The term "headshaping"[16] is used here to denote modification of cranial form for cultural reasons. It should be differentiated from malformations and deformations due to genetic causes, trauma, or pathological conditions. Headshaping can come about either as an unintentional effect of cultural infant care practices—such as cradleboarding, or laying babies to sleep

16. See n. 1.

consistently in a certain position on a hard surface—or as an intentional effect of manipulation of the infant head using cradleboards or particular types of devices.[17] In the former case where headshaping occurs as an unintentional secondary effect, this does not rule out its subsequent intentional usage for sociocultural differentiation, following the recognition of the modification. Thus, the term headshaping denotes here both intentional and unintentional modification of the cranial shape, of cultural origin.

There are at least three main headshaping types (Fig. 5.2): the circumferential or circular type, the antero-posterior type, and the post-bregmatic type. There is great variation in headshaping types and, at times, there is some overlap between their forms.[18] The circumferential or circular type includes all modifications involving bandaging the head circumferentially. Subtypes include the differential shapes that result from this process; for example, there are circular-erect, circular-oblique, and two-band circumferential types.[19] The post-bregmatic type is so far clearly attested only in Cyprus and consists of the superior flattening of the calvarium produced presumably by securing particular kinds of headgear on the superior aspect of the growing infant head.[20] Antero-posterior headshaping denotes the flattening or modification of the occipital part of the cranium, the frontal part of the cranium, or both parts; but without a circumferential involvement of the whole cranial vault. The various types and subtypes of antero-posterior modification can be brought about by the use of cradleboards, with or without additional accessories, or by securing particular kinds of headgear on the growing infant head.

Given the overall rarity of headshaping and the complexity of its expressions, it is important to be able to differentiate between pathological conditions, unintentional positional causes, and intentional causes for altered head shapes. The following information can be employed:

1. Alterations of cranial shape due to environmental causes (including headshaping) rather than growth disturbances and genetic deficiencies do not result in reduction of the brain case volume. Thus, crania with greatly reduced volumes may be attributed to pathological causes.[21] These individuals are often nonviable and die in infancy or childhood, and so these conditions are evidenced more often in immature skeletal remains.
2. Pathological premature suture closure, for example in the lambdoidal or sagittal sutures, results in different morphological expressions (see discussion below) than unintentional or intentional occipital flattening, which affects the same sutures.[22]
3. Unintentional environmental alterations of the cranial form are often asymmetrical and do not occur uniformly throughout a population. Positional plagiocephaly is an example of this phenomenon.[23]
4. Intentional headshaping aims at symmetrical shapes (cross-culturally attested) and may be uniform or patterned within a population. Cranial features indicating intentional cranial shape modification include pad and binding impressions, and particular features characteristic of various types of headshaping.[24]

17. Lorentz 2003a, 2003b.
18. Lorentz 2003b.
19. Dembo and Imbelloni 1938; Anton 1989.
20. Lorentz 2003b; Schulte-Campbell 1983b; Angel 1972.
21. Anton 1989.
22. Anton 1989; White 1996; Bennett 1965; El-Najjar and Dawson 1977.
23. Littlefield, Pomatto, and Kelly 2000.
24. Buikstra and Ubelaker 1994; Anton 1989.

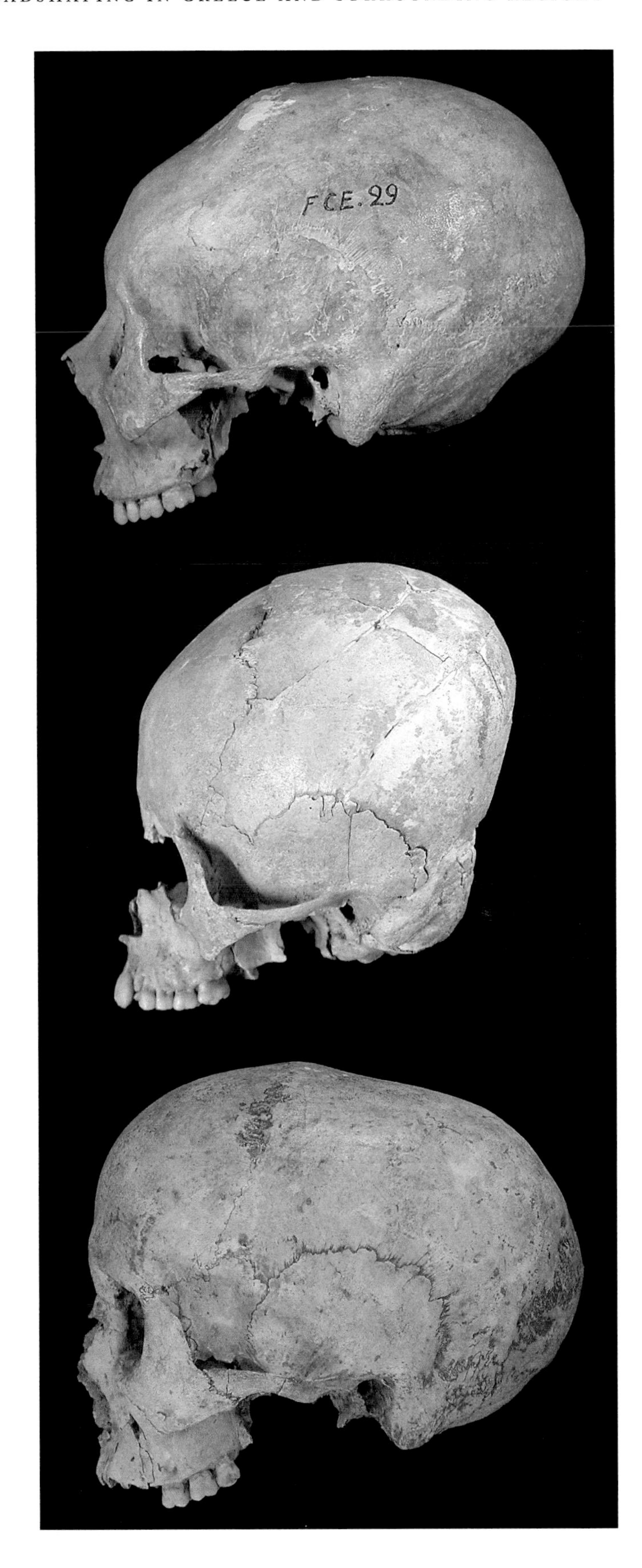

Figure 5.2. Types of cranial modification in Cyprus: circumferential (*top,* from Enkomi), antero-posterior (*middle,* from Khirokitia), and post-bregmatic (*bottom,* from Hala Sultan Tekke). Scale 1:2. Photos K. O. Lorentz, courtesy Department of Antiquities, Cyprus

Figure 5.3. Depictions of cradleboards, Bronze Age Cyprus: inv. no. CM A15 1935, H. 18.9 cm *(left)*; inv. no. CM A26 1935, H. 14.9 cm *(right)*. Photos K. O. Lorentz, courtesy Department of Antiquities, Cyprus

5. Intentional headshaping may involve differential degrees of severity of the alteration and/or different types of alterations within the same population.[25] When the degree of headshaping is employed to signal social difference, there are likely to be classes of severity (gradation) rather than a continuum.
6. Headshaping as an unintentional secondary effect of culturally specific infant care practices, such as cradleboarding,[26] is likely to be expressed universally within a population, with a continuum of degree of severity without marked cutoff points or a large number of extreme cases (Fig. 5.3).

Thus, it emerges that it may be possible to differentiate between the altered head shapes caused by pathological, environmental, positional, and cultural (unintentional and/or intentional) causes, provided that there is a large enough sample population available. This necessitates attention to detail on two different levels: the individual cranium and the population. At the population level, the degree of severity of headshaping, the variation expressed in this degree of severity, the nature of its expression (gradation or continuum), and the prevalence of headshaping all help to distinguish cultural from noncultural head shape modifications.[27]

Analysis of altered crania also entails some specialized procedures. In addition to complete descriptions, detailed data on the following features that are typically associated with headshaping are important:

1. Extrasutural bones. The etiology of extrasutural bones (sutural bones, supernumerary bones, and ossicles) has been debated. The main focus of the debate has been the extent to which the

25. Lorentz 2003b, 2004, 2005.
26. Karageorghis 1991; Lorentz 2004, 2005.
27. Lorentz 2003b.

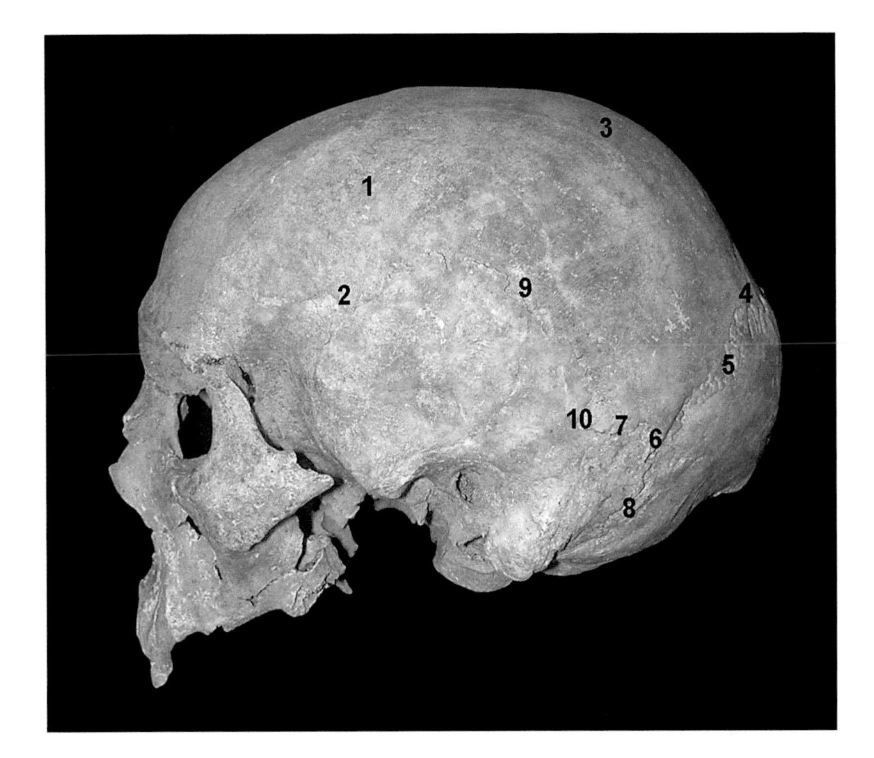

Figure 5.4. Lateral view of a cranium, showing the sutural locations for recording extrasutural bone presence and frequency (1, coronal; 2, pteronic; 3, sagittal; 4, apical; 5, lambdoid; 6, asteric; 7, parieto-mastoid; 8, occipito-mastoid; 9, squamosal; 10, parietal notch), according to the method presented by O'Loughlin (2004). Photo K. O. Lorentz

presence and frequency of extrasutural bones is attributable to genetic factors, environmental factors (such as headshaping), or both. Previous studies of the relationship between extrasutural bones and headshaping have provided ambiguous results, likely due their failure to investigate different headshaping types separately. The most recent study by O'Loughlin[28] indicates that both antero-posterior and circumferential types of headshaping affect the frequency of some types of extrasutural bones. In particular, these headshaping types exhibited significantly greater frequencies of lambdoid ossicles in O'Loughlin's study. Presence and frequency of extrasutural bones can be assessed by recording the number of extrasutural bones in the different cranial sutures, using the method presented by O'Loughlin.[29] These include extrasutural bones at the coronal, pterionic, sagittal, apical, lambdoid, asterionic, parietomastoid, occipitomastoid, squamosal, and parietal notch regions (see Fig. 5.4). Hauser and Stefano[30] present even more detailed divisions of the cranial sutures, allowing scoring of fragmentary cranial material.

2. Cranial synostoses. Craniosynostosis involves the premature fusion of one or more of the calvarial sutures.[31] Following the method presented by White[32] premature cranial synostosis can be scored as present or absent. Premature synostosis is identified by a marked differential fusion between one suture (or sutural region) and the other major vault sutures (i.e., the suture scored as displaying premature synostosis must be obliterated both endocranially and ectocranially when the other sutures are still open).

28. O'Loughlin 2004.
29. O'Loughlin 2004.
30. Hauser and De Stefano 1989.
31. Cohen 1986.
32. White 1996, pp. 401–402.

3. Sutural interdigitation. Interdigitation refers to another aspect of relative complexity of the cranial sutures, in addition to the presence and frequency of extrasutural bones. As in the case of extrasutural bones, it is debated whether the extent of sutural interdigitation is attributable to genetic factors, environmental factors (mechanical loading, such as headshaping), or both.[33] A study by Anton, Jaslow, and Swartz[34] indicates that different types of headshaping may have differential effects on the extent of sutural interdigitation. However, their results are ambiguous, and more studies are needed. The researchers quantified measures of sutural complexity (interdigitation and number and size of extrasutural bones) from digitized tracings of 13 sutures and compared these among three groups of crania showing antero-posterior headshaping, circumferential headshaping, or no modification. This method requires complete crania, however, and as such is not applicable to fragmentary materials, including series like Tharrounia. The degree of interdigitation can be assessed in incomplete crania by scoring sixteen separate portions of the coronal (six, three on each side), sagittal (four), and lambdoidal (six, three on each side) sutures, according to the method presented in Hauser and Stefano.[35] Using this method potentially allows for the comparative assessment of sutural complexity on fragmentary crania, where obtaining interdigitation data on the full length of a suture may not be possible.
4. Asymmetries. Some researchers have referred to a possible increase in cranial asymmetry in connection to antero-posterior headshaping achieved by cradleboarding.[36] Kohn, Leigh, and Cheverud explored the vault modification of Hopi crania using finite element scaling, and found significant correlation of size asymmetry with direction of modification in the cranial vault.[37] However, few satisfactory metric or morphometric methods[38] have been developed allowing quantitative analysis of incomplete crania. The method presented by Zonenshayn, Kronberg, and Souweidane[39] may provide a way forward. It is based on semiautomated, computerized analysis of digital images of the superior aspect of the cranium, surrounded by a headband with nasion and inion indicated. Cranial index of symmetry (CIS) can be calculated based on the shape and area of each hemisphere. This method is applicable to both complete and incomplete crania, as well as living populations, potentially allowing quantitative comparisons of cranial asymmetry between archaeological and living populations exhibiting cranial shape modifications (such as headshaping, or positional plagiocephaly).

Various other methods of shape analysis can also be applied to the analysis of headshaping. One of the recent methods is presented by Friess and Baylac.[40] It uses elliptic Fourier analysis of Procrustes aligned outlines. The problem with this method, as with many other shape analysis methods, is that it is only applicable to relatively complete crania.[41]

33. Anton, Jaslow, and Swartz 1992.
34. Anton, Jaslow, and Swartz 1992.
35. Hauser and De Stefano 1989.
36. Kohn, Leigh, and Cheverud 1995.
37. Kohn, Leigh, and Cheverud 1995.
38. Kohn, Leigh, and Cheverud 1995.
39. Zonenshayn, Kronberg, and Souweidane 2004.
40. Friess and Baylac 2003.
41. The Tharrounia series, like many of the Mediterranean and Near Eastern series, includes mainly incomplete crania.

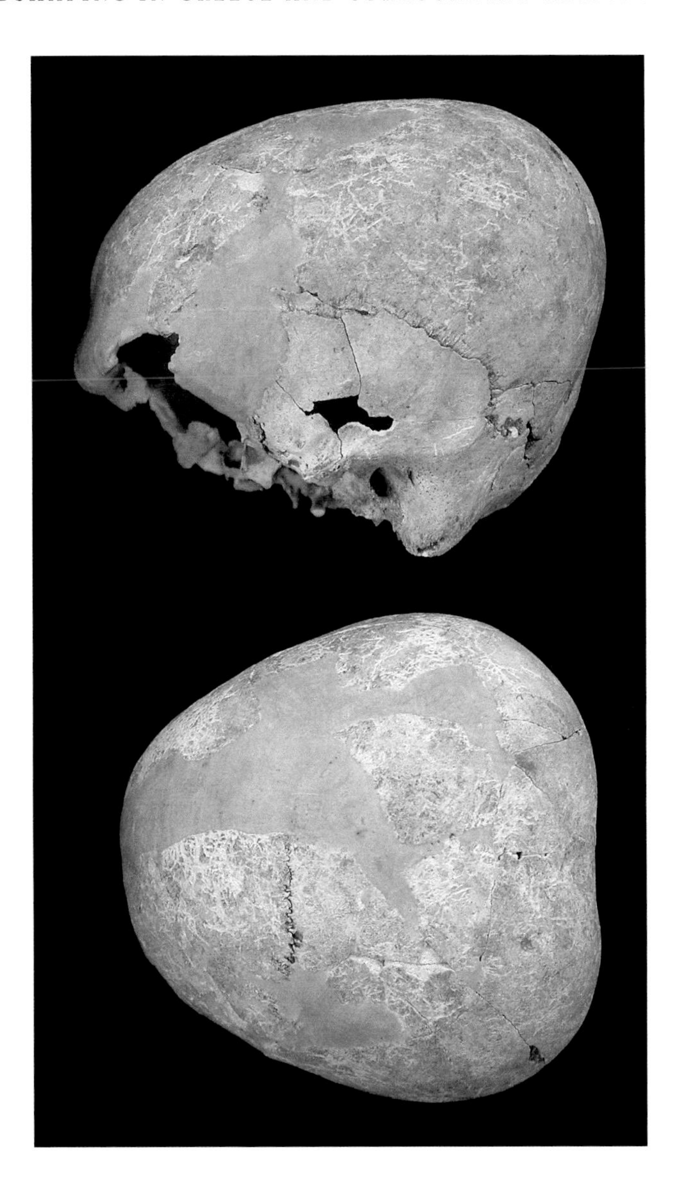

Figure 5.5. Antero-posterior headshaping at Tharrounia. Cranium 9, left lateral *(top)* and superior *(bottom)* views. Scale 1:2. Photos K. O. Lorentz

RESULTS

All of the crania within the Tharrounia series that are sufficiently complete for headshaping assessment—11 out of a total of 13—display some extent of antero-posterior headshaping.[42] The subtype of antero-posterior headshaping present at Tharrounia can be characterized as occipital flattening, with occasional slight involvement of the frontal (for example, in crania 8, 9, 10, and 12).[43]

Tharrounia cranium 9 is a typical example of an antero-posteriorly modified cranium within the series (Fig. 5.5). This cranium has been modified by pronounced headshaping. The curvature of the parietals is clearly increased. The superior aspect of the cranium appears sphenoid, and the sagittal suture is depressed, especially in the posterior half. The occipital squama curvature is decreased, and there is a depressed area on the

42. Lorentz and Manolis, forthcoming.
43. Lorentz and Manolis, forthcoming.

Figure 5.6. Posterior view of a cranium from Khirokitia with an extrasutural bone band in the lambdoid suture. Scale 1:2. Photo K. O. Lorentz, courtesy Department of Antiquities, Cyprus

occipital forming a clear concavity, a possible pad or binding impression. The frontal involvement is not clear. There is a small extrasutural bone at lambda (apical extrasutural bone), and at least two, but possibly three or four, joining extrasutural bones in a row in the right part of the lambdoidal suture, where fragmentation complicates further analysis.

The expression of these extrasutural bones is similar to the conjoining extrasutural bones in the extrasutural bone bands occurring in the antero-posteriorly shaped crania in the Neolithic Khirokitia series in Cyprus (Fig. 5.6).[44] The width of the probable band of extrasutural bones on cranium 9 is ca. 27 mm. There is no clear post-coronal depression.

Due to the incomplete nature of the crania available for analysis from Tharrounia, only broad estimates of age and sex could be performed (see Table 5.2). It seems that headshaping was practiced universally so that all individuals received the treatment in infancy at Neolithic Tharrounia. However, there are clear differences in the intensity or extent of headshaping among the crania.

Table 5.3 gives details of the extrasutural bone presence and number in the lambdoidal suture in the Tharrounia crania. It should be noted that full assessment of extrasutural bone presence in all sutures of all crania was also not possible due to the incompleteness of the crania.

Within the Tharrounia series, premature cranial synostosis is present in three examples, crania 8, 9, and 10. The premature synostosis in cranium 8 is asymmetrical, involving the left coronal suture only. Slight cranial asymmetry accompanies this asymmetrical synostosis. Crania 9 and 10 both display premature sagittal synostosis, assessed by the discrepancy between the fused and almost obliterated sagittal suture and the other cranial sutures, which are still open. In the case of cranium 10, the lambdic part of the lambdoid suture is also fused. It is interesting to note that both of these crania display pronounced headshaping, occupying positions 1 and 3 in the headshaping extent seriation respectively. The extent of

44. See Lorentz 2003b.

TABLE 5.2. EXTENT OF HEADSHAPING BY AGE AND SEX

Position in Series	*Cranium*	*Age Estimate**	*Sex Estimate***
1	9	A	F
1	7	SA (ca. 4–5 yrs)	
1	12	SA (+7 yrs)	
2	3	A	
3	10	A	F
3	6	A	
4	1	A	
5	5	A	Possibly M
6	11	SA (4–5 yrs)	
7	8	A	M
8	2	A	Possibly F

* The incompleteness of the adult crania and the possible effects of headshaping in cranial suture closure do not allow the use of the cranial suture closure method for adult aging. The lack of secure associations between post-cranial elements and the crania prevent use of aging methods based on post-cranial elements. Assignation of individual crania to the broad category of "adult" is based on general morphological features (Schwartz 1995) and the fusion of the spheno-basilar synchondrosis. Sub-adult age estimates are based on morphology and the developmental stage of the dentition (Smith 1991; Moorrees, Fanning, and Hunt 1963; Ubelaker 1989; Lorentz and Manolis, forthcoming). Abbreviations: A = adult, SA = sub-adult.

** Sex estimations are based on cranial morphology (Schwartz 1995; Lorentz and Manolis, forthcoming). The lack of secure associations between post-cranial elements and the crania prevented use of sex estimation methods based on post-cranial elements. Abbreviations: F = female, M = male.

headshaping (Table 5.2) was qualitatively determined by visual inspection, following O'Loughlin.[45] Several crania occupy the same position in the seriation as to extent of headshaping: crania 7, 9, and 12, for example, all display equally pronounced antero-posterior headshaping. Position 1 in the seriation denotes the most pronounced antero-posterior headshaping present in Tharrounia, while position 8 represents the least pronounced. Crania 4 and 13 could not be placed in the seriation, as they are not complete enough for assessment. The lack of male individuals high up in the seriation could be taken as an indication of potential gender differentiation as to the extent of headshaping. However, due to the problems with sex estimation (the number of subadults and the fragmentary nature of the cranial remains) and the small sample size, it is not possible to arrive at conclusions on this.

Pathological premature cranial synostosis can cause modification of the head shape, but in such cases the sagittal suture is often raised,[46] not depressed as in the Tharrounia examples. Further, the lack of cranial synostosis in the other similarly modified crania within the Tharrounia series negates the possibility that premature cranial synostosis would have caused the type of modifications seen in the sample. Instead, the fusion visible in crania 9 and 10 may be the consequence of tensile forces on cranial sutures caused by headshaping.[47] White[48] suggests that headshaping devices create tensile forces on the sagittal suture during the peak growth period of the parietals, and that these forces may induce an adaptive response that

45. O'Loughlin 2004, 1996.
46. Anton 1989.
47. White 1996.
48. White 1996.

TABLE 5.3. EXTRASUTURAL BONE PRESENCE AND NUMBER IN THE LAMBDOIDAL SUTURE

Cranium Number	*Age Estimate**	*Sex***	*Extrasutural Bones*
1	A		
2	A	Possibly F	None
3	A		
4	A		
5	A	Possibly M	None
6	A		
7	SA (ca. 4–5 yrs)		
8	A	M	2 apical 1 right lambdic 1 right asteric
9	A	F	1 apical 2 right lambdic 1 left asteric
10	A	F	None
11	SA (4–5 yrs)		1 right lambdic 1 left lambdic
12	SA (+7 yrs)		1 left asteric
13	A	Possibly F	

Note: For scoring, the lambdoidal suture is divided into sections following Hauser and DeStefano 1989, p. 88: apical part (at lambda); lambdic part (generally more complicated and extending from the lambda towards a more or less marked angle in the suture); intermediate part (generally more complicated part extending laterally from the more or less marked angle in the suture); and asteric part (more simple sutural pattern, near asterion). Data are not given where missing regions prevent assessment.

* For ageing methods used see first note to Table 5.2. Abbreviations: A = adult, SA = sub-adult.

** For sex estimation methods used see second note to Table 5.2. Abbreviations: F = female, M = male.

contributes to premature sagittal synostosis. The asymmetry of the cranial vault of cranium 10 may be the result of cradleboarding and induced premature sagittal synostosis.

The etiology of craniosynostosis is unclear. O'Loughlin[49] has posited a multifactorial etiology involving genetics, biochemical abnormalities, and environmental constraints. Further research into the possible connection between premature cranial synostosis and headshaping is needed.

Table 5.4 details the interdigitation assessment conducted according to the method published by Hauser and Stefano.[50] No assessments of interdigitation could be conducted on crania 7 and 13 due to high fragmentation, and Tharrounia cranium 6 allowed only one assessment. The rest of the crania allow between 5 and 11 assessments each, and in no cranium is a complete assessment possible.

There does not seem to be any correlation between the distribution of the combination interdigitation scores[51] and the severity of headshaping.[52] However, it should be noted that the amount of missing data prevents statistical analyses. Qualitative assessment of the sutural regions most often present in the Tharrounia crania does not show any patterning according to headshaping severity.

49. O'Loughlin 1996.
50. Hauser and De Stefano 1989.
51. Hauser and De Stefano 1989.
52. Lorentz and Manolis, forthcoming.

TABLE 5.4. SUTURAL INTERDIGITATION SCORES

Cran No.	*Cor-breg L*	*Cor-breg R*	*Cor-com L*	*Cor-com R*	*Cor-tem L*	*Cor-tem R*
1				5-3l-2		
2	oblit	2-2l-1	oblit	oblit	oblit	oblit
3	3-2l-1	3-2l-1		4-3l-2		
4		2-2l-1		4-3l-4		
5						
6						
7	partial		partial			
8	3-2l-1	3-2l-1	5-3l-4			
9	3-2l-1			4-3l-4		
10	3-2l-1	3-2l-1	4-3l-3	4-3d-3		3-2l-1
11				4-2d-1		1-1-1
12	1-1-1	1-1-1				3-2l-1

Cran No.	*Sag-breg*	*Sag-vert*	*Sag-obel*	*Sag-lam*
1	3-2l-1	6-3l-2	3-2l-1	4-3l-2
2	5-3l-1	5-3l-1	5-3l-1	
3	3-2l-1	4-2l-1	5-2l-1	1-1-1
4				
5	3-2l-1	4-2l-1	5-3l-1	4-2l-1
6			3-2l-1	
7	partial			
8	3-2l-1	4-2l-1	4-2l-1	1-1-1
9	fused	fused	fused	5-2l-1
10	fused (3-2l-1)?	fused	fused (5-2l-1)?	fused
11			4-2l-1	3-2l-1
12	2-2l-1	3-2l-1		

Cran No.	*Lam-lam L*	*Lam-lam R*	*Lam-int L*	*Lam-int R*	*Lam-ast L*	*Lam-ast R*
1		partial		partial		partial
2	4-2l-2	5-2l-1	5-3l-4	4-2l-1	4-2l-2	3-2l-1
3	partial	partial	partial	partial		
4						
5	4-2l-1	4-2l-1	5-2l-2	5-2l-1		
6						
7						
8	5-2l-1	5-3l-2		5-2l-1		3-2l-1
9	5-2l-2	2-2l-1	5-3l-3	5-3l-3	2-2l-1	
10	fused	fused	4-2l-2	5-2l-1	4-2l-3	3-2l-1
11		3-2d-1		4-2l-1		1-1-1
12						

Note: The first number in the series of three denotes Hauser and DeStefano's (1989, p. 88) first criterion (maximal sutural shape extension), followed by their second criterion (basic configurations), and ending with their third criterion (secondary protrusions). The second criterion is further qualified by a letter: d = dentate; l = looped. Where data are not given, it is because missing regions prevent assessment. This includes all sutures of cranium 13, which is omitted from the table. Cor-breg = Coronal suture, bregmatic part; Cor-com = Coronal suture, complicated part (pars complicata); Cor-tem = Coronal suture, temporal part; Sag-breg = Sagittal suture, bregmatic part; Sag-vert = Sagittal suture, vertex part; Sag-obel = Sagittal suture, obelic (interforaminal) part; Sag-lam = Sagittal suture, lambdic part; Lam-lam = Lambdoid suture, lambdic part; Lam-int = Lambdoid suture, intermediate part; Lam-ast = Lambdoid suture, asteric part; L = left, R = right; oblit = obliterated.

Comparison of the Tharrounia and the Khirokitia (Cyprus, Neolithic)[53] series crania reveals significant similarities in headshaping type and prevalence, both showing antero-posterior type headshaping with universal prevalence. The Tharrounian series does not, however, contain such pronounced cases of antero-posterior headshaping as those found at Khirokitia, where some of the crania display almost fully concave occipital squama.[54]

DISCUSSION

Other Evidence of Headshaping from Greece

Is Tharrounia the only site in the area of modern Greece with evidence of headshaping? This is unlikely, but a lot of basic research is still needed to locate series that might potentially show evidence of headshaping. Further, due to the fragmentary nature of many of the skeletal series in Greece, restorations of at least the parietal regions are needed to enable the assessment of the presence or absence of headshaping. In the following section, some notes on potential headshaping evidence from other sites and skeletal series from Greece are discussed. It should be stressed that what follows is by no means a final, comprehensive, or conclusive list.

At least one cranium from the Athenian Agora shows modification.[55] Cranium 13AA[56] dating from the Roman period displays clear antero-posterior headshaping, with slight plagiocephaly (the right side protrudes slightly more posteriorly).[57] There is no evidence of premature suture closure, as all of the cranial sutures are of a similar fusion stage (mostly visible and unfused). The posterior flattening centers on lambda, but there are no clear localized depressions in this region, nor is there any clear frontal involvement, apart from slight asymmetry corresponding to the posterior asymmetry mentioned above. The posterior third of the parietals show lateral expansions, making the superior aspect of the cranium appear sphenoid. There is a clear post-coronal depression. There are no extrasutural bones present. The extent of antero-posterior modification is such that it would have been clearly visible with a certain kind of hairstyle. Cranium 14AA, belonging to an adult male of same date, also displays "clear cut lambdoid flattening"[58] according to Angel. Angel[59] also mentions that another cranium, cranium 27AA,[60] of Neolithic date and from the same series, has "a sloping forehead, flat lambda region, and rising vertex profile" with "definite skull and face broadening." This description seems consistent with the shape modifications present at Tharrounia and begs the question of whether or not there is some antero-posterior modification present. Unfortunately, it has not yet been possible to locate this cranium. These preliminary findings would warrant the complete survey of the Athenian Agora series to assess the presence of headshaping within this important skeletal sample.

Some of the crania collected from modern Athenian cemeteries seem to show signs of shape modification, possibly related to midwifery practices of binding the infant's head. This is consistent with the evidence for the

53. Angel 1953, 1961.

54. Lorentz 2003b.

55. I am grateful to John Camp for providing the opportunity to view this cranium.

56. Roman period I, ca. A.D. 450. Adult female. Find context: section Y, well at 9/Xita. Angel 1945, p. 315.

57. Angel describes this cranium as "a very short brachycrane, sphenoid cranium, with narrow forehead and almost vertically flat lambdoid-occipital region" (Angel 1945, p. 315).

58. Angel 1945, p. 313. Roman period I, ca. A.D. 450. Adult male (age estimate by Angel: ca. 38 years at death). Find context: section Y, well at 9/Xita.

59. Angel 1945.

60. Angel provides the following age and sex estimates: "a young adult male, probably between 30 and 35 at death" (Angel 1945, p. 291). 27AA was found in section E, shaft grave 2.

circumferential type of headshaping during recent times from the village of Yeranica[61] and the district of Zalka (Tiflis).[62] Ethnographic and oral history research is urgently needed to investigate these more recent practices while elderly informants are still living. Sporadic and vague mentions of headshaping in Crete, both in recent and in archaeological populations, have also been made by Von Fürst,[63] Kurth and Röhrer-Ertl,[64] and Dingwall.[65] It is likely that the current paucity of secure evidence for headshaping from Greece is an artifact of the current research situation, rather than a real absence of the practice itself. However, if the practice really is unusual, the Tharrounia material is remarkable and would require a very specific kind of cultural interpretation.

Headshaping in the Mediterranean and the Near East

Table 5.5 presents some of the archaeologically known evidence for headshaping from the Near East. The areas with current unpublished and published evidence for headshaping from this wide region include Turkey, Iran, Iraq, Syria, Lebanon, and the southern Levant, as well as Cyprus, Greece, and Crete. It should be noted that it is only for Cyprus that we so far have a headshaping sequence researched in sufficient detail to begin to understand the cultural contexts in which the different types and combinations of types of headshaping were employed.[66]

In Cyprus,[67] headshaping occurs from the earliest known Neolithic mortuary series to at least the Early Iron Age. Interestingly, until the Late Bronze Age, and specifically until the Late Cypriot II period, there seems to be only one type of headshaping present in Cyprus, the so-called antero-posterior type. However, during the Late Cypriot, the headshaping practices become much more prolific, with as many as three different types of headshaping (antero-posterior, post-bregmatic, circumferential), as well as non-modified individuals.

It is conceivable that headshaping in Greece or in Cyprus did not arise in isolation from practices in surrounding regions and links between the information about Near Eastern practices shown in Table 5.5 and evidence from the Mediterranean should be explored. As the overall extent of skeletal evidence for headshaping in Greece is as yet poorly understood, the following discussion focuses on Cyprus and its interconnections as an example of potential avenues for future research. Although regions such as Iran may seem very far away from Cyprus, it should be noted that the Ubaid interaction sphere was vast, and many of the sites close enough to Cyprus to have an impact in some form, even if secondarily, were connected to this Ubaid sphere. The Ubaid phenomenon occurred in Mesopotamia between about 6000 B.C. or earlier, and 3800 B.C. It can be divided into six different chronological phases according to differences in pottery. Ubaid pottery is found in settled villages along the great rivers of Mesopotamia, the Tigris and Euphrates, but it seems that members of these communities traveled far in search of new lands and natural resources. Some went north, following the river systems to their sources. Others went downstream (south) as far as the Arabian Peninsula. By 4000 B.C. the influence of the Ubaid

61. Hasluck 1947.
62. Dingwall 1931.
63. Von Fürst 1933.
64. Kurth and Röhrer-Ertl 1981.
65. Dingwall 1931.
66. Lorentz 2003b, 2004, 2005.
67. Lorentz 2003b, 2004, 2005; Schulte-Campbell 1979, 1983a, 1983b, 1986; Angel 1953, 1961, 1972; Fox Leonard 1997.

TABLE 5.5. ARCHAEOLOGICAL EVIDENCE FOR HEADSHAPING FROM THE NEAR EAST

Site (Country)	*Period*	*Type of Headshaping*	*Reference*
Kurban Höyük (Turkey)	Halaf	Type not clear	Alpagut 1986
Degirmentepe (Turkey)	Ubaid (second half of 5th millennium B.C. uncal)	Circumferential	Özbek 1984, 1986, 2001
Şeyh Höyük (Turkey)	Ubaid	Circumferential	Senyürek and Tunakan 1951
Karataş (Turkey)	Early Bronze Age	Antero-posterior; post-bregmatic?	Angel 1968, 1970, 1975
Gordion (Turkey)	Phrygian	Circumferential	Unpublished
Tepe Ghenil (Iran)	Late 8th–early 6th millennium B.C.	Circumferential?	Meiklejohn et al. 1992
Ganj Dareh (Iran)	ca. 7500–6500 B.C. uncal	Circumferential	Lambert 1979; Meiklejohn et al. 1992
Tepe Abdul Hosein (Iran)	First half of 7th millennium B.C.	Circumferential	Lorentz, forthcoming
Ali Kosh (Iran)	7th millennium B.C.	Circumferential	Lambert 1979; Meiklejohn et al. 1992
Seh Gabi (Iran)	Chalcolithic	Circumferential	Lambert 1979; Meiklejohn et al. 1992
Chaga Sefid (Iran)	ca. 7500–5000 B.P. uncal	Circumferential	Hole 1977
Qumrud (Iran)	5th millennium B.C.	Circumferential	Lorentz, forthcoming
Choga Mish (Iran)	4500–4000 B.C.	Circumferential	Ortner 1996
Tepe Sialk (Iran)	5th–4th millennium B.C.	Antero-posterior	Soto-Heim 1986
Ghalecoti (Iran)	250 B.C.– 200 A.D.	Antero-posterior	Soto-Heim 1986
Bolghasian (Iran)	622–700 A.D.	Antero-posterior	Pardini 1968; Soto-Heim 1986
Shanidar (Iraq)	Mousterian; 9000–8500 B.C. uncal	Circumferential? (debated)	Trinkaus 1982; Meiklejohn et al. 1992
Eridu (Iraq)	Late Ubaid	Circumferential	Meiklejohn et al. 1992
Tell Arpachiyah (Iraq)	Ubaid	Circumferential	Molleson and Campbell 1995
Telul-eth Thalathat (Iraq)	Ubaid	Type not stated	Egami 1959
Tell Madhur (Iraq)	Ubaid	Circumferential?	Downs 1984
Tell Ramad (Syria)	Neolithic	Type not stated	Ferembach 1957, 1985; Arensburg and Hershkovitz 1988
Bouqras (Syria)	6500–5500 B.C.	Type not clear	Meiklejohn et al. 1992
Ras Shamra (Syria)	18th–13th century B.C.	Type not stated	Vallois 1960; Vallois and Ferembach 1962; Soto-Heim 1986
Minet el Beida (Syria)	14th–13th century B.C.	Type not stated	Vallois and Ferembach 1962; Soto-Heim 1986
Byblos (Lebanon)	Second half of 4th millennium B.C.	Circumferential	Vallois 1937; Özbek 1974a, 1974b, 1976
Sidon (Lebanon)	4th–5th century A.D.	Circumferential	Özbek 1974a, 1974b
Phoenician (sites not specified)	Phoenician	Type not stated	Lortet 1884; Arensburg and Hershkovitz 1988
Jericho (Southern Levant)	Aceramic	Antero-posterior	Kurth 1959, 1980; Kurth and Röhrer-Ertl 1981; Meiklejohn et al. 1992
Tell Duweir (Southern Levant)	8th–7th century B.C.	Circumferential	Risdon 1939; Özbek 1974a, 1974b
En Gedi (Southern Levant)	Hellenistic–Byzantine	Antero-posterior	Arensburg and Hershkovitz 1988
Yavne Yam (Southern Levant)	Hellenistic–Byzantine	Antero-posterior	Arensburg and Hershkovitz 1988
Jericho (Southern Levant)	Hellenistic–Byzantine	Antero-posterior	Arensburg and Hershkovitz 1988

Note: It should be noted that there are many more sites in the Near East with skeletal remains that have not been adequately assessed for presence or absence of headshaping, and there is scope for future work (Lorentz, forthcoming).

sphere on temple architecture and probably mythology and language had spread northward as far as Syria and Iran, with Ubaid eventually replacing the long-established Halaf and Samarran cultures. Whether this spread involved an actual movement of peoples or mimicry of material culture is less certain. Ubaid pottery is found as far afield as Arabia and Iran.[68] The evidence detailed in Table 5.5 seems to suggest that in addition to pottery and temple architecture, people living within the Ubaid sphere had another thing in common: circumferential headshaping. The evidence from sites such as Eridu in Iraq, the early Iranian sites, and the sites situated in the southeast of modern Turkey (Değirmentepe and Şeyh Höyük) seems to be consistent with this hypothesis, warranting its more detailed investigation in the future.[69] The evidence from Byblos, at a reasonable distance from Cyprus, falls close to the latter part of this period (second half of the 4th millennium B.C.). The first archaeologically detected evidence for circumferential headshaping from Cyprus dates from the Late Cypriot period (1600–1050 B.C.), a couple of thousand years later than the very similar Ubaid related circumferential headshaping. This begs the question of a possible connection between these occurrences of the same form of headshaping in Cyprus and on the mainland. In order to fully understand whether there are cultural transfers involved or whether it is rather a question of parallel innovation, it is paramount that the skeletal evidence for headshaping from sites dating to the intervening periods is analyzed in more detail in the future (e.g., Early Bronze Age evidence from Karataş in Turkey, and Ras Shamra and Minet el Beida in Syria). In this context it is interesting to note the iconographic similarities noted by many researchers between the anthropomorphic figurines of the Ubaid period[70] and the later "Birdfaced" figurines found in Cyprus and dated to the Late Cypriot Bronze Age.[71]

Other potential geographical and temporal patterns of interest emerging from Table 5.5 include the early evidence for circumferential headshaping from the area of Iran and the possible cessation of circumferential headshaping by the Hellenistic–Byzantine period, although evidence for antero-posterior headshaping continues to be found. The above discussion and inferences are necessarily preliminary in nature due to the lack of detail in publications mentioning headshaping evidence in the region.

Sociocultural Context

Headshaping is inherently tied up with sociocultural practices related to the care of children, which may be seen as traditional and slow to change. It thus has a heightened potential to serve as a cultural marker when investigating regional connections and, better still, to act as an indicator of the nature and intensity of culture contact, migration and population movement, cultural diffusion, and acculturation. Display artifacts (artifacts used to display socioeconomic and/or political prestige and power), such as daggers, earrings, and pots, are adopted and move much more easily between populations and areas than does headshaping. Further, the study of differential headshaping practices has the potential to aid our understanding of

68. Yoffee and Clark 1993, p. 265; Oates 1993, pp. 409–410, 414–415; Moorey 1994, p. 154.
69. Lorentz, forthcoming.
70. Molleson and Campbell 1995.
71. Karageorghis 1993.

emerging complex societies, and the ways in which past societies employed bodily markers and modifications for denoting hierarchical positions, as well as other forms of sociocultural difference.

How do cultural practices of body modification come about? How did it occur to people to begin modifying cranial shape? When a group of people adopt the practice of headshaping, two explanations can be invoked: culture contact and original innovation.

First, let us consider the situation where a population adopts the practice of body modification through culture contact with a population that is already practicing headshaping. Due to the length of time needed to bring about a permanent modification of the head shape[72] and the need to appreciate the rather narrow window of opportunity presented by the progress of cranial growth, it is improbable that headshaping was adopted through sporadic culture contact. Some kind of "apprenticeship," or a prolonged period of observation at least, would likely have been needed. So far, we have considered only the practical aspects of the acquisition of the knowledge of how to bring this body modification about. We also need to consider the intricacies involved in the conceptual acceptance, and the cultural context that would have allowed or indeed spurred the adoption of such a practice. Sociocultural practices related to the care of children rarely change abruptly, and often retain traditional aspects, even when other realms of culture show considerable malleability.[73] In addition, many village-based societies are highly prescriptive in terms of individual appearance and dress. It is unlikely that headshaping was begun as a whim or fashion (thus, it is unlike many modern body modifications). It is conceivable that a cultural demand for a highly visual marker of sociocultural difference would be needed to motivate the adoption of this rather labor-intensive and time-consuming modification, which requires detailed knowledge. This kind of demand could arise, for example, in a situation where the population already practicing headshaping occupies a higher social status in cultural exchanges, or its sociocultural (including ritual and religious) and/or economic practices, material culture, or technology are seen as desirable and linked to headshaping practices.[74] Headshaping has been used in various historical and archaeological contexts to denote status.[75]

Further, headshaping is done to one in infancy by others. It is not the kind of body modification that can be brought about at will later in life or instigated by the individual concerned. This aspect draws attention to the fact that it is the caretakers, and those who would have held power over the forms of care children received, who decided whether and what kind and intensity of modification was introduced. Thus, if headshaping was used for denoting status, it would have not been possible for an individual seeking power to bring about this modification later in life. This indicates a close link between the use of headshaping for denoting status and hereditary or ascribed status. Likewise, if headshaping was used to denote membership of a sociocultural group, be it ethnic or another kind of a group, it would not have been possible for an individual to instigate such modification on themselves upon joining such a group. This may indicate a certain rigidity for sociocultural groups using headshaping as a marker of difference. Headshaping would have been a barrier to change

72. Littlefield, Pomatto, and Kelly 2000; Lorentz 2003b.

73. Whiting 1963; Zelizer 1985.

74. See Dingwall 1931 on American Indian practices involving differentiating the free and the enslaved.

75. Dingwall 1931; Tiesler Blos 1998; Lorentz 2007, 2008.

in such sociocultural constellations. Thus, headshaping is unlike aspects of dress or bodily modifications that can be performed in adulthood (such as scarifications or plastic surgery) in that it has to be planned and instigated in infancy, by others.

However, how does headshaping originally come about? It is unlikely that there was a single area of origin for the varied headshaping practices known from all over the world and from different time periods. The practice could have arisen independently at different time periods and locations. A significant cognitive leap is required to move from the detailed observation of the modificatory power of devices (for example, cradleboards) and practices (swaddling, laying babies to sleep on relatively hard surfaces in consistent positions), to the conscious and intentional exploitation of the observed plasticity of the growing human head. Further, only particular cultural conditions would have conceivably allowed the widespread adoption and elaboration of intentional modification of the head shape by a whole population or significant groups within it.

It seems probable that the use of cradleboards to transport and protect infants, as attested ethnographically, caused the first significant modifications of the head form. These effects were subsequently observed and then intentionally exploited by caretakers. Certainly in Cyprus, the earliest known type of headshaping, the antero-posterior type, is consistent with this hypothesis. The earliest known evidence comes from the Kissonerga-Mylouthkia wells.[76] The cranium found at this site is dated to the Cypro-PPNB period.[77] It shows a moderate extent of antero-posterior headshaping that could have come about simply as a secondary effect of cradleboarding. However, already during the Khirokitian Neolithic (7000/6500 B.C.–5800/5500 B.C.),[78] the intensity and variation in antero-posterior type headshaping within the skeletal series from the site of Khirokitia, together with large concavities on the posterior aspect of some of the crania, suggest clearly intentional modification of the head shape.[79] Much later in the Cypriot cultural sequence new headshaping types are introduced: Late Bronze Age II sees the introduction and adoption of at least two additional types, the so-called post-bregmatic (also known as the "Cypriot" type) and the circumferential type. Indications of the different type of headshaping devices used to achieve these forms can be gained through the detailed analysis of morphological changes within the different types.[80] The antero-posterior type features are consistent with the use of cradleboards: the occipital flattening is most likely a result of a consistent supine position while attached to a relatively hard, flat surface. The frontal impressions visible in some of the antero-posterior modified crania most likely relate to accessories used to bind the head to the cradleboard. It should be noted, however, that it is enough to swaddle the infant body tightly on the board to result in occipital flattening. Use of such devices is documented ethnographically as well as archaeologically.[81] Cypriot coroplastic art also depicts cradleboards. The dating of these depictions of freestanding cradleboards is somewhat problematic due to their provenance from illicit excavations, but Karageorghis states the Early Cypriot period as a probable date for the freestanding cradleboards and the Middle Cypriot period for the depictions of plank figurines holding cradleboarded infants.[82]

76. Peltenburg et al. 2001.
77. Peltenburg et al. 2001.
78. Peltenburg 1989; Peltenburg et al. 2001.
79. Lorentz 2003b.
80. See Lorentz 2003b.
81. Dellinger 1936; Holliday 1993; Hudson 1966.
82. Karageorghis 1991.

It is conceivable that the two other general types of headshaping in Cyprus, the post-bregmatic and the circumferential types, were achieved with the use of a freestanding device, attached to the head only and unrelated to cradleboarding. This would be consistent with the changes in the anthropomorphic depictions of infants throughout time as well. Such freestanding devices could include bands, boards, or other artifacts secured to the head by bands.[83] Thus, we see a trajectory from the use of a large artifact used also for other purposes—the cradleboard—to the use of freestanding devices attached to the head with the sole purpose of modifying the head form, known archaeologically by the Late Cypriot II and continuing at least until the Iron Age. The fact that headshaping occurs in very rich Late Cypriot Bronze Age tombs, for example in Enkomi and at Kalavasos-Ayios Dhimitrios,[84] indicates the possibility that headshaping during the Late Cypriot period was related to social status and illustrates the very crucial position of the young[85] in realizing the socioculturally prestigious body form for privileged families.

Due to the so far unique nature of the Tharrounia finds from Greece, and the fragmentary condition of the skeletal material, it is difficult to make far reaching conclusions on the beginnings of this practice in Greece. However, the type of headshaping present in the Tharrounia skeletal series is consistent with the use of cradleboards. All of the Tharrounia crania that can be assessed for the presence of headshaping display evidence of antero-posterior modification, indicating that all individuals at Tharrounia were affected to some extent by headshaping practices. There is no artifactual evidence for headshaping devices from Tharrounia, but some of the features of morphology on the modified crania point to possibly intentional attempts to shape the infant head. These include the small concavities on the occipital bone and the lambda region. The pronounced extent of antero-posterior headshaping on crania 9 and 10 points to the possible intentionality of the practice at Tharrounia. However, until larger series of skeletal material are available, the evidence may be viewed as inconclusive although consistent with the hypothesis positing existence of intentional headshaping during the Greek Neolithic. Further research systematically assessing a wide sample of skeletal collections in Greece, starting from collections of similar time periods to Tharrounia and moving to the preceding and succeeding periods and surrounding regions, is needed to illuminate the full cultural context of headshaping in Tharrounia and in Greece.

CONCLUSIONS

The skeletal series from Neolithic Tharrounia displays the earliest substantial evidence for headshaping in Greece. The type of headshaping attested is antero-posterior, and the subtype is occipital flattening. The occurrence of headshaping at Tharrounia is likely universal, with all individuals receiving treatment that resulted at least in a slight modification of the head form. Several crania show a pronounced form of modification. It is difficult to ascertain whether the practice of headshaping is intentional at Tharrounia, but several corroborative factors point to this possibility. These include the very pronounced extent of headshaping in crania 7, 9, and 12, and the

83. For photographs of actual headshaping devices constructed of boards and string, see Imbelloni 1932.

84. Lorentz 2003b.

85. Headshaping has to be instigated in infancy.

potential connection between gender and headshaping extent.[86] Due to the number of sub-adults included within the series, DNA analyses would be needed to address the latter issue.

The need to further investigate headshaping practices in Greece becomes clear when one considers that headshaping is known from various time periods and from all the surrounding regions, including Anatolia, the Balkans, Levant, Cyprus, and Italy. The analysis of the Tharrounia skeletal series shows that headshaping was practiced also in prehistoric Greece. It is suggested that the previous lack of evidence for headshaping on the Greek mainland and islands is more a function of the lack of research into such practices rather than the result of an actual absence of this cultural practice in Greece.

The presence of headshaping at Tharrounia also indicates the existence of special kinds of infant care practices, relating both to the sleeping arrangements of infants and to the intentional modification of their bodies. This type of intentional modification may have been used as a marker of social difference to denote differences in gender, status, ethnicity, and/or other social groups. The important discoveries at Tharrounia should spur us on to investigate the cultural employment of the human body and its modification in prehistoric Greece.

86. Lorentz and Manolis, forthcoming.

REFERENCES

Alpagut, B. 1986. "The Skeletal Remains from Kurban Hüyük (Urfa)," *Anatolica* 13, pp. 149–174.

Angel, J. L. 1945. "Skeletal Material from Attica," *Hesperia* 14, pp. 279–363.

———. 1953. "The Human Remains from Khirokitia: Appendix II," in *Khirokitia,* ed. P. Dikaios, London, pp. 416–430.

———. 1961. "Neolithic Crania from Sotira: Appendix I," in *Sotira,* ed. P. Dikaios, Philadelphia, pp. 223–229.

———. 1968. "Human Skeletal Remains at Karataş," *AJA* 72, pp. 258–263.

———. 1970. "Human Skeletal Remains at Karataş," *AJA* 74, pp. 253–259.

———. 1972. "Late Bronze Age Cypriotes from Bamboula: The Skeletal Remains," in *Bamboula at Kourion,* ed. J. L. Benson, Philadelphia, pp. 148–165.

———. 1975. "Excavations in the Elmali Area, Lycia, 1975. Appendix: Early Bronze Karataş People and Their Cemeteries," *AJA* 80, pp. 385–391.

Anton, S. C. 1989. "Intentional Vault Deformation and Induced Changes of the Cranial Base and Face," *American Journal of Physical Anthropology* 79, pp. 253–267.

Anton, S. C., C. R. Jaslow, and S. M. Swartz. 1992 "Sutural Complexity in Artificially Deformed Human *(Homo Sapiens)* Crania," *Journal of Morphology* 214, pp. 321–332.

Arensburg, B., and I. Hershkovitz. 1988. "Cranial Deformation and Trephination in the Middle East," *Bulletins et Mémoires de la Société d'Anthropologie de Paris* 14, pp. 139–150.

Bennett, K. A. 1965. "The Etiology and Genetics of Wormian Bones," *American Journal of Physical Anthropology* 23, pp. 255–260.

Brown, P. 1981. "Artificial Cranial Deformation: A Component in the Variation in Pleistocene Australian Aboriginal Crania," *Archaeologia Oceania* 16, pp. 156–167.

Buikstra, J. E., and D. H. Ubelaker, eds. 1994. *Standards for Data Collection from Human Skeletal Remains,* Fayetteville, Ark.

Cheverud, J. M., L. A. P. Kohn, S. R. Leigh, and S. C Jacobs. 1993. "Effects of Annular Cranial Vault Modification on the Cranial Base and Face," *American Journal of Physical Anthropology* 90, pp. 147–168.

Cheverud, J. M., L. A. P. Kohn, L. W. Konigsberg, and S. R. Leigh. 1992. "Effects of Fronto-Occipital Artificial Cranial Vault Modification on the Cranial Base and Face," *American Journal of Physical Anthropology* 88, pp. 323–345.

Cheverud, J. M., and J. E. Midkiff. 1992. "Effects of Fronto-Occipital Cranial Reshaping on Mandibular Form," *American Journal of Physical Anthropology* 87, pp. 167–171.

Cohen, M. M., Jr. 1986. "The Etiology of Craniosynostosis," in *Craniosynostosis: Diagnosis, Evaluation and Management,* ed. M. M. Cohen Jr., New York, pp. 59–79.

Dellinger, S. C. 1936. "Baby Cradles of the Ozark Bluff Dwellers," *AmerAnt* 3, pp. 197–214.

Dembo, A., and J. Imbelloni. 1938. *Deformaciones intentionales del cuerpo humano de caracter étnico,* Buenos Aires.

Dingwall, E. J. 1931. *Artificial Cranial Deformation: A Contribution to the Study of Ethnic Mutilations,* London.

Downs, D. 1984. "The Human Skeletal Remains," *Sumer* 43, p. 127.

Egami, N. 1959. *Telul eth-Thalathat* 1: *The Excavation of Tell II, 1956–1957,* Tokyo.

El-Najjar, M. Y., and G. L. Dawson. 1977. "The Effect of Artificial Cranial Deformation on the Incidence of Wormian Bones in the Lambdoidal Suture," *American Journal of Physical Anthropology* 46, pp. 155–160.

Ferembach, D. 1957. "À propos du crâne trépané trouvé à Timna, origine de certaines tribus Berbères," *Bulletins et Mémoires de la Société d'Anthropologie de Paris* 8, pp. 244–275.

———. 1985 "Quelques coutumes et modes préhistoriques intéressant le crâne," *Arquelogia* 12, pp. 47–56.

Fox Leonard, S. C. 1997. "Comparative Health from Paleopathological Analysis of the Human Skeletal Remains Dating to the Hellenistic and Roman Periods, from Paphos, Cyprus, and Corinth, Greece" (diss. University of Arizona).

Friess, M., and M. Baylac. 2003. "Exploring Artificial Cranial Deformation Using Elliptic Fourier Analysis of Procrustes Aligned Outlines," *American Journal of Physical Anthropology* 122, pp. 11–22.

Hasluck, M. 1947. "Head-Deformation in the Near East," *Man* 47, pp. 143–144.

Hauser, G., and G. F. DeStefano. 1989. *Epigenetic Variants of the Human Skull,* Stuttgart.

Hole, F. 1977. *Studies in the Archaeological History of the Deh Luran Plain: The Excavation of Chaga Sefid* (Memoirs of the Museum of Anthropology, University of Michigan, 9), Ann Arbor.

Holliday, D. Y. 1993. "Occipital Lesions: A Possible Cost of Cradleboards," *American Journal of Physical Anthropology* 90, pp. 283–290.

Hudson, C. 1966. "Isometric Advantages of the Cradleboard," *American Anthropologist* 68, pp. 470–474.

Imbelloni, J. 1932. "Sobre un ejemplar mimetico de deformacion craneana," in *Anales del Museo Nacional de Historia Natural de Buenos Aires* 33:74, pp. 193–204.

Karageorghis, V. 1991. *Coroplastic Art of Cyprus* I: *Chalcolithic–Late Cypriot I,* Nicosia.

———. 1993. *Coroplastic Art of Cyprus* II: *Late Cypriot II–Cypro-Geometric III,* Nicosia.

Kiszely, I. 1978. *The Origins of Artificial Cranial Formation in Eurasia from the Sixth Millennium* B.C. *to the Seventh Century* A.D. (*BAR-IS* 50), Oxford.

Kohn, L. A. P., S. R. Leigh, and J. M. Cheverud 1995. "Asymmetric Vault Modification in Hopi Crania," *American Journal of Physical Anthropology* 98, pp. 173–195.

Kohn, L., S. Leigh, S. Jacobs, and J. Cheverud. 1993. "Effects of Annular Cranial Vault Modification on the Cranial Base and Face," *American Journal of Physical Anthropology* 90, pp. 147–168.

Kurth, G. 1959. "Anthropologische Beobachtungen von der Jerichograbung 1955–1958," *Homo* 6, pp. 115–130.

———. 1980. "Beiträge zur anthropologie und populationsbiologie des Nahen Osten aus der Zeit vom Mesolithikum bis zum Chalkolithikum. Ein exemplarischer versuch anhand der serien vom Tell es Sultan/Jericho, Khirokitia/Cypern, Byblos/Libanon, Eridu/Irak und Sialk/Iran," *Bonner Hefte zur Vorgeschichte* 21, pp. 31–203.

Kurth, G., and O. Röhrer-Ertl. 1981. "On the Anthropology of the Mesolithic to Chalcolithic Human Remains from the Tell Es-Sultan in Jericho, Jordan: Appendix B," in *Excavations at Jericho* 3: *The Architecture and Stratigraphy of the Tell,* ed. K. M. Kenyon, London, pp. 407–499.

Lambert, P. J. 1979. "Early Neolithic Cranial Deformation at Ganj Dareh Tepe, Iran," *Canadian Review of Physical Anthropology* 1, pp. 51–54.

Littlefield, T. R., J. K. Pomatto, and K. M. Kelly. 2000. "Dynamic Orthotic Cranioplasty: Treatment of the Older Infant: Report of Four Cases," *Neurosurgery Focus* 9, pp. 1–4.

Lorentz, K. O. 2003a. "Cultures of Physical Modifications: Child Bodies in Ancient Cyprus," *Stanford Journal of Archaeology* 2, http://www.stanford.edu/dept/archaeology/journal/newdraft/2003_Journal/lorentz/paper.html (accessed June 8, 2008).

———. 2003b. "Minding the Body: The Growing Body in Cyprus from the Aceramic Neolithic to the Late Bronze Age" (diss. University of Cambridge).

———. 2004. "Age and Gender in Eastern Mediterranean Prehistory: Depictions, Burials and Skeletal Evidence," *Ethnographisch-Archäologische Zeitschrift* 45, pp. 297–315.

———. 2005. "Late Bronze Age Burial Practices: Age as a Form of Social Difference," in *Cyprus: Religion and Society from the Late Bronze Age to the End of the Archaic Period,* ed. V. Karageorghis, H. Matthaus, and S. Rogge, Mohnesee-Wamel, pp. 41–55.

———. 2007. "From Life Course to Longue Durée: Headshaping as Gendered Capital?" in *Gender through Time in the Near East,* ed. D. Bolger, Walnut Creek.

———. 2008. "From Bodies to Bones and Back: Theory and Human Bioarchaeology," in *Between Biology and Culture,* ed. H. Schutkowski, Cambridge.

———. Forthcoming. "Ubaid Headshaping: Negotiations of Identity through Physical Appearance," in *The Ubaid and Beyond: Exploring the Transmission of Culture in the Developed Prehistoric Societies of the Middle East,* ed. R. Carter and G. Philip, Durham.

Lorentz, K. O., and S. Manolis, forthcoming. *First Evidence for Headshaping in Greece: The Neolithic Population of Tharrounia, Euboea Island, Greece.*

Lortet, L. 1884. "Cause des déformations que présentent les crânes des Syro-Phéniciens," in *Bulletin de la Société d'Anthropologie de Lyon* 3, pp. 30–40.

Meiklejohn, C., A. Agelarakis, P. A. Akkermans, P. E. L. Smith, and R. Solecki. 1992. "Artificial Cranial Deformation in the Proto-Neolithic and Neolithic Near East and Its Possible Origin: Evidence from Four Sites," *Paléorient* 18, pp. 83–97.

Molleson, T., and S. Campbell. 1995. "Deformed Skulls at Tell Arpachiyah: The Social Context," in *Archaeology of Death in the Ancient Near East* (Oxbow Monograph 51), ed. S. Campbell and A. Green, Oxford, pp. 45–55.

Moorey, P. R. S. 1994. *Ancient Mesopotamian Materials and Industries: The Archaeological Evidence,* Oxford.

Moorrees, C. F. A., E. Fanning, and E. E. J. Hunt. 1963. "Formation and Resorption of Three Deciduous Teeth in Children," *American Journal of Physical Anthropology* 21, pp. 205–213.

Oates, J. 1993. "Trade and Power in the Fifth and Fourth Millennia B.C.: New Evidence from Northern Mesopotamia," *WorldArch* 24, pp. 403–422.

O'Loughlin, V. D. 1996. "Comparative Endocranial Vascular Changes Due to Craniosynostosis and Artificial Cranial Deformation," *American Journal of Physical Anthropology* 101, pp. 369–385.

———. 2004. "Effects of Different Kinds of Cranial Deformation on the Incidence of Wormian bones," *American Journal of Physical Anthropology* 123, pp. 146–155.

Ortner, J. 1996. "Artificial Cranial Deformation of a Human Skull from Choga Mish," in *Choga Mish* 1: *The First Five Seasons of Excavations, 1961–1971,* ed. P. Delougaz and H. J. Kantor, Chicago, pp. 319–322.

Ortner, D. J., and W. G. J. Putschar. 1981. *Identification of Pathological Conditions in Human Skeletal Remains,* Washington, D.C.

Ossenberg, N. S. 1970. "The Influence of Artificial Cranial Deformation on Discontinuous Morphological Traits," *American Journal of Physical Anthropology* 33, pp. 357–372.

Özbek, M. 1974a. "Étude de la déformation crânienne artificielle chez les chalcolithiques de Byblos (Liban)" (diss. Université Paris VIII).

———. 1974b. "Etude de la déformation crânienne artificielle chez les chalcolithiques de Byblos (Liban)," *Bulletins et Mémoires de la Société d'Anthropologie de Paris* 1, pp. 455–481.

———. 1976. "Hommes de Byblos: Etude comparative des squelettes des ages des métaux au Proche-Orient" (diss. Université de Bordeaux).

———. 1984. "Etude anthropologique des restes humains de Hayaz Huyuk," *Anatolica* 11, pp. 155–169.

———. 1986. *Değirmentepe Insanlarının Antropologik Acidan Analizi,* Ankara.

———. 2001. "Cranial Deformation in a Subadult Sample from Değirmentepe (Chalcolithic, Turkey)," *American Journal of Physical Anthropology* 115, pp. 238–244.

Pardini, E. 1968. "Il Sistan e un area di deformazione crania?," *Quaderni di Scienze Anthropologische* 1, pp. 188–194.

Peltenburg, E., ed. 1989. *Early Society in Cyprus,* Edinburgh.

Peltenburg, E., P. Croft, A. Jackson, A., C. McCartney, and M. A. Murray. 2001. "Well-Established Colonists: Mylouthkia 1 and the Cypro-Pre-Pottery Neolithic B," in *The Earliest Prehistory of Cyprus: From Colonisation to Exploitation* (CAARI Monograph Series 2), ed. S. Swiny, Boston, pp. 61–94.

Risdon, D. L. 1939. "A Study of the Cranial and Other Human Remains from Palestine Excavated at Tell Duweir (Lachish) by the Welcome Marston Archaeological Research Expedition," *Biometrika* 31, pp. 99–166.

Sampson, A. 1993. *Skotini, Tharrounia: The Cave, the Settlement and the Cemetery,* Athens.

Schendel, S. A., G. Walker, and A. Kamisugi. 1980. "Hawaiian Craniofacial Morphometrics: Average Mokapuan Skull, Artificial Cranial Deformation, and the 'Rocker' Mandible," *American Journal of Physical Anthropology* 52, pp. 491–500.

Schulte-Campbell, C. 1979. "Preliminary Report on the Human Skeletal Remains from Kalavasos-Tenta" in *Vasilikos Valley Project: Third Preliminary Report,* ed. I. A. Todd, *JFA* 6, pp. 298–299.

———. 1983a. "A Late Bronze Age Cypriot from Hala Sultan Tekke and Another Discussion of Artificial Cranial Deformation: Appendix V," in *Hala Sultan Tekke 8* (*SIMA* 45), ed. P. Åström, Göteborg, pp. 249–252.

———. 1983b. "The Human Skeletal Remains from Palepaphos-Skales: Appendix XII," in *Palaepaphos-Skales: An Iron Age Cemetery in Cyprus,* ed. V. Karageorghis, Konstantz, pp. 439–451.

———. 1986. "Human Skeletal Remains," in *Vasilikos Valley Project I: The Bronze Age Cemetery in Kalavasos Village* (*SIMA* 71), ed. I. Todd, Göteborg, pp. 168–178.

Schwartz, J.H. 1995. *Skeleton Keys: An Introduction to Human Skeletal Morphology, Development, and Analysis,* Oxford.

Senyürek, S., and S. Tunakan. 1951. "Şeyh Höyük iskeletleri" *Belleten* 60, pp. 431–445.

Smith, B. H. 1991. "Standards of Human Tooth Formation and Dental Age Assessment" in *Advances in Dental Anthropology,* ed. M. A. Kellen and C. S. Larsen, Wilmington, Del., pp. 143–168.

Soto-Heim, P. 1986. "Déformation crânienne artificielle dans l'Iran ancien," *Bulletins et Mémoires de la Société d'Anthropologie de Paris: Séries XIV* 3, pp. 105–116.

Stravopodi, H. 1993. "An Anthropological Assessment of the Human Findings from the Cave and the Cemetery: Appendix IV," in *Skotini, Tharrounia: The Cave, the Settlement and the Cemetery,* ed. A. Sampson, Athens pp. 378–391.

Tiesler Blos, V. 1998. *La costumbre de la deformacion cefalica entre los antiguos mayas: Aspectos morfologicos y culturales*, Mexico City.

Trinkaus, E. 1982. "Artificial Cranial Deformation in the Shanidar 1 and 5 Neanderthals," *CurrAnthr* 23, pp. 198–200.

Ubelaker, D. H. 1989. *Human Skeletal Remains: Excavation, Analysis, Interpretation,* Washington, D.C.

Vallois, H. V. 1937. "Note sur les ossements humains de la nécropole énéolithique de Byblos," *BMusBeyr* 1, pp. 23–33.

———. 1960. "Vital Statistics in Prehistoric Populations as Determined from Archaeological Data," in *The Application of Quantitative Methods in Archaeology,* ed. R. F. Heizer and S. F. Cook, Chicago, pp. 186–204.

Vallois, H. D., and D. Ferembach. 1962. "Les restes humains de Ras Shamra et de Minet el Beida: Étude anthropologique," *Ugaritica* IV, pp. 565–630.

Von Fürst, C. M. 1933. *Zur kenntnis der anthropologie der prähistorischen bevolkerung der insel Cypern,* Lund.

White, C. D. 1996. "Sutural Effects of Fronto-Occipital Cranial Modification," *American Journal of Physical of Anthropology* 100, pp. 397–410.

White, T., and P. Folkens. 1991. *Human Osteology,* San Diego.

Whiting, B. B., ed. 1963. *Six Cultures: Studies of Child Rearing,* New York.

Yoffee, N., and J. J. Clark, eds. 1993. *Early Stages in the Evolution of Mesopotamian Civilization: Soviet Excavations in Northern Iraq,* Tucson.

Zelizer, V. A. 1985. *Pricing the Priceless Child: The Changing Social Value of Children,* New York.

Zonenshayn, M., E. Kronberg, and M. M. Souweidane 2004. "Cranial Index of Symmetry: An Objective Semiautomated Measure of Plagiocephaly," *Journal of Neurosurgery* (*Pediatrics* 5) 100, pp. 537–540.

CHAPTER 6

Skeletal Evidence for Militarism in Mycenaean Athens

by Susan Kirkpatrick Smith

Militarism was a central concern to the people of the Mycenaean culture of Late Bronze Age Greece. The Mycenaeans left behind heavily fortified buildings; artwork with militaristic themes; Linear B references to chariots, weapons, and soldiers; and burials that contained weaponry and armaments.[1] While militarism in Mycenaean Greece appears to be an established fact, there has not yet been any in-depth study of the persons who lived this military life. Human remains from Mycenaean burials are an untapped reserve that can shed light on Mycenaean Greek military activity. We should expect to find skeletal remains from Mycenaean cemeteries with evidence of traumatic injuries caused by military activity. An analysis of the presence or absence of these injuries, coupled with an analysis of burial treatment, will let us begin to understand how warriors or soldiers may be recognized.[2]

Traumatic injuries in archaeological skeletons can be used to infer particulars of life experiences for individuals and for populations. Common patterns of trauma in a population may point to specific activities performed by group members. Age, sex, and skeletal location of wounds can help determine whether injuries were received from interpersonal conflict, large animals, or accidents.[3] While it may be difficult to make specific inferences based on the analysis of trauma in one individual, patterns of trauma can be quite informative. Stirland, for example, found occupational markers consistent with activities practiced by men in the British navy.[4] In a study of the skeletons from the *Mary Rose,* Henry VIII's flagship that sank in the 16th century, she determined that many of these men (primarily aged 15–25) exhibited arthropathies and enthesopathies that she related to their naval battle activities and weapon use.

Warriors in Mycenaean Greece cannot be found in a single repository as in the *Mary Rose*, but have been identified in Linear B tablets, on

1. See Vermeule 1972 and Dickinson 1994 for discussions of the role of militarism in Mycenaean Greece.

2. Funding for this work was provided by the Wiener Laboratory at the American School of Classical Studies at Athens, a Sigma Xi Grant-in-Aid of Research, and the Ruth Hindman Foundation. I would like to thank James Whitley for sending me a copy of an interesting paper relevant to my topic. My thanks also go to the anonymous reviewer who provided me with helpful comments.

3. Jurmain 2001.

4. Stirland 2000.

artwork, and by their burials with weaponry. The evidence points to at least two major classes of military personnel for Mycenaean culture: a "warrior aristocracy"[5] based on weapons and other rich goods being found together in graves, and more ordinary soldiers[6] who have been described in Linear B tablets. Linear B tablets also clearly indicate high-status leaders with likely military roles (*lawagetas* and possibly *e-qe-ta*[7]), and lower status soldiers, charioteers, and archers. To date, it seems that only the warrior aristocracy has been identified in burials. Driessen and MacDonald, discussing Mycenaean Knossian burials, state:

> The graves furnished with swords at Knossos are not poor burials and none are likely to represent the lower echelons of society nor the rank and file of the Knossian military. *Whether the latter could be identified in other graves is a matter for further research.* (Emphasis mine)[8]

Existing evidence from many Mycenaean sources gives a picture of a society that was heavily invested in warfare, its preparations, and equipment. This study will present the information from the human skeletal remains from nine Mycenaean burials at the Athenian Agora, to show how military and combat activities can be documented in individual warriors and soldiers.

The skeleton is a rich source of information about military and conflict activity, and a source that has not been previously investigated in detail for Mycenaean culture. The Mycenaean burials from the Athenian Agora are an instructive sample of human remains that shows how we might further examine the evidence for warfare or interpersonal conflict activity on a population-wide basis in the context of the existing framework of archaeological interpretation. This kind of study has been undertaken in many other archaeological contexts and has proven to be informative and enlightening.[9]

MYCENAEAN ATHENS

Mycenaean Athens has the same kinds of evidence for social stratification and militaristic activity that are typical of Mycenaean centers in the Argolid.[10] Stratification is suggested by the range of burial types (chamber tombs, cist graves, and pit graves) and the wide variation of the grave goods found in the burials.[11] There are traces of enclosure walls on the Acropolis similar to those at the citadels of Mycenae and Tiryns that would have surrounded a well and allowed for extended defense of the Acropolis.[12]

5. Fortenberry 1990, p. 304. A warrior aristocracy has also been proposed at Knossos. See, for example, Sandars 1963, pp. 127, 128, and Driessen and MacDonald 1984, pp. 56, 58.

6. Driessen and MacDonald 1984, p. 56.

7. Ventris and Chadwick 1973, pp. 122, 408.

8. Driessen and MacDonald 1984, p. 58.

9. See, for example, Jurmain 2001; Kilgore, Jurmain, and Van Gerven 1997; Milner 1995.

10. See Smith 1998 for a discussion of the social status divisions in Mycenaean Athens.

11. *Agora* XIII is a catalog and extensive description of all finds from the Mycenaean period at the Athenian Agora.

12. Iakovides 1983, p. 90.

While the remains at Athens from the Mycenaean period are relatively few, the remains from the Acropolis suggest that Athens was an important site during this time.[13] We can assume, therefore, that Athenian social structure would have been similar to that from Mycenaean centers for which we have more archaeological evidence. We should expect there to have been high and low status individuals,[14] and, in particular, military personnel of various ranks.[15]

Mycenaean Athens in the area of the Agora is primarily represented by burials (see Fig. 6.1).[16] Forty-four separate tombs or graves, 34 of which contained human skeletal remains, have been excavated. The majority of the tombs date to the LH III period (1400–1125 B.C.), with three tombs dating perhaps into the LH I–LH II (1550–1450 B.C.). Three burials are not datable using current methods.

The burials vary in style, size, and contents. The pit and cist graves and chamber tombs each have examples of single and multiple burials. Some are barely large enough for one person. Others have room for multiple individuals to be laid out at once. The grave goods vary from simple pots to gold ornaments and bronze weapons. It is clear that social distinctions were being made in the burial treatment of Mycenaean Athenians.[17] The question for this study is whether we can determine if military activity is one of the social dimensions reflected in the burials.

THE SAMPLE

Burials or skeletons with evidence of military activity are considered in this study (see Fig. 6.1 and Table 6.1). Six Mycenaean Athenian burials contain weapons, which include bronze rapiers, swords, razors,[18] arrowheads, knives, spears, and obsidian arrowheads (Table 6.2). It is not currently possible to determine how these weapons were used, whether for hunting, warfare, or as status symbols. The knives may have been used as offensive weapons[19] or as hunting weapons or razors. There is disagreement about the use of a large spear found in tomb XL. It may have been a weapon for offensive use against another person or it may have been used for hunting.[20]

13. McDonald and Thomas 1990, p. 342; *Agora* XIII, p. 153.

14. Smith 1998, 1999.

15. Smith 1995.

16. The graves are published in *Agora* XIII and the skeletons in Smith 1995, 1998, 1999. The skeletons were aged and sexed by the author according to standard osteological protocols, and without reference to Angel's previous reports. Likewise, all pathologies were described and analyzed by the author, other than AA 28 (see note 29). Two additional Mycenaean chamber tombs were excavated at the Athenian Agora during the summers of 1998 and 1999 (preliminarily published by Camp 2003, p. 254). These tombs each contain several bronze weapons. The skeletal remains have not been published in detail, but Maria Liston has conducted a preliminary analysis. This indicates that tomb J–K 2:2 contained three adults (two males and one female) and one child. Tomb K 2:5 contained at least four adults and three children. One of the adults, a male, survived injuries to a collarbone and rib. Final publication of the human remains will show whether the burials in tombs J–K 2:2 and K 2:5, which appear to have been from a different cemetery than the burials in this study, will exhibit the same patterns of trauma.

17. Smith 1998 has a complete discussion of the nature of the social statuses that were present in Mycenaean Athens.

18. There is some disagreement about whether the items identified as "razors" were in fact knives or other offensive weapons. *Agora* XIII, p. 106 and Blegen 1937 pp. 347–348.

19. Blegen 1937, pp. 347–348.

20. *Agora* XIII, pp. 103, 245.

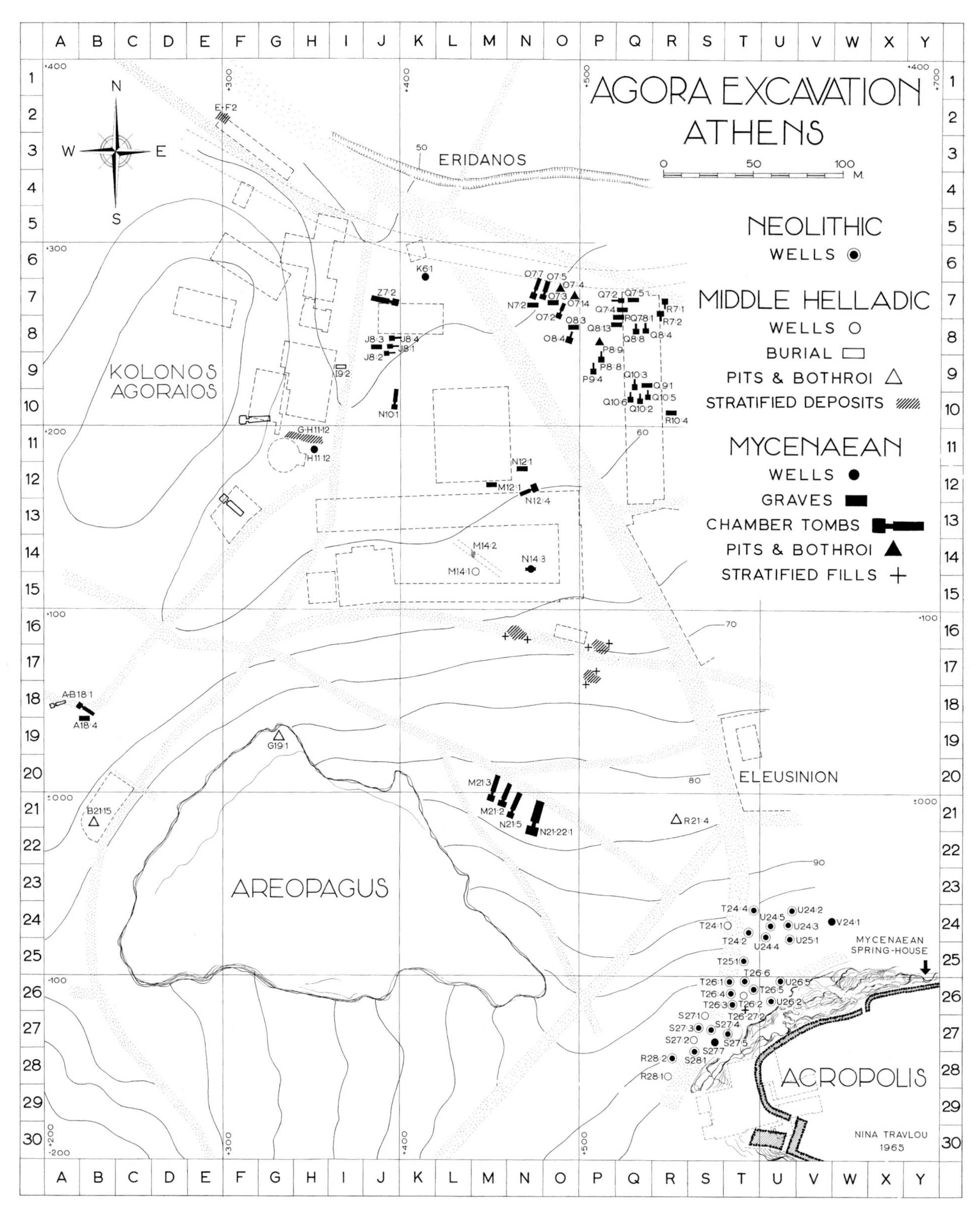

Figure 6.1. Plan of the prehistoric deposits from the Agora excavations in Athens. *Agora* XIII, pl. 91. Courtesy Trustees of the American School of Classical Studies at Athens

TABLE 6.1. TOMB LOCATIONS

Tomb	*Burial*	*Period*	*Map Location*
III	AA 41	LH IIIA2	M 21:2
IV	AA 45	LH IIIA1–LH IIIA2	M 21:3
VII	AA 113	LH IIB–LH IIIA1, LH IIIA2	J 7:2
XIV	AA 118	LH IIIA1–LH IIIA2 early	O 7:5
XV	AA 132	LH IIIA2	O 7:2
XVII	AA 124	LH IIIA–LH IIIB	O 7:3
XVIII	AA 134	LH IIIA2	O 8:4
XXXVII	AA 28	LH IIIA1	M 12:1
XL	AA 300	LH IIIA1	N 12:4

TABLE 6.2. TYPES OF WEAPONRY

Tomb	*Burial*	*Period*	*Disturbed in Antiquity*	*Grave Goods, Weaponry*	*Grave Goods, Other*	*Age*	*Sex*	*Additional Skeletons*
III	AA 41	LH IIIA2	no	rapier, short sword	painted pottery, bronze bowl, ivory and gold rosettes, stone buttons	35+	M	none
VII	AA 113	LH IIB–LH IIA1, LH IIIA2	slight	razor, thrusting weapon, obsidian and bronze arrowheads	painted pottery, ivory comb, bone pin, necklace of quartz and glass paste, steatite bead, buttons	35+	M	juvenile
XIV	AA 118	LH IIIA1–LH IIIA2 early	no	knife or razor (or offensive weapon per Blegen)*	painted pottery, gold bead, paste bead, stone buttons	40+	M	2 adult females
XXVII	AA 124	LH IIIA–LH IIIB	no	bronze knife	painted pottery	15	M	none
XL	AA 300	LH IIIA1	no	thrusting spear	painted pottery, buttons, grave marker, amber bead, carnelian sealstone, scrap of ivory	17–20	M	none
XV**	AA 132	LH IIIA2	extensive	dagger/razor	painted pottery	40+	F	juvenile

* Immerwahr in *Agora* XIII, 1971, has suggested that this tool is a razor, but Blegen (1937) suggested it was a knife.

** This tomb has been eliminated from the analysis because it was extensively disturbed in antiquity and no skeletal remains from a male were found in the tomb.

All but one of the tombs that held weapons contained at least one teenage or adult male burial.[21] The six tombs with weapons date from the same time period, LH IIIA (1400–1325 B.C.). Tomb VII might be from slightly earlier, from LH IIB (1450 B.C.), but the dating on that tomb is not secure.[22]

Three of the skeletons from this cemetery have evidence of trauma that might have been the result of military activity (Table 6.3). An adult male of 35+ years (inv. no. AA 28) was buried in tomb XXXVII. This tomb contained painted pottery when it was excavated, but it had been disturbed in antiquity. Tomb XVIII contained a teenage male (AA 134) along with two adult females. This tomb was also disturbed in antiquity, and

21. Tomb XV, which was disturbed extensively in antiquity, contained the remains of an adult female of at least 40 years and a juvenile of 5–9 years. Because it was disturbed, it is certainly possible that an adult male had been buried in the tomb as well.

22. *Agora* XIII, pp. 183–184.

TABLE 6.3. SKELETAL PATHOLOGIES

Tomb	*Burial*	*Age*	*Sex*	*Weapons Present*	*Skeletal Pathologies and Anomalies*
III	AA 41	35+	M	Y	Several robust muscle attachments (medial head of right gastrocnemius; patellar ligament exostoses on both patellae; humerus, clavicle); disk herniation on many thoracic vertebrae; possible fusion of C4 and C5 at articular facets; body expansion on C5–7, T1–5, S1
VII	AA 113	35+	M	Y	Eburnation on superior articular facet (right) of right C4; some disk herniation
XIV	AA 118	40+	M	Y	Ligamentous ossification of trapezoid ligament on right clavicle, not present on left
XVII	AA 124	mid-teens	M?	Y	Probable healed cribra orbitalia
XL	AA 300	17–20	M	Y	Healed cribra orbitalia
IV	AA 45	adult	M	N	Healed fractures to left metacarpals I and IV
XVIII	AA 134	17–19	M	N	Healed sharp trauma to left frontal; healed sharp trauma to left pterion area; perimortem blunt trauma to right parietal; robust muscle attachments on upper body
XXXVII	AA 28	35+	M	N	Weapon wound to left shoulder; healed fracture of left radius; heavily muscled humeri, scapulae, innominates, and femora

Note: The individuals discussed in the table are from tombs with weapons or who have evidence of possible combat-related trauma.

contained painted pottery and a stone button. Tomb IV, containing an adult male (AA 45), was extensively disturbed in antiquity. There were multiple adult male and female burials in the tomb, and the only grave goods present in the tomb were pottery fragments. All three of these burials with skeletal pathologies date to LH IIIA (1400–1325 B.C.).

RESULTS

The skeletons from the burials with weapons are alike in that none has any evidence of trauma or injury. Four out of the five males from these burials are older adults (35+ years old). The fifth is in his middle teens. Two of the five have fairly robust muscle attachment areas, but there is no consistency in location. Also, because both of these individuals are over 35 years old, it is likely that age is the primary reason for this robusticity.[23]

AA 41 is the only skeleton from a burial with weapons that had an injury that might have resulted from combat. He has a healed fracture of one of the middle ribs on the left side. An injury to the left side of the body is consistent with an injury from a right-handed assailant. However, most rib fractures are the result of falls.[24] At this point, the more parsimonious explanation is a fall.

The remaining four males from tombs with weapons have no evidence of any kind of trauma or pathology that would suggest any form of injury. AA 113 has some eburnation on his fourth cervical vertebra (C4) and some disk herniation, but this is most likely a result of age. AA 118 has

23. Jurmain 1999.

24. Sirmali et al. (2003) examined 548 cases of rib fracture in a modern population. Three hundred seventy-two cases were the result of either auto or industrial accidents. Of the remaining, 122 fractured ribs were caused by falls and only 54 by assault.

TABLE 6.4. BURIALS WITH POSSIBLE COMBAT-RELATED INJURIES

Tomb	*Burial*	*Period*	*Disturbed in Antiquity*	*Grave Goods, Weaponry*	*Grave Goods, Other*	*Age*	*Sex*	*Additional Skeletons*
IV	AA 45	LH IIIA1–LH IIIA2	Extensive	None	Pottery fragments	adult	M	multiple adult males and females
XVIII	AA 134	LH IIIA2	Yes	None	Painted pottery, stone button	17–19	M	2 adult females
XXXVII	AA 28	LH IIIA1	Yes	None	Painted pottery	35+	M	no

extensive ligamentous ossification of the trapezoid ligament of the right clavicle, and a lesion on the inferior surface of the right clavicle at the costoclavicular ligament attachment. Neither of these conditions is present on the left. While the presence of heavy use of the right arm is obvious, the specific nature of his activity cannot be determined.[25] Most people are right handed, so it is not surprising that heavy use of the right arm would be present in a male skeleton.

AA 124 and AA 300 are younger than the other three males buried with weapons and have no pathologies at all, except for some possible healed cribra orbitalia (both) and dental caries (AA 124). The cribra orbitalia, which is not extensive, is not prevalent in this population.[26] The cribra orbitalia in this population is most likely the result of childhood illnesses or parasitic infections,[27] or the result of iron sequestering by the body to combat viral and/or bacterial infections[28] and would not have had any effect on his adult functioning. The dental carries on AA 124 were not extensive and did not produce abscesses or antemortem tooth loss. It is unlikely that they contributed to his death or even to much discomfort.

The final burial with weapons, tomb XV, contained an adult female and a juvenile. This burial was heavily disturbed in antiquity and could have had additional bodies buried in it. There is no archaeological, written, or artistic evidence to suggest that women engaged in military activity during this time, so this tomb will be excluded from further analysis.

Two males from the cemetery, AA 28 and AA 134 (Table 6.4), do have evidence of weapon wounds, though they are buried in tombs that contained no weapons. Both of these males were found in burials disturbed in antiquity, so we do not know if their lack of weaponry is the result of plundering or original absence. AA 28 is a middle-aged male from tomb XXXVI who has a healed "rounded wound depression in the right posterior rim of the joint socket of the right shoulder-blade [that] might easily result from arrow or spear thrust from behind."[29] In addition, this male has a healed fracture of the right radius, which Angel[30] suggests might be the result of a direct force being applied during a fight.

While the scapula injury is most likely the result of a weapon wound, the radius injury may have been caused by a fall with an outstretched arm. Injury to the radius from a fall usually occurs more distally on the radius than it did on this man (11 cm from the distal end of the radius), but defensive

25. Jurmain 1999.

26. Smith 1998, p. 154.

27. Kent 1991; Stuart-Macadam 1982.

28. Weinberg 1984 and Kent, Weinberg, and Stuart-Macadam 1990.

29. Angel 1945, p. 297. This skeleton is no longer housed at the Agora Museum. It was lost several decades ago. The only evidence for it now is Angel's thorough description in this article.

30. Angel 1945, p. 297.

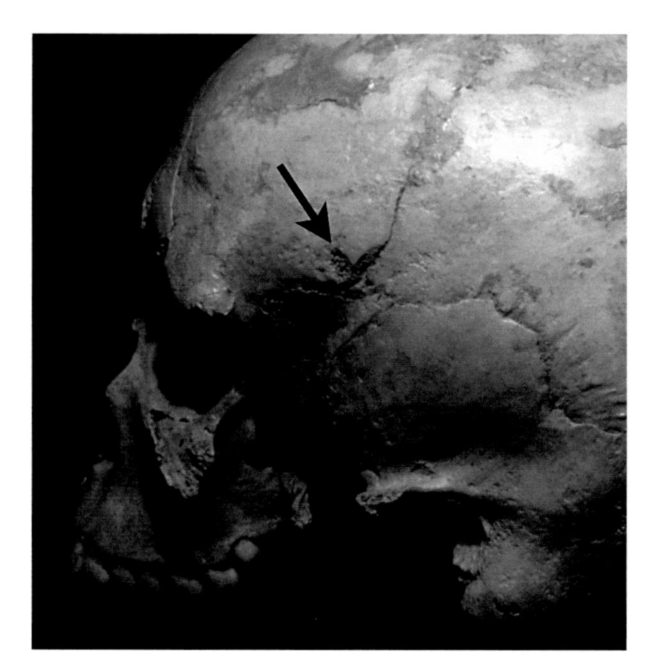

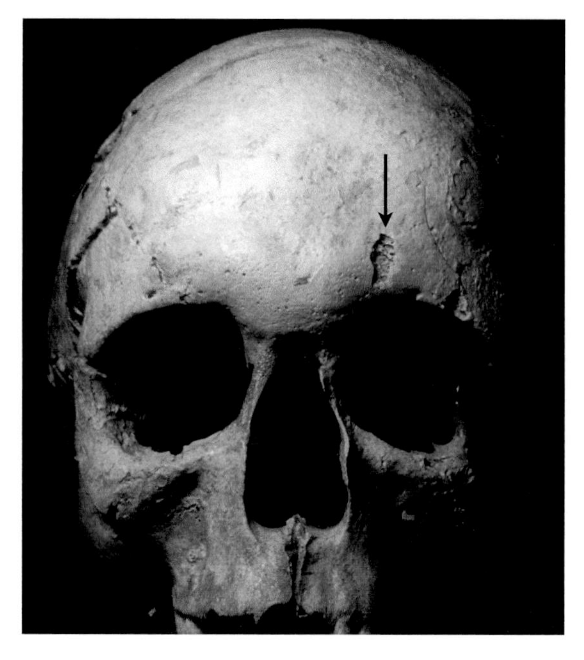

Figure 6.2 ***(left).*** **Agora skeleton AA 134: healed wound at left pterion.** Photo S. K. Smith

Figure 6.3 ***(right).*** **Agora skeleton AA 134: healed wound on left frontal.** Photo S. K. Smith

wounds of the arm usually result in the fracture of the ulna rather than the radius as the arm is held up in a defensive posture.[31]

AA 134 is a 17–19 year old male.[32] The skeleton is fairly complete. He exhibits three head wounds, two with evidence of healing (Figs. 6.2, 6.3) and one that is perimortem and likely the cause of death (Fig. 6.4). The healed fractures, on the left frontal superior to the superior margin of the orbit, and at the left pterion region, have smooth margins with no evidence of infection. The third fracture on the right parietal at the coronal suture is a circular depression with no evidence of healing. Small bone fragments remain in place indicating the fracture occurred while soft tissue was still intact.[33] Lime concretions are adhering to the bone, but this was deposited after the body skeletonized and is therefore not responsible for the small bones being held in place.

The skeletal evidence indicates that this young man engaged in combat at least twice in his life. The healed wounds are consistent with sword or dagger wounds, both weapons that are found at the Agora from this time. His third wound, the circular depression, is consistent with an injury from a sling stone.[34] Sling stones have not been found at the Agora, but there are representations of them in artwork from the period.[35]

One other skeleton exhibits signs of trauma that might be combat related. A commingled collection of adult males and females (designated AA 45) from tomb IV has healed fractures to left metacarpals I and IV. It is probable that these are both from the same male individual, but this is not definite. Angel attributed fractures of the hand in the Mediterranean to hand-to-hand combat.[36] These injuries do not have any evidence of a weapon wound, so it is not possible to say that is how they were broken and hand fractures are among the most common fractures.[37] These metacarpals could not be firmly grouped with other skeletal elements. However, no other trauma is present in any of the other bones in this tomb.

31. Ortner and Putschar 1985, p. 76.
32. Smith 1995.
33. Ortner and Putschar 1985, p. 72.
34. Merbs 1994, p. 174.
35. Vermeule 1972, p. 101, pl. XIV.
36. Angel 1974, p. 17.
37. Landin (1983) analyzed 8,682 fractures in children aged 0–16 years. He found that the bones of the hand (carpals, metacarpals, and phalanges) were the second most common to be broken, after the distal forearm bones. Playing and sports were the most common causes of the fractures. The metacarpals of AA 45 were completely healed and could have been broken in childhood or adulthood. Due to the frequency of fractures in this area caused by accidents, it is not possible to say with any certainty how the injuries of AA 45 were sustained.

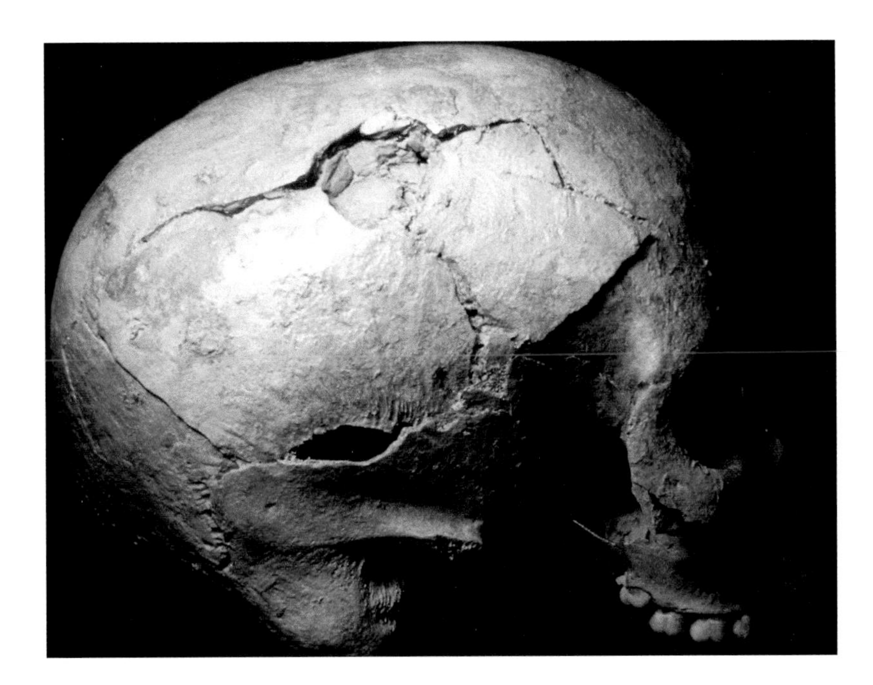

Figure 6.4. Agora skeleton AA 134: perimortem blunt trauma to right parietal. Photo S. K. Smith

DISCUSSION AND CONCLUSIONS

The evidence from burials at the Mycenaean Athenian Agora points to two very different potential pictures of military or combat activity. The archaeological data point to a scenario of five males ranging in age from middle teens to over 40 years, all of whom had evidence of military or fighting weaponry.[38] In contrast, skeletal data present three completely different males ranging in age from late teens to over 35 years of age with evidence for wounds consistent with combat or interpersonal violence. How are we to reconcile these two data sets?

Angel says of skeleton AA 28 that "without much doubt this Athenian was a warrior."[39] This statement is at first glance problematic, but on second glance provocative. The initial problematic issue is that it is impossible to verify that the shoulder and arm injuries were sustained during combat. The other, more provocative, issue is whether we agree with Angel that a man buried without the weapons signifying a warrior status might be considered a "warrior."

"Warriors buried without weapons" is perhaps not as outlandish as it seems. While "'warrior graves' and a 'warrior aristocracy' have . . . become an integral part of the furniture of Aegean archaeology,"[40] it is important to differentiate between who was doing the fighting and who was being lauded in death as a warrior, whether he participated in any fighting or not. It may be that those men buried with weapons at the Mycenaean Athenian Agora were not the soldiers. They may have had the weapons as signs of their high status or leadership roles, rather than as signs of their role as soldiers.

The inverse relationship between presence of weapons in burials and presence of injuries consistent with weapon wounds has been documented at other archaeological sites. Early Anglo-Saxon burials in England have

38. *Agora* XIII lists only tomb III as containing materials similar to a warrior aristocracy. However, Deger-Jalkotzy 2006, has suggested that the presence of weaponry alone may be indicative of warrior burials.

39. Angel 1945, p. 296.

40. Whitley 2002, p. 208.

shown this pattern of large, elaborate metal weapons being buried with adolescents, too young to be the rightful users of the weapons, while skeletal analysis has documented warfare trauma in individuals buried without these trappings of warriorhood.[41]

In such cases, where weapons do not evince warrior status, evidence of who the soldiers themselves were needs to come from their skeletons. If this was true at Athens, then what we may be seeing with AA 28 and AA 134 (Table 6.4) are examples of the rank and file of the Mycenaean military, the men who did the actual fighting.

This fits with the observation that Mycenaean culture may have been characterized by a simple structure rather than a hierarchical system.[42] There would have been an elite group, but otherwise the population would not have been differentiated in terms of their social standing.

Mee and Cavanagh[43] have also suggested that tomb size and grave goods (not burial type) are the important indicators for social status. In the case at Athens, then, the bronze and obsidian tools may be meaningful status differentiators only among high social and/or military ranks. This might reveal some divisions in the elite group itself, which is in fact what is observed. The only Mycenaean burial in the Athenian Agora to be described as belonging to a member of the "warrior aristocracy"[44] is tomb III, which contained a bronze rapier, a short bronze sword, ivory and gold rosettes, a bronze bowl, stone buttons, and painted pottery. The other burials with bronze or obsidian weapons may represent a lower segment of the elite group, or other interpretations of how to bury a "warrior."

Greek prehistorians have many questions in common with the interests of physical anthropologists. These questions can be answered by analysis of the skeletons in conjunction with the artifacts and features of the archaeological record. Due to the small number of "warrior graves" that have been found to date in Mycenaean cemeteries, it is clear that we must look for the men engaged in warfare in burials other than these "warrior graves."[45] We must work with the archaeological and physical remains in conjunction with each other in order to document who the warriors of Mycenaean Greece were.

41. Härke 1990.
42. Mee and Cavanaugh 1984, p. 56.
43. Mee and Cavanaugh 1984.
44. *Agora* XIII, p. 105.
45. Deger-Jalkotzy 2006, p. 152.

REFERENCES

Agora XIII = S. A. Immerwahr, *The Neolithic and Bronze Ages,* Princeton 1971.

Angel, J. L. 1945. "Skeletal Material from Attica," *Hesperia* 14, pp. 279–363.

———. 1974. "Patterns of Fractures from Neolithic to Modern Times," *Anthropologia Keozlemenyek* 18, pp. 9–18.

Blegen, C. W. 1937. *Prosymna: The Helladic Settlement Preceding the Argive Heraeum,* Cambridge, Mass.

Camp, J. McK. II. 2003. "Excavations in the Athenian Agora, 1998–2001," *Hesperia* 72, pp. 241–280.

Deger-Jalkotzy, S. 2006. "Late Mycenaean Warrior Tombs," in *Ancient Greece: From the Mycenaean Palaces to the Age of Homer,* ed. S. Deger-Jalkotzy and I. S. Lemos, Edinburgh, pp. 151–180.

Dickinson, O. T. P. K. 1994. *The Aegean Bronze Age,* Cambridge.

Driessen, J., and C. McDonald. 1984. "Some Military Aspects of the Aegean in the Late 15th and Early 14th Centuries BC." *BSA* 79, pp. 49–74.

Fortenberry, C. D. 1990. "Elements of Mycenaean Warfare" (diss. Univ. of Cincinnati).

Härke, H. 1990. "'Warrior Graves?' The Background of the Anglo-Saxon Weapons Burial Ritual," *Past and Present* 126, pp. 22–43.

Iakovides, S. 1983. *Late Helladic Citadels on Mainland Greece,* Leiden.

Jurmain, R. 1999. *Stories from the Skeleton: Behavioral Reconstruction in Human Osteology,* Amsterdam.

———. 2001. "Paleoepidemiological Patterns of Trauma in a Prehistoric Population from Central California," *American Journal of Physical Anthropology* 115, pp. 13–23.

Kent, S. 1991. "The Shift to Sedentism as Viewed from a Recently Sedentary Kalahari Village," *Nyame Akuma* 35, pp. 2–10.

Kent, S., E. Weinberg, and P. Stuart-Macadam. 1990. "Dietary Prophylactic Iron Supplements: Helpful or Harmful?" *Human Nature* 1, pp. 55–81.

Kilgore, L., R. Jurmain, and D. Van Gerven. 1997. "Paleoepidemiological Patterns of Trauma in a Medieval Nubian Skeletal Population," *International Journal of Osteoarchaeology* 7, pp. 103–114.

Landin, L. A. 1983. "Fracture Patterns in Children: Analysis of 8,682 Fractures with Special Reference to Incidence, Etiology and Secular Changes in a Swedish Urban Population 1950–1979," *Acta Orthopaedica Scandinavica Supplementum* 202, pp. 1–109.

McDonald, W. A., and C. G. Thomas. 1990. *Progress into the Past: The Rediscovery of Mycenaean Civilization,* Bloomington, Ind.

Mee, C. B., and W. G. Cavanagh. 1984. "Mycenaean Tombs as Evidence for Social and Political Organisation," *OJA* 3, pp. 45–64.

Merbs, C. F. 1994. "Trauma," in *Reconstruction of Life from the Skeleton,* ed. M. Y. İşcan and K. A. R. Kennedy, New York, pp. 161–189.

Milner, G. R. 1995. "An Osteological Perspective on Prehistoric Warfare," in *Regional Approaches to Mortuary Analysis,* ed. L. A. Beck, New York, pp. 221–244.

Ortner, D. J., and W. G. J. Putschar. 1985. *Identification of Pathological Conditions in Human Skeletal Remains,* 2nd ed., Washington, D.C.

Sandars, N. K. 1963. "Later Aegean Bronze Swords," *AJA* 67, pp. 117–153.

Sirmali, M., H. Türüt, S. Topçu, E. Gülhan, Ü. Yazici, S. Kaya, and I. Tastepe. 2003. "A Comprehensive Analysis of Traumatic Rib Fractures: Morbidity, Mortality, and Management," *European Journal of Cardio-Thoracic Surgery* 24, pp. 133–138.

Smith, S. K. 1995. "Head Trauma and Occupational Markers in a Late Bronze Age Warrior from Athens, Greece," *American Journal of Physical Anthropology* Supplement 20, p. 198 (abstract).

———. 1998. "A Biocultural Analysis of Social Status in Mycenaean (Late Bronze Age) Athens, Greece" (diss. Indiana Univ.).

———. 1999. "Skeletal and Dental Evidence for Social Status in Late Bronze Age Athens," in *Paleodiet in the Aegean,* ed. S. J. Vaughan and W. D. E. Coulson, Oxford, pp. 105–113.

Stirland, A. J. 2000. *Raising the Dead: The Skeleton Crew of Henry VIII's Great Ship, the* Mary Rose, Chichester.

Stuart-Macadam, P. 1982. "A Correlative Study of a Paleopathology of the Skull" (diss. Univ. of Cambridge).

Ventris, M., and J. Chadwick. 1973. *Documents in Mycenaean Greek,* 2nd ed., Chicago.

Vermeule, E. 1972. *Greece in the Bronze Age,* Chicago.

Weinberg, E. 1984. "Iron Withholding: A Defense against Infection and Neoplasia," *Physiological Review* 64, pp. 65–102.

Whitley, J. 2002. "Objects with Attitude: Biographical Facts and Fallacies in the Study of Late Bronze Age and Early Iron Age Warrior Graves," *CAJ,* 12, pp. 217–232.

CHAPTER 7

Patterns of Trauma in a Medieval Urban Population (11th Century A.D.) from Central Crete

by Chryssi Bourbou

The record of traumatic incidents imprinted upon a skeleton may contain a wealth of information about a lifetime of encounters with the environment and fellow humans. Because they are probably the most easily diagnosed and among the most common types of skeletal trauma, fractures have traditionally garnered the most interest from paleopathological investigators. Although case studies usually predominate in the paleopathological literature, systematic work on fracture prevalence was applied as early as 1910.[1] During the last decades, a number of papers stimulated wider interest in population studies of fractures.[2] This report describes patterns of fractures recorded in an urban population from central Crete. The investigation assesses cranial and postcranial fractures recorded in these remains, dating to the Middle Byzantine period (11th century A.D.) in order to (1) present the raw data, frequencies, and general description of the fractures observed among the population; (2) assess the prevalence of fractures for male and female individuals; (3) investigate any evidence of complications during the healing process or possible treatment; and (4) compare the distribution of fractures between the urban site in question and two rural sites in western Crete, Pemonia and Stylos, that date to the same period.

MATERIALS AND METHODS

The Archaeological Context and Skeletal Sample

In 2003, during the restoration of the Venetian church of St. Peter at the site of Kastella in the modern city of Heraklion in central Crete, a Middle Byzantine cemetery came to light. A total of 32 excavated burials contained single and multiple inhumations of 59 individuals (35 adult and 24 subadult individuals). The typology of the relatively few accompanying goods (mainly pottery and bronze jewels) permitted assigning the exact chronology of these burials to the 11th century A.D. In the history of the continuous occupation of the densely populated modern city of Heraklion, the discovery of this Middle Byzantine cemetery is considered as the "missing chronological link" for understanding the urban life of the city during

1. Wood-Jones 1910.
2. E.g., Lovejoy and Heiple 1981; Merbs 1989; Walker 1989; Roberts 1991; Berger and Trinkaus 1995; Grauer and Roberts 1996; Smith 1996; Stirland 1996; Kilgore, Jurmain, and Van Gerven 1997; Jurmain and Bellifemine 1997; Lambert 1997; Judd and Roberts 1999; Neves, Barros, and Costa 1999; Jurmain 2001; Judd 2004; Domett and Tayles 2006; Djuri et al. 2006; Mitchell, Nagar, and Ellenblum 2006; Brickley 2006. For a thorough bioarchaeological analysis on the history of violence, see Walker 2001.

the Middle Ages. It is a unique, up-to-date, and thorough excavation of a Middle Byzantine site in Crete, revealing traces of occupational activity and mortuary evidence that essentially contribute to the reconstruction of life patterns during a generally unknown period of the Middle Ages.

The study sample consists of the skeletal remains of the 35 adult individuals: 15 males, 8 females, and 12 individuals of unknown sex. Adult age at death was estimated using the pubic symphysis, the auricular surface of the ilium,[3] and dental wear.[4] Sex was determined using dimorphic aspects of the pelvis and skull.[5]

Method of Fracture Analysis

Evaluation of fractures included only adult individuals and was based on macroscopic analysis, supplemented by radiography. All elements with macroscopically visible lesions were radiographed. Each skull bone (frontal, parietal, temporal, occipital, zygomatic, nasal, maxilla, and mandible) was scrutinized for fractures, following the definitions given by Lovell.[6] All skull elements were distinguished by side with the exception of the nasal bones, which were counted as one bone only. In the postcranial skeleton, the presence of fracture was determined for the long bones (clavicle, humerus, ulna, radius, metacarpals, femur, tibia, fibula, metatarsals). Each long bone was identified as present (90% present), incomplete (50%–90% present), fragmentary (<50% present), or absent.[7] Both incomplete and complete long bones with fractures formed the observable corpus. Primarily, the following information was documented for each long bone lesion: side and position of bone (proximal, middle and distal diaphyses; proximal and distal epiphyses), as well as type of fracture, following the definitions summarized by Lovell.[8] Also following Lovell,[9] long bone fracture description included recording of length, apposition (shift), rotation, and angulation (alignment). Finally, aspects of fracture healing, such as duration of healing (i.e., presence of complete or partial callus formation), as well as complications of healing, were also assessed. Fractures in the postcranial skeleton also included injuries to the thorax and the vertebral column.

RESULTS

Kastella Fracture Pattern and Frequencies

The demographic and fracture profile of the Kastella adults is given in Table 7.1. The fracture frequency for individuals was 28.5%, as 10 of the 35 individuals sustained one or multiple fractures. Of the ten traumatized persons, seven were males (70%) and three (30%) were females. Individuals who died between 35 to 44 years of age accounted for the majority of the traumatic lesions (5/35 = 14%), and all of them were males. Table 7.2 provides a categorized summary of the total bones observed and adult fractures from the Kastella sample. A total of 228 long bones was examined and 13 healed fractures were recorded, resulting in a long bone fracture prevalence of 5.70%. Males had ten injuries (77%) and females accounted for the remaining three lesions (23%).

3. As cited in Buikstra and Ubelaker 1994.
4. Brothwell 1981.
5. As cited in Buikstra and Ubelaker 1994.
6. Lovell 1997, pp. 149–150.
7. Judd and Roberts 1999.
8. Lovell 1997, pp. 141–144.
9. Lovell 1997, p. 150.

TABLE 7.1. FREQUENCY OF FRACTURES BY AGE AND SEX IN THE KASTELLA SAMPLE

Age at Death	*Male Fractures*			*Female Fractures*			*Total Fractures*		
(in years)	*N*	*n*	%	*N*	*n*	%	*N*	*n*	%
18–24	1	0	–	1	0	–	2	0	–
25–34	2	1	50	2	1	50	4	2	50
35–44	5	5	100	1	0	–	6	5	83
45+	2	0	–	2	1	50	4	1	25
Adult	5	1	20	2	1	50	7	2	29

N = total number of individuals; *n* = individuals exhibiting fractures; % = $n/N \times 100$.

TABLE 7.2. FREQUENCY OF FRACTURED BONES AT KASTELLA

	Right Fractures			*Left Fractures*			*Total Fractures*		
Bone	*N*	*n*	%	*N*	*n*	%	*N*	*n*	%
Parietal	12	1	8.3	5	1	20	17	2	11.8
Occipital	14	1	7.1	5	0	–	19	1	5.7
Clavicle	6	0	–	18	2	11.1	24	2	8.3
Scapula (body)	8	0	–	12	1	8.3	20	1	5
Humerus	26	2	7.7	11	1	9.1	37	3	8.1
Radius	12	0	–	22	2	9.1	34	2	5.9
2nd mcp	10	0	–	20	1	5	30	1	3.3
5th mcp	15	0	–	20	2	10	35	2	5.7
Ribs	30	1	3.3	80	6	7.5	110	7	6.4
5th lumbar	–	–	–	–	–	–	14	1	7.1
Tibia	12	0	–	25	2	8	37	2	5.4
Fibula	10	0	–	18	1	5.5	28	1	3.6

N = total number of bones observed; *n* = number of bones with fractures; % = $n/N \times 100$; 2nd mcp = second metacarpal; 5th mcp = fifth metacarpal.

Only two cranial fractures were recorded in the sample, both involving male individuals aged 30–35 years and 39 years, respectively. Skeleton 002 exhibited a diagonal cut from the left parietal to the central part of the occipital bone (Fig. 7.1). The sharp-edged defect was 10.1 cm long with a maximum width of 1.5 cm. The internal edges were sharp with no evidence of healing. Skeleton 014 presented an oval depressed skull fracture (1.1 × 0.5 cm) on the right parietal, just above the lambdoidal suture (Fig. 7.2).

Three individuals experienced multiple fractures. Skeleton 001, a female ca. 25 years old, suffered from a compression fracture on the fifth lumbar vertebra (Fig. 7.3). Protrusion to the spinal cord and ossified nodules were observed on the superior body, while the inferior body was normal. Slight marginal osteophytosis was also present on the superior body. Five left ribs (fifth to eighth and eleventh) presented transverse fractures toward the neck (Fig. 7.4). Callus formation was visible in all cases. Involvement of the pleura sheath in the healing process was noted in fifth to seventh ribs and woven bone formation on the eighth rib. A transverse fracture was also observed on the distal third of the left tibia (Fig. 7.5). Callus formation

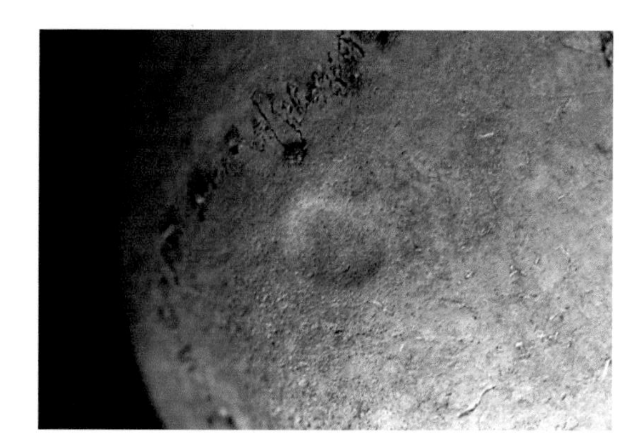

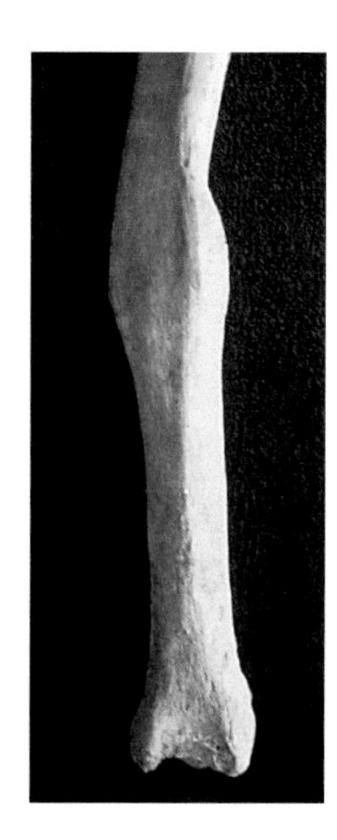

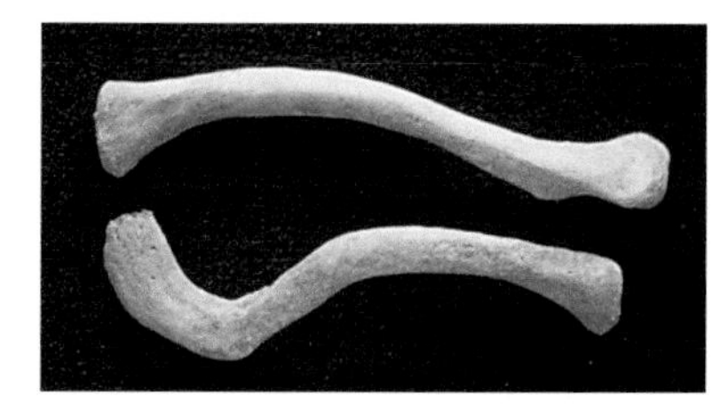

Figure 7.1 *(top left)*. **Kastella skeleton 002: a diagonal cut from the left parietal to the central part of the occipital bone.** Photo C. Bourbou

Figure 7.2 *(top right)*. **Kastella skeleton 014: oval depression fracture on the right parietal.** Photo C. Bourbou

Figure 7.3 *(middle left)*. **Kastella skeleton 001: compression fracture on the fifth lumbar vertebra.** Scale 1:2. Photo C. Bourbou

Figure 7.4 *(above left)*. **Kastella skeleton 001: transverse fracture toward the neck of the eighth left rib.** Scale 1:2. Photo C. Bourbou

Figure 7.5 *(above center)*. **Kastella skeleton 001: transverse fracture on the distal third of the left tibia.** Scale 1:3. Photo C. Bourbou

Figure 7.6 *(above right)*. **Kastella skeleton 001: transverse fracture at the distal end of the left clavicle.** Scale 1:3. Photo C. Bourbou

was visible and the affected bone was shorter than the contralateral (L = 34 cm, R = 36 cm). Secondary osteoarthritis at the ankle joint developed as evidenced by the osteophytic formation on several tarsal bones: on the plantar surface of the talus, the posterior articular surface of the calcaneus, and the plantar surface of the navicular. Finally, a transverse fracture was also present at the distal end of the left clavicle (Fig. 7.6). The affected clavicle was shorter than the contralateral (L = 10.1 cm, R = 11 cm) and the bone exhibited angular alignment.

Skeleton 011a, a male of 44 years, sustained a transverse fracture on the mid-shaft of the fifth left metacarpal (Fig. 7.7). Slight callus formation was visible. In addition, the left scapula presented a sharp cut on the body (Fig. 7.8). Complications in the healing process included a displacement of the bone fragments in the lateral border and ossified nodules due to possible involvement of the infraspinatus muscle. Skeleton 017, a male of 35–40 years, also presented multiple fractures. Transverse fractures were observed at the mid-shaft of the right humerus, the distal third of the left radius, the mid-shaft of the left fifth metacarpal, the distal third of the left fibula, the distal end of the left tibia (Fig. 7.9), and in two right and one left fragmentary ribs. Callus formation was present in all cases and complications were noted in the fractured tibia: ossified nodules were

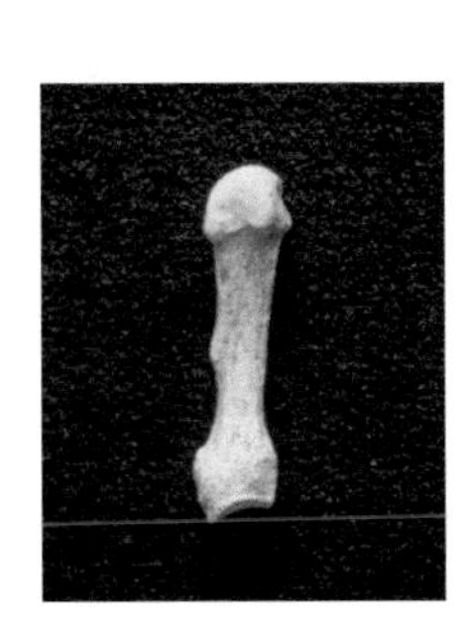

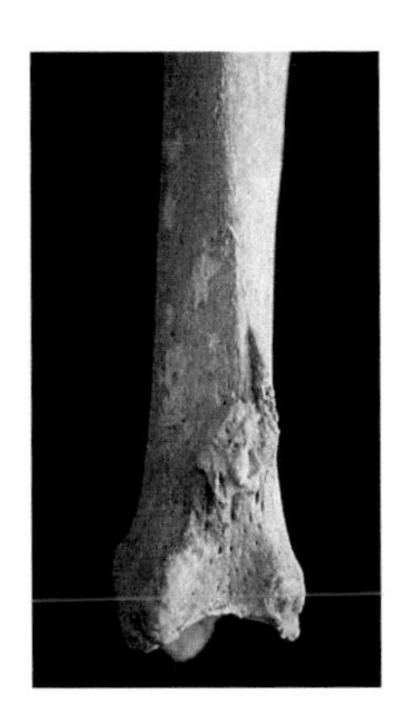

Figure 7.7 *(left)*. **Kastella skeleton 011a: transverse fracture on the mid-shaft of the fifth left metacarpal.** Scale 1:2. Photo C. Bourbou

Figure 7.8 *(center)*. **Kastella skeleton 011a: sharp cut on the body of the left scapula.** Scale 1:3. Photo C. Bourbou

Figure 7.9 *(right)*. **Kastella skeleton 017: transverse fracture at the distal end of the left tibia.** Scale 1:3. Photo C. Bourbou

present on the insertion of the interosseous ligament, and there was osteoarthritis of the ankle joint (marginal osteophytosis at the distal end of the tibia and the talus).

Skeleton 008b, a mature adult male, presented a transverse fracture, well remodeled, at the third proximal diaphysis of the right humerus. Skeleton 013, a male of 44 years, suffered from a Galleazi's fracture at the distal end of the left radius. Angular alignment was observed, as well as osteoarthritis on the wrist joint in the presence of marginal osteophytosis and slight eburnation on the radius. Skeleton 022, a male of 35 years, presented a transverse fracture at the distal end of the left clavicle. Angular alignment and shortening of the bone were noted (L = 14 cm, R = 14.5 cm). Skeleton 023, a female of 50–60 years, sustained a transverse fracture on the mid-shaft of the left second metacarpal. Callus formation was present and the affected bone appeared shorter than the contralateral (L = 5 cm, R = 6 cm). Finally, skeleton 025, an adult female, exhibited a transverse fracture at the mid-shaft of the left humerus. Callus formation was visible, as was additional shortening of the bone (L = 30 cm, R = 32 cm).

COMPARISON WITH OTHER SITES

It is only during the last few years that human skeletal material from highly ignored time periods (i.e., the Byzantine and the post-Byzantine) has contributed to the construction of past health patterns in Greece.[10] For Middle Byzantine Crete, data on human skeletal remains are available for two rural sites in the western part of the island, Pemonia and Stylos.[11]

Pemonia is a village in the region of Apokoronas, some 30 km from Khania and at an altitude of 210 m. In 1994 and 1995, excavations at the interior of the church of Ayios Georgios revealed five tombs. The skeletal remains came from two cist graves (tomb 1 and tomb 5). Tomb 1 contained the commingled remains of secondary burials (minimum number of individuals = 15: 11 adults and 4 subadults), and tomb 5 contained an in situ burial of an adult, as well as the remains of two secondary burials (one adult and one subadult). Although this is a small sample, numerous fractures were recorded for the adult individuals (Table 7.3): a right ulna presented a parry fracture (Fig. 7.10), a right radius presented a Galleazi's fracture, a left and right humerus exhibited transverse fractures on the mid-shafts,

10. Garvie-Lok 2001; Bourbou 2004.

11. Bourbou 2003.

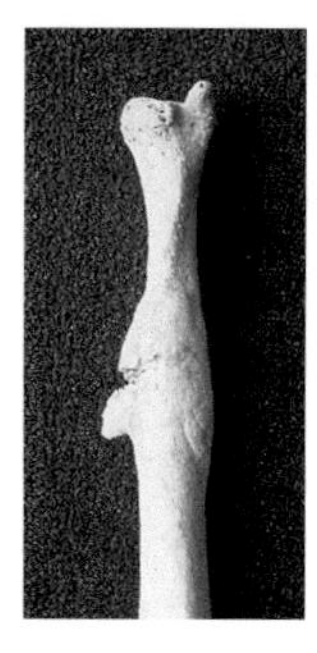

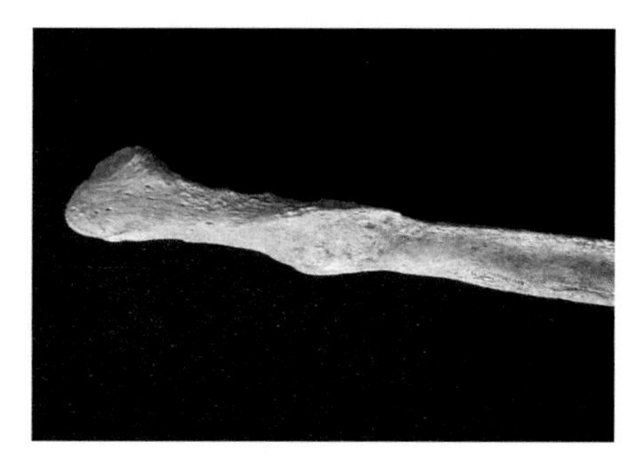

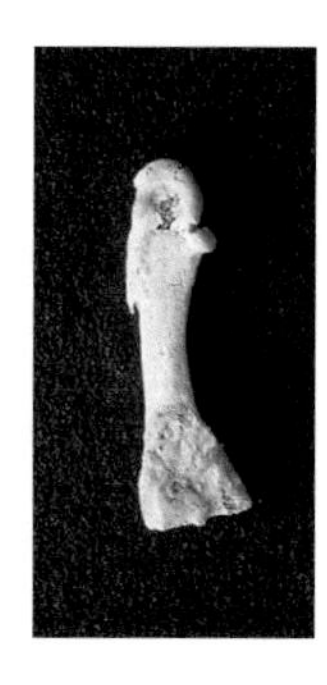

Figure 7.10 ***(left).*** **Pemonia, tomb 1 (commingled remains): parry fracture on a right ulna.** Scale 1:3. Photo C. Bourbou

Figure 7.11 ***(center).*** **Pemonia, tomb 5 (commingled remains): transverse fracture on the midshaft of a left fibula.** Scale 1:3. Photo C. Bourbou

Figure 7.12 ***(right).*** **Pemonia, tomb 5 (commingled remains): transverse fracture on the proximal end of a third right metatarsal.** Scale 1:3. Photo C. Bourbou

TABLE 7.3. FREQUENCY OF FRACTURED BONES AT PEMONIA

	Right Fractures			*Left Fractures*			*Total Fractures*		
Bone	*N*	*n*	*%*	*N*	*n*	*%*	*N*	*n*	*%*
Ulna	7	1	14.3	7	0	–	14	1	7.1
Radius	7	1	14.3	9	0	–	16	1	6.2
Humerus	8	1	12.5	10	1	10	18	2	11.1
Fibula	9	1	11.1	6	1	16.6	15	2	13.3
3rd mtt	5	1	20	6	0	–	11	1	9

N = total number of bones observed; n = number of bones with fractures; % = $n/N \times 100$; 3rd mtt = third metatarsal.

a left fibula exhibited a transverse fracture on the mid-shaft (Fig. 7.11), and finally, a right third metatarsal exhibited a transverse fracture on the proximal end (Fig. 7.12).

During a number of excavation seasons, extended work took place in the church of Ayios Ioannis Theologos at Stylos. The excavations started in 1988 and continued, with intervals, until 1997. Although a large number of burials were excavated, only a small sample of human remains is available. The majority were cist or tile graves. The practice of secondary burial was commonly followed, thus the majority of graves contained the disarticulated remains of both adults and subadults. The adult sample consisted of 29 individuals: 9 males, 11 females, and 9 individuals of unknown sex. The number of fractures recorded in the sample is shown in Table 7.4. One distal left tibia of an adult (burial 26a, 1989) exhibited a transverse fracture (Fig. 7.13). Gross callus formation is present and is characterized by thick layers of lamellar bone covering the entire surface of the distal end. The fracture does not seem to have healed well due to the slight deformation of the bone, which has an inclination towards the medial aspect. The absence of a cloaca excludes the differential diagnosis of osteomyelitis. The distal right ulna of an adult (burial 1, 1990) presented a well-remodeled parry fracture. Two adult individuals from burial 28, 1997, presented fractures: one had a transverse fracture on the mid-shaft of the left clavicle and the other exhibited a Galleazi's fracture at the distal end of the left radius.

The distribution of fractures in the adult samples is presented in Table 7.5. The Kastella sample exhibits the higher frequency of fractures, and in all three samples the majority of fractures affected the upper limbs (i.e., Kastella, 11/25 or 44%; Pemonia, 4/6 or 67%; Stylos, 3/4 or 75%). Studies

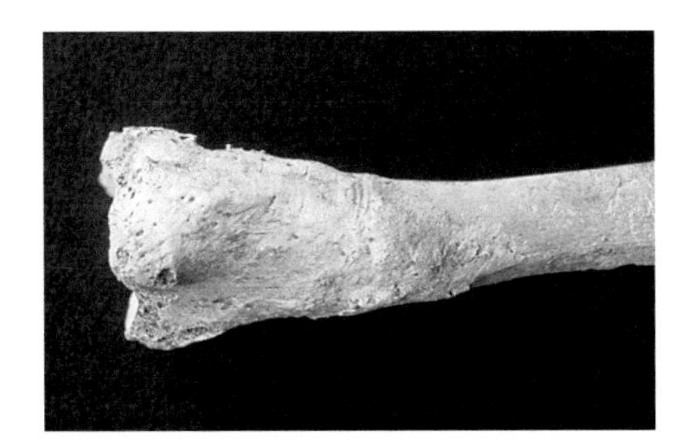

Figure 7.13. Stylos, burial 26a/1989: transverse fracture on the left tibia of an adult. Scale 1:3. Photo C. Bourbou

TABLE 7.4. FREQUENCY OF FRACTURED BONES AT STYLOS

	Right Fractures			*Left Fractures*			*Total Fractures*		
Bone	*N*	*n*	*%*	*N*	*n*	*%*	*N*	*n*	*%*
Clavicle	6	0	–	10	1	10	16	1	6.2
Ulna	13	1	7.7	11	0	–	24	1	4.2
Radius	11	0	–	13	1	7.7	24	1	4.2
Tibia	11	0	–	11	1	9	22	1	4.5

N = total number of bones observed; *n* = number of bones with fractures; % = *n*/*N* × 100.

TABLE 7.5. DISTRIBUTION OF FRACTURES IN THE SAMPLES

Bone Affected	*Kastella*	*Pemonia*	*Stylos*
Cranium	2	–	–
Upper limbs	11	4	3
Lower limbs	3	3	1
Thorax	8	–	–
Vertebrae	1	–	–
Total	25	7	4

comparing the frequencies of fractures between urban and rural sites[12] in medieval Britain suggested that rural environments were more hazardous than urban centers. However, this argument cannot be supported by the results of the current study, because the higher frequency of fractures is noted at the urban settlement of Kastella and not in the rural sites. It is also noteworthy that the urban males of Kastella sustained a greater number of fractures at a variety of anatomical locations and especially on the upper limbs, possibly reflecting riskier and more diverse activities when compared to females.

In all three samples most of the fractures can be attributed to accidental injuries. For example, the fractures of the lower radial shaft (Galleazi's fracture), and associated subluxation/dislocation of the distal radio-ulnar joint are typically produced by a fall on the hand with rotation force. Regarding the fractured clavicles, the mechanism of injury is usually a fall onto the outstretched hand or a fall on the shoulder. However,

12. E.g., Grauer and Roberts 1996; Judd and Roberts 1999.

the possibility of injuries due to conflict must also be considered in some cases. Injuries most frequently associated with interpersonal violence and observable in archaeological remains include cranial injuries attributed to direct blows,[13] multiple lesions from habitual or severe assault, and distal ulna shaft fractures (parry fracture) resulting from defending a blow to the head.[14] Although traditionally, parry fractures are associated with interpersonal violence, and especially with female abuse,[15] it is not always clear whether all or any of the injuries actually resulted from parrying a blow.[16] According to Lovell,[17] fractures of the skull (especially the nasal and zygomatic bones and the mandible), posterior rib fractures, vertebral spinous process fractures, and fractures of the hand and foot bones, which can result from the direct trauma of punches and kicks, are considered to have a high specificity for a clinical diagnosis of assault.

In the case of the cranial fracture of skeleton 002, such sharp, clearly defined edges could not be produced postmortem in a dry skull. There is no evidence of remodeling or secondary infection. It seems that the individual had not survived long enough for the healing process to begin, thus the injury appears to have been perimortem. The sharp edge, as well as the lack of microfracture, suggests a long, straight, sharp-edged weapon, such as a blade or sword. The depressed fracture observed on the skull of skeleton 014 is not typical of those from falls from high places, or those involving striking the head against a large mass, such as a rock or a hard surface, which would result in a large comminuted or linear fracture.[18] Most probably, this fracture pattern indicates a blow with a small weapon (i.e., slingshot, small club, implement used as a tool).

In the paleopathological record, fractures to the scapular body are relatively rare, most probably because this bone does not usually survive intact. The type of fracture recorded for the scapula of skeleton 011a, a sharp cut, most probably can be also attributed to a case of interpersonal violence, where again a long, straight, sharp-edged weapon, such as a blade or sword, was used. In the cases of multiple fractures in the same individual, the possibility of physical abuse can be safely excluded because the fractures were simultaneous and did not present different stages of healing, thus representing an accidental trauma event rather than a continuous abusive behavior.

Finally, the observed fractures provide no evidence that they were the result of an underlying disease. Callus formation was present in most of the cases, but no direct evidence of any possible treatment was observed. In several cases, complications were noted. The most common included shortening and realignment of the bone, as well as periosteal reaction due to bone surface infection or remnants of the healing process, and secondary osteoarthritis.

CONCLUSIONS

The purpose of this study was to present fracture data for the urban population of Middle Byzantine Kastella in central Crete, and to compare the fracture pattern with two rural sites (Pemonia and Stylos) dating to the same time period. As evidenced by the raw fracture data, the majority of

13. E.g., Powers 2005; Mays 2006; Paine et al. 2007.

14. E.g., Walker 1989, 2001; Smith 1996; Bridges 1996; Anderson 1996; Hutchinson 1996; Jurmain and Bellifemine 1997; Jurmain 2001; Judd 2004.

15. E.g., Wood-Jones 1910; Wells 1964.

16. Mafart 1991; Grauer and Roberts 1996; Lovell 1997, p. 165.

17. Lovell 1997, p. 166.

18. Galloway 1999, pp. 67–68.

cases occurred at the urban site of Kastella. In all three samples, the highest frequency of traumatic lesions was observed among the upper limbs, and males were more frequently affected than females, possibly reflecting sex-based differences in activity or risk of fracture. The majority of the fractures were recorded in the age group of 35–44 years. A small portion of individuals in the Kastella sample experienced more than one fracture, but the absence of different stages of healing suggests that the fractures were simultaneous and associated perhaps with an accidental trauma rather than sustained physical abuse. Most fractures are attributable to accidental injuries (e.g., falls), and only some to interpersonal violence. None appeared to be a result of an underlying pathological condition. Complications were noted in some cases, resulting in infections (periosteal reaction), degenerative changes of the joint (osteoarthritis), deformations (loss of normal shape of bone), and involvement of associated muscles or ligaments; however, no evidence of medical treatment was observed. With the exception of the cranial trauma of skeleton 014 from Kastella that most possibly was fatal, all the other fractures, albeit exhibiting complications, demonstrated healing, indicating that the individual survived the incident.

REFERENCES

Anderson, T. 1996. "Cranial Weapon Injuries from Anglo-Saxon Dover," *International Journal of Osteoarchaeology* 6, pp. 10–14.

Berger, D. T., and E. Trinkaus. 1995. "Patterns of Trauma among the Neanderthals," *JAS* 22, pp. 841–852.

Bourbou, C. 2003. *Interpreting Life and Death in Middle-Byzantine Greece (9th–13th Centuries A.D.).* Unpublished Report, Wiener Laboratory.

———. 2004. *The People of Early Byzantine Eleutherna and Messene (6th–7th Centuries A.D.): A Bioarchaeological Approach,* Athens.

Brickley, M. 2006. "Rib Fractures in the Archaeological Record: A Useful Source of Sociocultural Information?," *International Journal of Osteoarchaeology* 16, pp. 61–75.

Bridges, P. S. 1996. "Warfare and Mortality at Koger's Island, Alabama," *International Journal of Osteoarchaeology* 6, pp. 66–75.

Brothwell, D. R. 1981. *Digging Up Bones,* 3rd ed., Ithaca, N.Y.

Buikstra, J. E., and D. H. Ubelaker, eds. 1994. *Standards for Data Collection from Human Skeletal Remains,* Fayetteville, Ark.

Djuri, M. P., C. A. Roberts, Z. B. Rakoevi, D. D. Djondi, and A. R. Lei. 2006. "Fractures in Late Medieval Skeletal Populations from Serbia," *American Journal of Physical Anthropology* 130, pp. 167–178.

Domett, K. M., and N. Tayles. 2006. "Adult Fracture Patterns in Prehistoric Thailand: A Biocultural Interpretation," *International Journal of Osteoarchaeology* 16, pp. 185–199.

Galloway, A. 1999. "Fracture Patterns and Skeletal Morphology: Introduction and the Skull," in *Broken Bones: Anthropological Analysis of Blunt Force Trauma,* ed. A. Galloway, Springfield, Ill., pp. 63–80.

Garvie-Lok, S. 2001. "Loaves and Fishes: A Stable Isotope Reconstruction of Diet in Medieval Greece" (diss. Univ. of Calgary).

Grauer, A. L., and C. A. Roberts. 1996. "Paleoepidemiology, Healing, and Possible Treatment of Trauma in the Medieval Cemetery Population of St. Helen-on-the-Walls, York, England," *American Journal of Physical Anthropology* 100, pp. 531–544.

Hutchinson, D. L. 1996. "Brief Encounters: Tatham Mound and the Evidence for Spanish and Native American Confrontation," *International Journal of Osteoarchaeology* 6, pp. 51–65.

Judd, M. A. 2004. "Trauma in the City

of Kerma: Ancient versus Modern Injury Patterns," *International Journal of Osteoarchaeology* 14, pp. 34–51.

Judd, M. A., and C. A. Roberts. 1999. "Fracture Trauma in a Medieval British Farming Village," *American Journal of Physical Anthropology* 109, pp. 229–243.

Jurmain, R. 2001. "Paleoepidemiological Patterns of Trauma in a Prehistoric Population from Central California," *American Journal of Physical Anthropology* 115, pp. 13–23.

Jurmain, R., and V. I. Bellifemine. 1997. "Patterns of Cranial Trauma in a Prehistoric Population from Central California," *International Journal of Osteoarchaeology* 7, pp. 43–50.

Kilgore, L., R. Jurmain, and D. Van Gerven. 1997. "Paleoepidemiological Patterns of Trauma in a Medieval Nubian Skeletal Population," *International Journal of Osteoarchaeology* 7, pp. 103–114.

Lambert, P. M. 1997. "Patterns of Violence in Prehistoric Hunter Gatherer Societies of Costal Southern California," in *Troubled Times: Violence and Warfare in the Past,* ed. D. W. Frayer and D. L. Martin, Amsterdam, pp. 77–109.

Lovejoy, C. O., and K. G. Heiple. 1981. "The Analysis of Fractures in Skeletal Populations with an Example from the Libben Site, Ottowa County Ohio," *American Journal of Physical Anthropology* 55, pp. 529–541.

Lovell, N. C. 1997. "Trauma Analysis in Paleopathology," *Yearbook of Physical Anthropology* 40, pp. 139–170.

Mafart, B.-Y. 1991. *Apport de l' étude des fractures osseuses pour la connaissance des populations anciennes* (Dossiers de Documentation Archéologique 14), Paris.

Mays, S. 2006. "A Possible Case of Surgical Treatment of Cranial Blunt Force Injury from Medieval England," *International Journal of Osteoarchaeology* 16, pp. 95–103.

Merbs, C. F. 1989. "Trauma," in *Reconstruction of Life from the Skeleton,* ed. M. Y. İşcan and K. A. R. Kennedy, New York, pp. 161–189.

Mitchell, P. D., Y. Nagar, and R. Ellenblum. 2006. "Weapon Injuries in the 12th Century Crusader Garrison of Vadum Iacob Castle, Galilee," *International Journal of Osteoarchaeology* 16, pp. 145–155.

Neves, W., A. M. Barros, and M. A. Costa. 1999. "Incidence and Distribution of Postcranial Fractures in the Prehistoric Population of San Pedro de Atacama, Northern Chile," *American Journal of Physical Anthropology* 109, pp. 253–258.

Paine, R. R., D. Mancinelli, M. Ruggieri, and A. Coppa. 2007. "Cranial Trauma in Iron Age Samnite Agriculturists, Alfedena, Italy: Implications for Biocultural and Economic Status," *American Journal of Physical Anthropology* 132, pp. 48–58.

Powers, N. 2005. "Cranial Trauma and Treatment: A Case Study from the Medieval Cemetery of St. Mary Spital, London," *International Journal of Osteoarchaeology* 15, pp. 1–14.

Roberts, C. A. 1991. "Trauma and Treatment in the British Isles in the Historic Period: A Design for Multidisciplinary Research," in *Human Paleopathology: Current Syntheses and Future Options,* ed. D. J. Ortner and A. C. Aufderheide, Washington, D.C., pp. 225–240.

Smith, M. O. 1996. "'Parry' Fractures and Female-Directed Interpersonal Violence: Implications from the Late Archaic Period of West Tennessee," *International Journal of Osteoarchaeology* 6, pp. 84–91.

Stirland, A. 1996. "Patterns of Trauma in a Unique Medieval Parish Cemetery," *International Journal of Osteoarchaeology* 6, pp. 92–100.

Walker, P. L. 1989. "Cranial Injuries as Evidence of Violence in Prehistoric Southern California," *American Journal of Physical Anthropology* 80, pp. 313–323.

———. 2001. "A Bioarchaeological Perspective on the History of Violence," *Annual Review of Anthropology* 30, pp. 573–596.

Wells, C. 1964. *Bones, Bodies and Disease,* London.

Wood-Jones, F. 1910. "Fractures Bones and Dislocations," in *The Archaeological Survey of Nubia Report for 1907–1908,* ed. G. E. Smith and F. Wood-Jones, Cairo, pp. 293–342.

CHAPTER 8

Investigating the Human Past of Greece during the 6th–7th Centuries A.D.

by Chryssi Bourbou and Agathoniki Tsilipakou

Archaeologists in Greece are currently incorporating more skeletal studies into their research designs by testing hypotheses and drawing inferences about diet and nutrition, health and disease, demography and physical behavior, as well as lifestyle in the past.[1] Current skeletal analyses in the country do not focus only on prehistoric populations; during the last few years, human skeletal material from highly ignored time periods (i.e., the Byzantine and the post-Byzantine) has also contributed to the reconstruction of health patterns in the past.[2]

In this study, we present the results of the analysis of human remains from the proto-Byzantine site (6th–7th centuries A.D.) of Sourtara Galaniou Kozanis in northern Greece.[3] Additionally, we intend this research to provide information on specific pathological conditions that are suggestive of stress—either environmentally or culturally induced—during a turbulent era of Greek history, and the impact of these factors on specific age groups, such as subadults. These results are expected to contribute to our relatively restricted knowledge of that era in northern Greece, and, when compared with analyses of other contemporaneous populations, help us to reconstruct the patterns of life and disease in the country.

MATERIALS

During 1998–1999, rescue excavations along the national road of Egnatia, at the section between the area of Kozani-Polymylos (6 km southwest of the current settlement of Ayios Dimitrios, and 4 km northwest of the settlement of Koilada), a proto-Byzantine cemetery came to light at Sourtara Galaniou Kozanis.[4] One hundred and nineteen single inhumations in simple pit graves were investigated. Some evidence was found for the mode of disposal of the deceased (fragments of wooden coffins or carriers) or features associated with burial ceremonies (hearths for the preparations of ceremonial feasts). The typology of the few, and generally poor, accompanying goods enables us to date the cemetery to the 6th–7th centuries A.D. The material in question (71 skeletons) is generally well preserved and comes from 70 burials.

1. Papathanasiou 2001; Triantaphyllou 2001.
2. Bourbou 2004; Garvie-Lok 2001; Tritsaroli 2006.
3. Part of the study was possible thanks to the generous funding of the Wiener Laboratory (ASCSA) through a J. Lawrence Angel Fellowship (2001–2002). We would like to especially thank the Director of the Lab, Sherry Fox, for her useful comments during the analysis of the collection, and the personnel of the 11th Ephorate of Byzantine Antiquities for curating a large part of the material.
4. Tsilipakou 1998, forthcoming; Petkos 2004.

METHODS

A thorough anthropological and paleopathological analysis was conducted. Adult age at death was estimated using the pubic symphysis, the auricular surface,[5] and dental wear.[6] Sex was determined in adults using dimorphic aspects of the pelvis and skull.[7] Age was estimated in subadults by dental eruption and development, as well as by measurements of long bone length.[8] No attempt was made in the present study to determine sex in immature individuals. Estimation of stature was based on the formulae given by Trotter.[9]

The study primarily focused on the impact of physiological disruption (stress) on the health status of the population.[10] To test how individuals were affected by historical and environmental changes, the collection was analyzed for the presence of dental diseases,[11] metabolic and hematopoietic disorders,[12] physical stress as evidenced in the frequencies of degenerative joint diseases (DJD),[13] infections, and trauma.[14] In order to delineate the role played by sanitary, social, and other environmental conditions in subadult mortality and survival, two age categories have been constructed: neonatal (after birth–4 years) and postnatal (5–9 years old).

RESULTS

Determination of sex was possible for 42 individuals (27 males and 15 females), while the sex of 29 individuals, 15 of which are subadults, remains unknown. Average age at death for males is 42 years, and for females it is 34 years (Fig. 8.1). Most individuals died between the ages of 35–39 years. Mean living stature is 1.70 cm for males and 1.52 cm for females.

Data on dental pathologies in the sample are presented in Table 8.1. The percentage of the recorded dental diseases is 39% (210 out of 538 teeth). The most striking pathological condition is antemortem tooth loss (24%, $n = 130$), followed by calculus (11%, $n = 57$) and carious lesions (4%, $n = 20$) (Fig. 8.2). Most dental diseases are observed in the age groups between 26–45 years old, and males (81.5%) seem to be more affected than females (66.7%).

Table 8.2 presents the hematopoietic disorders in the sample. Cribra orbitalia and porotic hyperostosis are the more commonly observed skeletal lesions. Two cases of cribra orbitalia (skeletons 046 and 047) and two cases of porotic hyperostosis (skeletons 019 and 047) are recorded for subadult individuals (Fig. 8.3). Where determination of sex was possible, all cases were male individuals (skeletons 018, 041, 050); in one case, the sex of the individual was not determined (skeleton 070). Metabolic disorders include a possible case of osteoporosis (skeleton 027, female, aged 50 years). Evident loss of bone mass (osteopenia) is attested for all preserved long bones of the skeleton, which appear much lighter than similar bones from other burials from the site. Although postmortem changes can lead to lighter bones, the age and sex of the individual, as well as a compression fracture on the eighth thoracic vertebra that is commonly seen in osteoporotic cases,[15] reinforces the possibility of osteoporosis.

5. As cited in Buikstra and Ubelaker 1994.
6. Brothwell 1981.
7. As cited in Buikstra and Ubelaker 1994.
8. Ubelaker 1989.
9. Trotter 1970.
10. Goodman et al. 1988.
11. Criteria after Brothwell 1981; Lukacs 1989.
12. Criteria after Stuart-Macadam 1991; Ortner 2003.
13. Criteria after Rogers et al. 1987; Rogers and Waldron 1995.
14. Aufderheide and Rodríguez-Martín 1998; Ortner 2003.
15. Ortner 2003, p. 411.

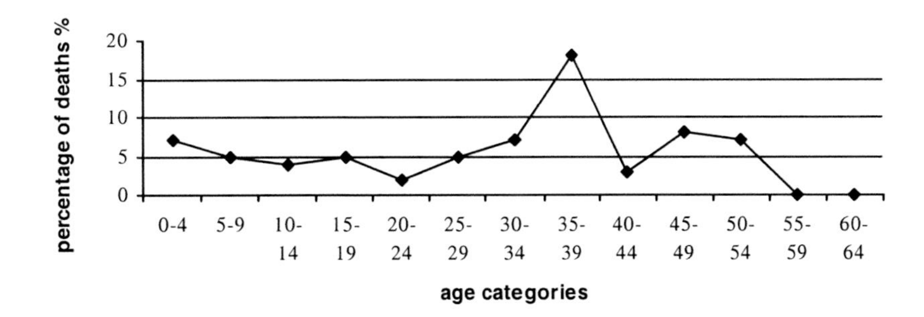

Figure 8.1. Mortality curve of the Sourtara sample.

TABLE 8.1. RECOVERED TEETH AND TEETH EXHIBITING DENTAL PATHOLOGIES IN THE ADULT SAMPLE FROM SOURTARA

	N	*Proportion of Total Possible Teeth (%)*
Total number of recovered teeth	538	30.5
Total number of recovered maxillary teeth	265	30.1
Total number of recovered mandibular teeth	273	31

	Max N	*Max %*	*Man N*	*Man %*	*Total*
Total number of teeth demonstrating pathologies	58	22	152	56	210
Total number of teeth lost antemortem	27	10	103	38	130
Total number of teeth with carious lesions	7	3	13	5	20
Total number of teeth with calculus	23	9	34	12	57
Total number of teeth with abscesses	1	0.4	–	–	1
Total number of teeth with enamel hypoplasias	–	–	2	1	2

Man = mandible; Max = maxilla.

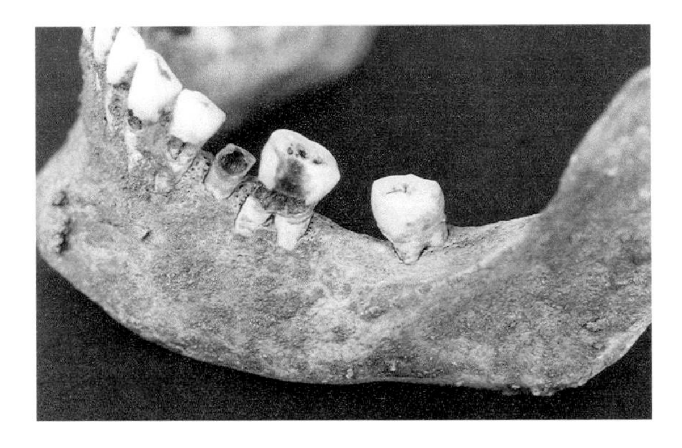

Figure 8.2. Soutara skeleton 009: large carious lesions on the mandibular teeth. Scale 1:2. Photo C. Bourbou

Figure 8.3. Soutara skeleton 019: porotic hyperostosis on a parietal fragment. Scale 1:2. Photo C. Bourbou

TABLE 8.2. HEMATOPOIETIC DISORDERS IN THE SOURTARA SAMPLE

Skeletal Lesion	*N*	*n*	%
Cribra orbitalia (orbits)	30	4	13
Porotic hyperostosis (parietals)	92	7	8
Porotic hyperostosis (occipitals)	83	4	5

N = total number of bones present; n = total number of bones exhibiting pathological conditions; % = $n/N \times 100$.

All reported cases of osteoarthritis for the upper and lower limbs refer to mature adult (over 46 years) males (Table 8.3). Osteophytic formations, pitting, and eburnation are manifested in all recorded cases (Fig. 8.4). In a total of 728 preserved vertebrae, the overall frequency of osteoarthritis is 11% (n = 80). Looking at the specific segments of the vertebral column, the breakdown for the involvement of different regions is cervical vertebrae 46% (n = 37), thoracic vertebrae 59% (n = 47), and lumbar vertebrae 21% (n = 17). Males (57%) are again more frequently affected than females. The overall frequency of Schmorl's nodes in the sample is also 11% (n = 81). They are more common in thoracic vertebrae (75%, n = 61) than in lumbar vertebrae (25%, n = 20) (Fig. 8.5), and are considerably more frequent in males (67%).

Skeleton 024 (a male individual aged 51 years) clearly demonstrated degenerations associated with diffuse idiopathic skeletal hyperostosis (DISH),

TABLE 8.3. OSTEOARTHRITIS IN THE SOURTARA SAMPLE

Joint Surface	*N*	*n*	%
Sternoclavicular	77	2	3
Distal ulna	44	1	2
Proximal femur	88	2	2
Acetabulum	82	2	2
Distal fibula	35	1	3
Proximal 1st metatarsal	42	1	2

N = number of joint surfaces observable; n = number of joint surfaces exhibiting degenerative lesions; % = $n/N \times 100$.

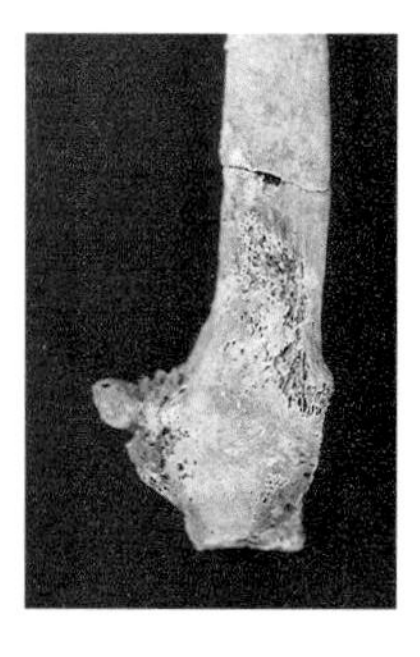

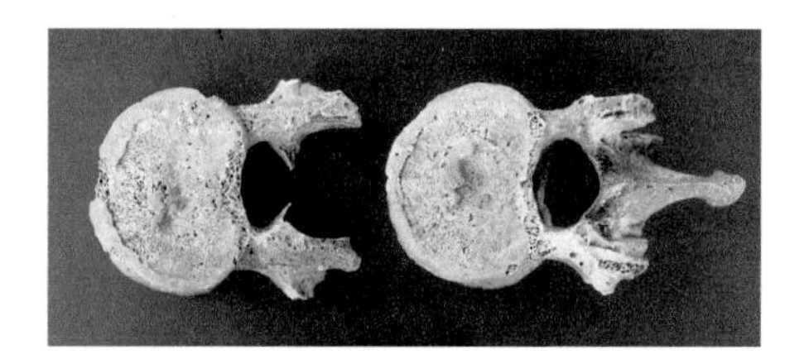

Figure 8.4 *(left)*. Soutara skeleton 016: osteoarthritis on the distal end of the fibula. Scale 1:2. Photo C. Bourbou

Figure 8.5 *(right)*. Soutara skeleton 032: Schmorl's nodes on lumbar vertebrae. Scale 1:3. Photo C. Bourbou.

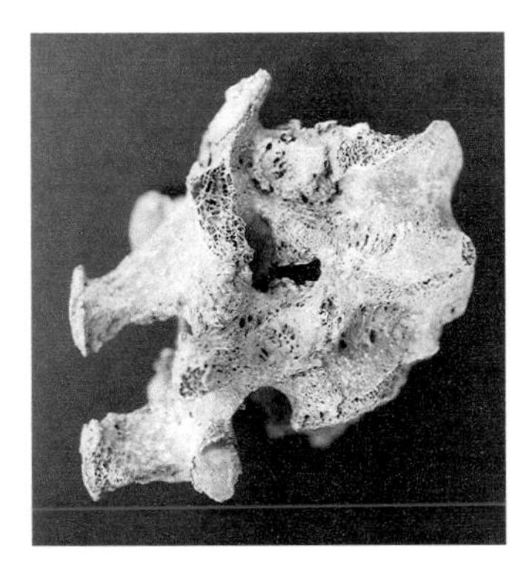

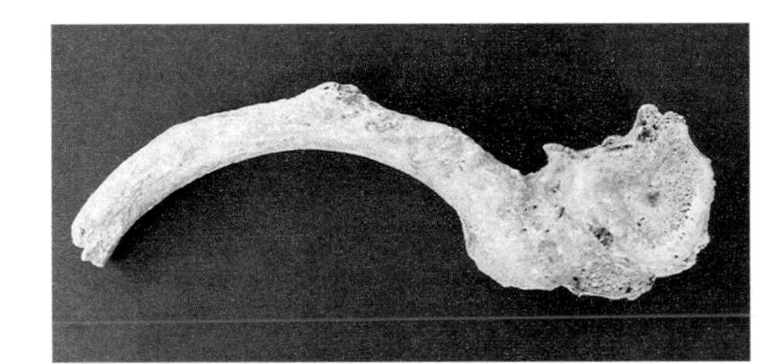

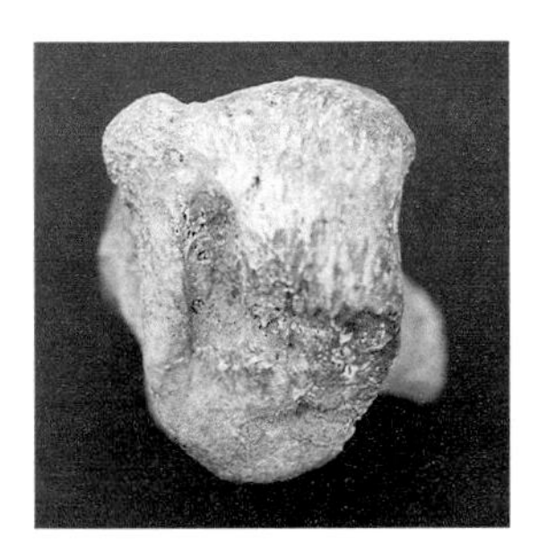

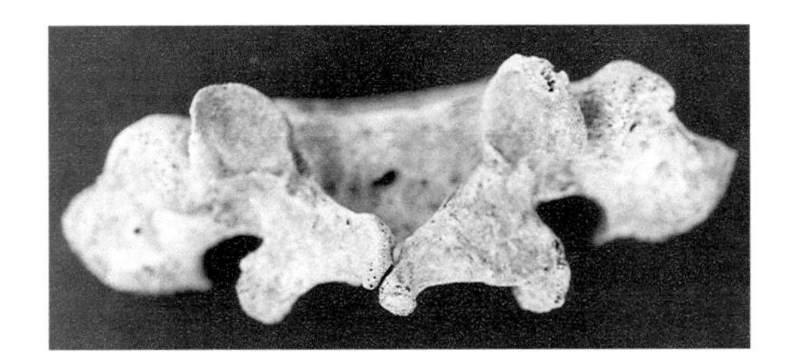

Figure 8.6 *(top left)*. Soutara skeleton 024: ankylosis of T10–T11 vertebrae (DISH). Scale 1:2. Photo C. Bourbou

Figure 8.7 *(top right)*. Soutara skeleton 024: ankylosis of T6 vertebra with associated rib (DISH). Scale 1:2. Photo C. Bourbou

Figure 8.8 *(above left)*. Soutara skeleton 010: trauma on posterior aspect of right calcaneus. Scale 1:2. Photo C. Bourbou

Figure 8.9 *(above right)*. Soutara skeleton 002: cleft neural arch on fifth lumbar vertebra. Scale 1:2. Photo C. Bourbou

such as ankylosis of the 10th–11th thoracic (Fig. 8.6) and 1st–2nd lumbar vertebrae, ankylosis of the 6th rib and associated vertebra (Fig. 8.7); numerous enthesophytes on the patellae, the olecranon processes of the ulnae, the iliac crests, ischial tuberosities and acetabula of the innominates, the linea aspera and trochanters of the femora, and the calcanei; additionally, osteophytic formation is noted on heads of ribs after their possible fracture.

Only two cases of periosteal reaction are included in the infectious conditions. Periostitis in the form of woven bone deposits is observed along the shaft of the left fifth metatarsal of skeleton 008 (1/31; 3.2%); lamellar bone formation is observed along the shaft of a right fibula (skeleton 016) (1/73; 1.4%). Few traumatic incidents were observed. These include a fracture, most likely induced by a blunt instrument on the posterior aspect of a right calcaneus (skeleton 010) (Fig. 8.8) (1/77; 1.3%) and a case of *myositis ossificans* at the distal end of a right radius (skeleton 021) (1/77; 1.3%). Finally, an interesting developmental defect is observed on the fifth lumbar vertebra of skeleton 002, a ca. 20-year-old male (Fig. 8.9). A failure of the two halves of the spinous process to coalesce resulted in a bifid or cleft neural arch.[16]

THE COMBINATION OF CULTURAL AND ANTHROPOLOGICAL DATA

At present, the information on the history and bioarchaeology of proto-Byzantine Greece is scattered and incomplete, because few records or major archaeological excavations refer to this turbulent era. From the available data, we know that it was generally a period of dramatic historical, social,

16. Barnes 1994, p. 119.

economic, and environmental changes. Proto-Byzantine Greeks suffered invasions, earthquakes, and the social upheavals associated with the introduction of Christianity.[17] The picture for northern Greece, and especially for Greek Macedonia—a highly strategic region at the crossroads between the north–south and east–west parts of the Byzantine Empire—is quite similar.[18] Most of our knowledge regarding the area derives from rescue excavations, surveys, and catalogs of scattered material. These sources provide important information on religious edifices, settlements, and defensive works.[19] From the 3rd century A.D. onward, the region suffered from extensive invasions and installations of various groups (i.e., Goths and Slavs). It was only during the 6th century A.D. that the Byzantine Empire, under the rule of Justinian, applied an extended plan of restoration and reinforcement of sensitive areas through the building of defensive settlements. This did not have a lasting effect. During the 6th–8th centuries A.D., further invasions and natural disasters following earthquakes resulted in the abandonment of numerous sites.[20] Some of these were repopulated when the danger diminished. The excavation of the proto-Byzantine cemetery at Sourtara suggests the existence of a so-far-unknown proto-Byzantine settlement in a strategic position along the road connecting the upper and lower parts of Greek Macedonia, and in a generally unexplored area in terms of Byzantine occupation.

Taking into account the hardships of the period, mean age at death at Sourtara (42 and 34 years for males and females, respectively) appears to have been considerably higher than in the preceding centuries. For example, for the Roman Imperial period in Greece, the average age at death was estimated at 38.8 years for males and 34.2 years for females.[21] Mortality rates among subadults for the Sourtara sample are higher from birth to 4 years and slowly decline thereafter. Pathological observations in the subadult individuals included cribra orbitalia and porotic hyperostosis. The high incidence of dental pathologies suggests a diet rich in sugar and carbohydrates. The presence of cribra orbitalia and porotic hyperostosis in the sample, diagnosed as possible iron-deficiency anemia, can be attributed to a low-protein diet and other synergistic factors (i.e., elevated environmental stressors and/or infections).

The presence of an older female (skeleton 027) with marked osteoporosis is of particular interest. Much clinical research is focused on the osteoporosis that results from the hormonal changes associated with aging and is particularly common in postmenopausal women.[22] Although in archaeological human skeletal populations the prevalence of the condition is less common because few people lived long enough to acquire the disease, some cases are reported in the paleopathological record.[23] According to Ortner,[24] "osteoporosis usually does not manifest itself before the fifth decade, and it is more frequent and severe in females than in males." The skeleton is not affected proportionately, and bones rich in spongiosa, such as the spine, ribs, sternum, and pelvis are most involved. Of the long bones, the femoral neck shows the most characteristic changes.[25] The vertebral bodies usually exhibit the earliest and most severe expression of the condition. Ortner[26] argues that "if the intervertebral disks retain their normal

17. Mango 1980; Cameron 1993; Bowersock, Brown, and Grabar 2001.

18. Karagianni 1999; Konstantokopoulou 1984; Louggis 1966.

19. Keramopoulos 1932, 1933a, 1933b, 1934, 1937; Michailidis 1965, 1971; Pelekanidis 1960; Tsigaridas and Loverdou-Tsigarida 1982; Tsilipakou 1995, 1997.

20. For a thorough discussion on the historical and archaeological context of the proto-Byzantine era, see Bourbou 2004, pp. 59–62.

21. Bisel and Angel 1985, table 4.

22. Ortner 2003, p. 410.

23. Dequeker et al. 1997; Roberts and Wakely 1992; Földes et al. 1994; Mays 1996; Mays, Lees, and Stevenson 1998; Brickley and Howell 1999.

24. Ortner 2003, p. 411.

25. Ortner 2003, p. 411.

26. Ortner 2003, p. 411.

turgor, they may contribute to multiple microfractures of the vertebral bodies, resulting in concave endplates and the biconcave appearance known as fish vertebra," which is reported for the case in question. Because there is no evidence of any underlying pathological condition associated with osteoporosis (i.e., trauma, infection, cancer, or *osteogenesis imperfecta*), the most probable diagnosis is that of senile osteoporosis.

It is also clear from the skeletal data that adult males in Sourtara have substantially higher frequencies of degenerative joint diseases. In addition, the data are more suggestive of physiological wear and tear on joints due to advanced age than other factors. The presence of Schmorl's nodes is most probably caused by degenerative changes associated with repetitive stress on the vertebral column. Regarding the cases of DISH, two other disorders were considered, although they were ultimately excluded: ankylosing spondylitis and spondylosis. Cases of DISH in Greece have been reported for preceding eras.[27] This disease affects the spine, but it has very characteristic bony abnormalities elsewhere in the body.[28] Males are affected more often than females, and 85% of the cases are aged over 50 years.[29] The etiology of DISH is unknown; however, the most recent suggestion is that DISH is a multisystem hormonal disorder.[30] Observations by Rogers,[31] Rogers et al.,[32] Bruintjes,[33] Bourbou,[34] Janssen and Maat,[35] and Jankauskas[36] indicate that high-status individuals are often affected by this disease. Waldron,[37] Rogers and Waldron,[38] and de la Rua and Orue[39] suggest a relationship between DISH and monastic life, considering that the explanation must lie in the daily activities of the monastery and the monks' diet. Nevertheless, the presence of DISH in a skeleton cannot be taken as an indication that the individual was either of high status or a monk. Usually, individuals demonstrating the disorder are significantly older than the rest of the population; their longevity alone might be suggestive of a better lifestyle and nutritional status.

The very low percentage of infectious conditions at Sourtara, considered in conjunction with the relatively high mean ages of death, is further suggestive of fairly good living conditions and nutritional status for the population. Moreover, the small number of fractures in the sample indicates the presence of only minor accidents in everyday life.

Finally, in the case of the individual with the neural arch defect, according to Barnes,[40] "a developmental delay resulting in hypoplasia or aplasia of one or both parts of the precursors of the pedicles, laminae, or spinous process can lead to failure of the two halves to coalesce and result in a bifid or cleft neural arch." Most probably, due to a minor delay in the development of the neural arches, the two halves came together but did not coalesce, forming a bifid neural arch with a small midline cleft or bifurcation of the spinous process. In general, symmetrical clefting is the most common type of developmental defect of the neural arch.[41] The tough fibrous tissue covering the cleft protects the underlying tissues as the missing bony part would, and thus is clinically insignificant.[42] It generally occurs in the border regions of the vertebral column, especially at the lumbosacral border, affecting only one or two vertebrae, although, in some cases, more vertebrae can be involved.

27. Fox Leonard 1997; Fox 2005; Bourbou 1998, 2005; Lagia 1999.
28. Aufderheide and Rodríguez-Martín 1998.
29. Julkunen, Heinonen, and Pyorala 1971.
30. Rogers and Waldron 2001.
31. Rogers 1982.
32. Rogers, Watt, and Dieppe 1985; Rogers et al. 1987.
33. Bruintjes 1989.
34. Bourbou 1998, 2005.
35. Janssen and Maat 1999.
36. Jankauskas 2003.
37. Waldron 1985.
38. Rogers and Waldron 2001.
39. de la Rua and Orue 1993.
40. Barnes 1994, pp. 118–119.
41. Barnes (1994, p. 119) notes that "clefting from developmental delay of the neural arches is quite common, with frequencies as high as 25%."
42. Saluja 1988.

TABLE 8.4. DATA ON HUMAN SKELETAL COLLECTIONS FROM PROTO-BYZANTINE SITES IN GREECE

	Total Number of Individuals				*Average Age at Death*		*Average Height (m)*	
Site	*M*	*F*	*?*	*Total*	*M*	*F*	*M*	*F*
Sourtara	27	15	29	71	42	34	1.70	1.52
Eleutherna (Bourbou 2003a, 2004)	52	21	78	151	40–45	30–35	1.69	1.60
Messene (Bourbou 2003a, 2004)	23	12	39	74	45	35	1.70	1.52
Gortyn (Mallegni 1988)	18	16	20	54	41	34	1.73	1.60
Knossos (Musgrave 1976)	9	12	10	35–50	46% died between ages 18–35		1.61–1.78	1.50–1.62
Corinth (Lerna) (Wesolowsky 1973)	54	43	67	164	6 individuals reached the 5th decade of life		1.63–1.73	1.60
Abdera (Polystylon) (Agelarakis and Agelarakis 1989)	16	5	19	40	Nobody over 60 years		1.68	1.56
Thassos (Aliki II) (Buchet and Sodini 1984)	22	1	124	147	40–50	30–40	1.62	1.50

A COMPARISON STUDY WITH CONTEMPORANEOUS POPULATIONS FROM GREECE

Skeletal reports that provide useful comparative data on proto-Byzantine populations are limited, because few refer to large samples and far fewer to the frequencies of the pathological conditions considered here. However, appropriate data are available for the sites of Eleutherna,[43] Messene,[44] Gortyn,[45] Knossos,[46] Corinth,[47] Abdera,[48] and Aliki.[49] A number of studies on human skeletal remains, although dating to the proto-Byzantine period, have been excluded from this survey, since their data are problematic either in terms of exact chronology or because they derive from case studies.[50]

Table 8.4 summarizes the data presently available for proto-Byzantine populations in Greece. As shown in the table, values for mean age at death and average stature for the Sourtara sample do not vary considerably from the other sites. Subadult mortality (Fig. 8.10) at Sourtara, as in Messene and Gortyn, involves highest mortality among infants and slowly declines thereafter.[51] However, in some sites where postnatal mortality exceeded

43. Bourbou 2003a, 2004.
44. Bourbou 2003a, 2004.
45. Mallegni 1988.
46. Musgrave 1976.
47. Wesolowsky 1973.
48. Agelarakis and Agelarakis 1989.
49. Buchet and Sodini 1984.
50. Angel, for example, studied the material from eight burials (at least four of them subadult burials) from the ancient Agora in Athens, but refers to them only as "Byzantine" or "medieval" without further chronological distinction (L. M. Little, pers. comm.). On the other hand, studies such as that of Pitsios (1985–1986) on three burials in Abdera refer to case studies identifying a specific cause of death (decapitation or death in childbirth).
51. It is commonly assumed that we would expect the highest proportion of deaths to be among the newborns in any mortality sample. For a thorough discussion, see Barker and Osmond 1986; Barker 1992; Barker and Martyn 1992.

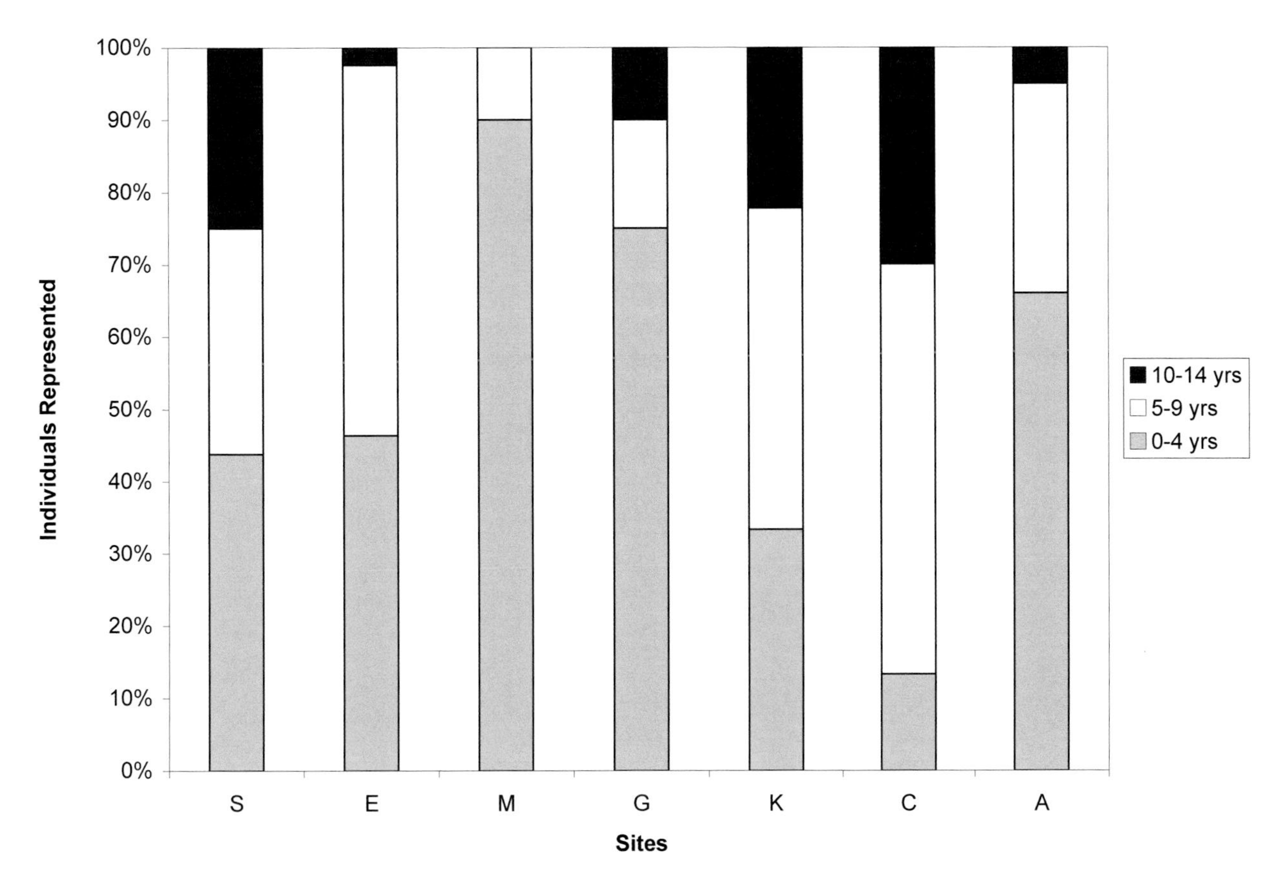

Figure 8.10. Subadult mortality patterns in the proto-Byzantine era. *Key:* **S = Sourtara; E = Eleutherna (Bourbou 2003a, 2004); M = Messene (Bourbou 2003a, 2004); G = Gortyn (Mallegni 1988); K = Knossos (Musgrave 1976); C = Corinth (Wesolowsky 1973); A = Aliki II (Buchet and Sodini 1984).**

neonatal mortality (i.e., Eleutherna, Knossos, Corinth), this discrepancy could be largely rooted in environmental factors, such as poor sanitation and nutrition. Furthermore, if general living conditions were poor and thus hindered a rapid recovery from diseases or other stresses, all children were likely at risk, especially those who had already been through the usual episodes of stress during their early years.

Cultural factors influencing dietary choices cannot be ruled out as a possible explanation for the differences in mortality rates seen in the subadult remains from these populations. Such cultural factors may include differences in weaning practices and in the selection of dietary elements, as well as in the proportions of those elements.[52] The age of weaning and the quality of the supplementary food may also contribute to the development of specific pathological conditions; for example, subadults at Eleutherna display more pathological conditions than those at Messene or Sourtara. Chronic bleeding and subsequent ossification at multiple sites on the skull

52. Textual evidence from the Roman Imperial period reveals interesting information about breastfeeding and weaning practices. Both Soranus and Galen (2nd century A.D.) recommended a gradual weaning process in which crumbled bread, moistened with liquid, was substituted for breast milk. The liquid used to moisten the bread could be milk (mainly from goats); *hydromel,* which is a mixture of honey and water; sweet wine; or honey mixed with wine (Jackson 1989). Actual weaning practices for the populations in question may have approximated those outlined by Soranus and Galen. For the direct consequences for the health of the infants by introduction of both goat's milk and honey, see Fairgrieve and Molto 2000; Arnon et al. 1979; Merenstein, Kaplan, and Rosenberg 1991.

associated with scurvy (vitamin C deficiency)[53] is the most likely reason for the bone lesions reported in the case from Eleutherna.[54]

Additionally, in the samples from both Eleutherna and Messene, periostitis was observed along the shafts of one or two subadult long bones, mainly the tibiae.[55] The skeletal tissue was loosely organized woven bone, suggesting that the lesions were still active at the time of death. The differential diagnosis could be a general ongoing infectious process that may have affected the individuals since birth or even before, and contributed directly to their early deaths; trauma is another possible diagnosis due to the localized nature of the lesion and the normal internal architecture of all other bones.[56]

Some general inferences can be drawn regarding the pathological lesions observed in the adult individuals of the samples. For Sourtara, dental data suggest a dependence on carbohydrates in the diet; for the other sites, most authors suggest that their dental data demonstrate a low carbohydrate diet, rich in plant tissues and fluorides that contribute to good oral hygiene.[57] For the interpretation of the hematopoietic disorders, such as cribra orbitalia and porotic hyperostosis observed in the samples of Sourtara, Eleutherna, Messene,[58] and Abdera,[59] an iron deficiency anemia of a multifactorial etiology (i.e., a poor and low protein diet and environmental constraints) is argued.[60] However, Angel's hypothesis[61] that these lesions may reflect thalassemia genotypes as a response to endemic malaria has been cited repeatedly in the skeletal reports from the proto-Byzantine era.[62] Because none of the authors mention any additional bone abnormalities that are pathognomonic of genetic anemias,[63] their assumptions—which contributed to a generalized hypothesis of a "Mediterranean malarial belt"—are questionable. Analyses of porotic hyperostosis in prehistoric material from Greece have thus far only succeeded to demonstrate the connection between anemic conditions, malaria and constrained environmental conditions, such as marshy areas; ongoing research is expected to explore the subject further.[64]

Data on the frequency of degenerative joint diseases are available for Eleutherna and Messene;[65] unfortunately, most other authors refer only vaguely to the presence of degenerative lesions. The overall frequency of degenerative diseases of the vertebral column in the sample of Eleutherna is 16% (*n* = 92 out of 566 preserved vertebrae) and 5% (*n* = 32 out of 695 preserved vertebrae) for Messene. Males are again more affected than females (Eleutherna: 60%, Messene: 17%). For Eleutherna the data on degenerative joint diseases are suggestive of high levels of physical stress in young and middle adult males and of differential male/female activity patterns. For Messene, however, as for Sourtara, the data are more suggestive of physiological wear and tear of joints due to advanced age. The overall frequency of Schmorl's nodes in the sample from Eleutherna is 10% (*n* = 59), and for Messene it is 8% (*n* = 53). Males have also substantially higher frequencies of Schmorl's nodes. Degenerative change associated with repetitive stress upon the vertebral column is the most plausible explanation for the presence of Schmorl's nodes in these samples. Last, erosive lesions are observed on the tarsometatarsal and metatarsophalangeal joints of skeletons 019 and 035 from Messene. These are attributed to

53. For a thorough review of the condition, see Stuart-Macadam 1989; Aufderheide and Rodríguez-Martín 1998; Ortner 2003. For a review of the pathognomic features associated with scurvy, see Ortner, Kimmerle, and Diez 1999; Ortner et al. 2001.

54. Bourbou 2003b; for cases in the paleopathological record, see Ortner 1984; Roberts 1987; Mogle and Zias 1995; Ortner and Ericksen 1997; Ortner, Kimmerle, and Diez 1999; Ortner et al. 2001.

55. Bourbou 2006.

56. Anderson and Carter 1994.

57. Wesolowsky 1973; Mallegni 1988; Bourbou 2003a, 2004.

58. Bourbou 2004.

59. Agelarakis and Agelarakis 1989.

60. For a discussion on the iron bioavailability in different sources, see Baynes and Brothwell 1990. For the contribution of various nondietary factors to the development of the condition, see Stuart-Macadam 1989.

61. Angel 1964, 1966.

62. Mallegni 1988, p. 385; Wesolowsky 1973, pp. 349–350.

63. Roberts and Manchester 1995, p. 170.

64. E. Stravopodi (pers. comm.).

65. Bourbou 2003a, 2004.

gouty arthritis.[66] The relatively young age (between 25–35 years) of both individuals demonstrating gouty arthritis most probably indicates a special type of idiopathic gout with an early onset.

Only two cases of periosteal reaction were recorded for the Sourtara sample. On the other hand, infectious conditions appear more frequent in the samples of Eleutherna and Messene.[67] Periostitis in these samples demonstrates most possibly active but chronic inflammations limited to a single bone of the skeleton. These appeared to be of presumably hematogenous origin or from a bony reaction to overlying skin ulcerations. However, minor trauma resulting from everyday activities, primarily performed by males, could explain the high prevalence of periostitis in individuals from Eleutherna and Messene.[68] A case of osteomyelitis reported in the sample of Messene possibly demonstrates an acute hematogenous osteomyelitis, because no evidence of fracture, injury or surgery is noted. The individual seemed to have survived long enough for the osteomyelitis to be chronic as evidenced by bone destruction and abscess formation, as well as draining sinuses (cloaca) and involucrum formation. A direct result of the interaction between the population and the environment (close contact with cattle resulting in poor sanitation) was the most probable explanation for a possible case of a bacterial infection (actinomycosis?) reported for Eleutherna.[69] The differential diagnosis of a fungal infection must be excluded, because bone infection in these cases constitutes only a very small percentage and morphologically the skeletal lesions are mainly lytic. Finally, reactive periostitis on the visceral aspects of 10 rib fragments from Messene may be the result of tuberculous pleuritis. This condition typically has an empyema formation, which is an extremely common aspect of active pulmonary tuberculosis. Rib lesions are also seen on individuals who died from chronic bronchitis, but they are quite different in texture, porosity, and location from those of pulmonary tuberculosis, the tentative diagnosis for this individual.[70]

As for infectious conditions, traumatic incidents were relatively rare in the Sourtara sample. Most reported traumatic incidences in the samples of Gortyn, Eleutherna, Messene, and Aliki II are presumably caused by minor accidents. Mallegni[71] attributed the bilateral Colles's fractures observed at Gortyn, together with compression fractures of two thoracic vertebrae, to a fall. The isolated case of parry fracture from Eleutherna may imply interpersonal violence, because it might have occurred while defending a blow against the head.[72] Analysis of the skeletal remains from Eleutherna found that two Colles's fractures and a fracture of a fifth metatarsal exhibited complications in the form of secondary osteoarthritis and loss of normal shape, respectively. Complications after fracture were also reported for Aliki: a male individual presented a fracture at the distal end of his tibia that probably resulted in an abnormal posture, as evidenced by severe osteophytic lesions on his lumbar vertebrae, sacrum, and pelvis.[73] Another male displayed secondary osteoarthritis on his elbow joint and similar degenerative lesions on both knee joints. All individuals with multiple degenerative lesions came from the same burial and the authors argue that they were most probably laborers. Of minor importance is the case of *myositis ossificans* reported for the sample from Messene, a condition

66. Bourbou 2001a.
67. Bourbou 2006.
68. Bourbou 2003a, 2006, 2004.
69. Bourbou 2006.
70. For a thorough discussion on recent advances in the diagnosis of tuberculosis in ancient populations, see Bourbou 2006.
71. Mallegni 1988.
72. Bourbou 2004; see also Bourbou this volume.
73. Buchet and Sodini 1984.

that may occur without obvious skeletal injury and after only trivial muscle trauma.[74] Finally, a case of *humerus varus* is reported for the proximal epiphysis of a subadult individual from Eleutherna. There is no evidence of underlying disease or recent trauma; however, all other epiphyses present have developed and/or fused normally, and this isolated deformity does not correspond to any established syndrome. The condition appears to have been caused by untreated intrauterine or early childhood trauma, resulting in a deformity moderate to severe in degree, with impingement upon the acromion.[75]

CONCLUSIONS

In this study, we presented the anthropological and paleopathological analysis of a proto-Byzantine cemetery population at Sourtara in northern Greece. The bioarchaeological data derived from this analysis add essential information to the generally poor record of the proto-Byzantine era in Greece, and especially in Greek Macedonia. The paleodemographic profile of the site, including the mean age at death, does not vary considerably from that of the other proto-Byzantine sites. Mortality rates among subadults are higher at birth–4 years and slowly decline thereafter, a pattern usually expected in most mortality samples and also attested in the sites of Gortyn, Messene, and Aliki. However, at sites where postnatal mortality exceeded neonatal mortality, as at Eleutherna, it could be rooted largely in environmental factors, such as poor sanitation and nutrition, that are thought to be characteristic of this time period.

A low-carbohydrate and plant-rich diet that has been proposed for the other populations (i.e., Eleutherna, Gortyn) cannot be postulated for Sourtara because of the high prevalence of dental pathology. Instead, we suggest there was a dependence on carbohydrates in the diet. A multifactorial etiology would seem the most appropriate model for interpreting the data on hematopoietic disorders, as also suggested for the sites of Eleutherna, Messene, and Abdera. The diagnosis of iron-deficiency anemia due to a poor, low-protein diet and other synergistic factors in the form of elevated environmental stressors and/or infections appears very probable. Osteoporosis is more common in females and especially postmenopausal ones due to the hormonal changes acquired with advanced age.

It is also clear from the skeletal data that Sourtara males have substantially higher frequencies of degenerative joint diseases than do females. In particular, the data are more suggestive of physiological wear and tear on joints due to advanced age, a pattern also attested for Messene. Regarding DISH, the most recent suggestion is that the condition is a multisystem hormonal disorder and might be suggestive of a better lifestyle and nutritional status for the individual in question. The very low percentage of infectious conditions in the sample suggests a prosperous everyday life with fairly good living conditions and nutritional status for the general population. In addition, the small number of fractures in the sample suggests the presence of only minor accidents in the everyday life.

74. Aufderheide and Rodríguez-Martín 1998.
75. Bourbou 2001b.

It is important to emphasize that the multiple and intensive socio-political and environmental phenomena of the proto-Byzantine era highlight the caution with which a bioarchaeological approach must be attempted. Thus, the elements that combine to form the image of a specific site need to be investigated on a case-by-case basis. However, one must bear in mind that adaptation to a continuously changing environment is the key characteristic of life in the past. The complex phenomena of the proto-Byzantine period had different effects on the populations. In general, two distinctive patterns of adaptation to strenuous conditions can be traced: for some groups, deteriorating living conditions and growing insecurity due to earthquakes and fear of invasions forced the population to abandon the settlement and look for a safer place to start a new life (i.e., Eleutherna, Gortyn). In other areas, despite the hard living conditions of the era, the population appears strong enough to resist a variety of pressures and even enjoy prosperity (i.e., Messene, Sourtara), until the proceeding centuries (9th–10th centuries A.D.) when cities started to flourish again. Despite the general current shortage of comparable bioarchaeological information available for proto-Byzantine Greece, future examination of other cemetery populations dating to the same period will help to assess the effects of stress, especially on individuals undergoing active growth.

REFERENCES

Agelarakis, A., and A. Agelarakis. 1989. "The Paleopathological Evidence at Polystylon, Abdera," *Byzantinische Forschungen* 14, pp. 9–25.

Anderson, T., and A. R. Carter. 1994. "Short Report: Periosteal Reaction in a New Born Child from Sheppey, Kent," *International Journal of Osteoarchaeology* 4, pp. 47–48.

Angel, J. L. 1964. "Osteoporosis: Thalassemia?," *American Journal of Physical Anthropology* 22, pp. 369–374.

———. 1966. "Porotic Hyperostosis, Anemias, Malarias, and the Marshes in the Prehistoric Eastern Mediterranean," *Science* 153, pp. 760–762.

Arnon, S. S., T. F. Midura, K. Damus, B. Thompson, R. M. Wood, and J. Chin. 1979. "Honey and Other Environmental Risk Factors for Infant Botulism," *Journal of Pediatrics* 94, pp. 331–336.

Aufderheide, A., and C. Rodríguez-Martín. 1998. *The Cambridge Encyclopedia of Human Paleopathology*, Cambridge.

Barker, D. J. P. 1992. "Fetal Growth and Adult Disease," *British Journal of Obstetrics and Gynaecology* 99, pp. 275–282.

Barker, D. J. P., and C. N. Martyn. 1992. "The Maternal and Fetal Origins of Cardiovascular Disease," *Journal of Epidemiology and Community Health* 46, pp. 8–11.

Barker, D. J. P., and C. Osmond. 1986. "Childhood Respiratory Infection and Adult Chronic Bronchitis in England and Wales," *British Medical Journal* 239, pp. 1271–1275.

Barnes, E. 1994. *Developmental Defects of the Axial Skeleton in Paleopathology*, Niwot, Colo.

Baynes, R. D., and T. H. Brothwell. 1990. "Iron Deficiency," *Annual Review of Nutrition* 10, pp. 133–148.

Bisel, S. C., and J. L. Angel. 1985. "Health and Nutrition in Mycenaean Greece: A Study in Human Skeletal Remains," in *Contributions to Aegean Archaeology: Studies in Honor of William A. McDonald*, ed. N. C. Wilkie and W. D. E. Coulson, Minneapolis, pp. 197–209.

Bourbou, C. 1998. "More Evidence on the Association of DISH and Upper Class Individuals from the Hellenistic Crete," *Paleopathology Association Newsletter* 101, pp. 7–10.

———. 2001a. "Pathological Conditions in the Lower Extremities of Two Skeletons from Early-Byzantine Greece," in *Proceedings of the XIIIth European Meeting of the Paleopathology Association, Chieti, September 18–23, 2000,* ed. M. La Verghetta and L. Capasso, Teramo, pp. 22–28.

———. 2001b. "A Proto-Byzantine Case of Unilateral *Humerus varus*" (paper, Chios 2001).

———. 2003a. "Health Patterns of Proto-Byzantine Populations (6th–7th Centuries A.D.) in South Greece: The Cases of Eleutherna (Crete) and Messene (Peloponnese)," *International Journal of Osteoarchaeology* 13, pp. 303–313.

———. 2003b. "The Interaction Between a Population and its Environment: Probable Case of Subadult Scurvy from Early-Byzantine Greece," *Eres* 11, pp. 105–114.

———. 2004. *The People of Early Byzantine Eleutherna and Messene (6th–7th Centuries A.D.): A Bioarchaeological Approach,* Athens.

———. 2005. "Biological Status in Hellenistic and Roman Elites in Western Crete (Greece)," *Eres* 13, pp. 87–110.

———. 2006. "Infectious Conditions Observed on Greek Proto-Byzantine (6th–7th Centuries A.D.) and Middle-Byzantine (11th Century A.D.) Skeletal Series," in *Actes du 8ème Journées Anthropologique de Valbonne, La Paléodémographie. Mémoire d'os, Mémoire d'hommes,* Valbonne June 5–8, 2003, ed. L. Buchet, C. Dauphin, and I. Séguy, Antibes, pp. 85–99.

Bowersock, G. B., P. Brown, and O. Grabar. 2001. *Interpreting Late Antiquity: Essays on the Postclassical World,* Cambridge.

Brickley, M., and P. Howell. 1999. "Measurements of Changes in Trabecular Bone Structure with Age in an Archaeological Population," *JAS* 26, pp. 151–157.

Brothwell, D. R. 1981. *Digging Up Bones,* 3rd ed., Ithaca, N.Y.

Bruintjes, T. J. D. 1989. "Diffuse Idiopathic Skeletal Hyperostosis (DISH): A 10th Century A.D. Case from the St. Servaas Church at Maastricht," in *Bones, Treasuries of Human Experience in Time and Space* 1, ed. W. R. K. Perizonius, Leiden, pp. 23–28.

Buchet, J. L., and J. P. Sodini. 1984. "Les tombes," in *Aliki II. La Basilique Double,* ed. J. P. Sodini and K. Kolokotsas, Athens, pp. 213–243.

Buikstra, J. E., and D. H. Ubelaker, eds. 1994. *Standards for Data Collection from Human Skeletal Remains,* Fayetteville, Ark.

Cameron, A. 1993. *The Mediterranean World in Late Antiquity A.D. 395–600,* London.

de la Rua, C., and J. M. Orue. 1993. "Health Conditions in a Monastic Community of the Basque Country (16th–17th Centuries)," *Journal of Paleopathology* 4, pp. 193–200.

Dequeker, J., D. Ortner, A. Stix, X.-G. Cheng, P. Brys, and S. Boonen. 1997. "Hip Fracture and Osteoporosis in a XIIth Dynasty Female Skeleton from Lisht, Upper Egypt," *Journal of Bone and Mineral Research* 12, pp. 8881–8888.

Fairgrieve, S., and J. E. Molto. 2000. "Cribra Orbitalia in Two Temporally Disjunct Population Samples from the Dakhleh Oasis, Egypt," *American Journal of Physical Anthropology* 111, pp. 319–331.

Földes, A., A. Moscovici, M. Popovtzer, P. Mogle, D. Urman, and J. Zias. 1994. "Extreme Osteoporosis in a Sixth Century Skeleton from the Negev Desert," *International Journal of Osteoarchaeology* 5, pp. 157–162.

Fox, S. 2005. "Health in Hellenistic and Roman Times: The Case Studies of Paphos, Cyprus, and Corinth, Greece," in *Health in Antiquity,* ed. H. King, London, pp. 59–82.

Fox Leonard, S. C. 1997. "Comparative Health from Paleopathological Analysis of the Human Skeletal Remains Dating to the Hellenistic and Roman Periods, from Paphos, Cyprus, and Corinth, Greece" (diss. Univ. of Arizona).

Garvie-Lok, S. 2001. "Loaves and Fishes: A Stable Isotope Reconstruction of Diet in Medieval Greece" (diss. Univ. of Calgary).

Goodman, A. H., R. Brooke-Thomas, A. C. Swedlund, and G. Armelagos. 1988. "Biocultural Perspectives on Stress in Prehistoric, Historical, and Contemporary Population Research," *Yearbook of Physical Anthropology* 31, pp. 169–202.

Jackson, R. 1989. *Doctors and Diseases in the Roman Empire,* Norman, Okla.

Jankauskas, R. 2003. "The Incidence of Diffuse Idiopathic Skeletal Hyperostosis and Social Status Correlations in Lithuanian Skeletal Materials," *International Journal of Osteoarchaeology* 13, pp. 289–293.

Janssen, H. A. M., and G. J. R. Maat. 1999. "Canons Buried in the 'Stiftskapel' of the Saint Servaas Basilica at Maastricht A.D. 1070–1521. A Paleopathological Study," *Barge's Anthropologica* 5, Leiden.

Julkunen, J., O. Heinonen, and K. Pyorala. 1971. "Hyperostosis of the Spine in an Adult Population: Its Relation to Hyperglycaemia and Obesity," *Annals of the Rheumatic Diseases* 30, pp. 605–612.

Karagianni, F. 1999. "Οι βυζαντινοί οικισμοί στη Μακεδόνια μέσα από τα αρχαιολογικά δεδομένα" (diss. Univ. of Thessaloniki).

Keramopoulos, A. D. 1932. "Ανασκαφαί και έρευναι εν τη άνω Μακεδονία," *ArchEph* 1932, pp. 81–87, 96–97, 99–100, 113–114.

———. 1933a. "Ανασκαφαί και έρευναι εν τη άνω Μακεδονία," *ArchEph* 1933, pp. 25–67.

———. 1933b. "Έρευναι εν Δυτική Μακεδονία," *Prakt* 1933, pp. 63–65.

———. 1934. "Ανασκαφαί και έρευναι εν τη άνω Μακεδονία," *Prakt* 1934, pp. 87–89.

———. 1937. "Ανασκαφαί και έρευναι εν τη άνω Μακεδονία," *Prakt* 1937, pp. 71–72.

Konstantokopoulou, A. 1984. "Ιστορική γεωγραφία της Μακεδονίας" (diss. Univ. of Ioannina).

Lagia, A. 1999. "Στοιχεία καθημερινού βίου. Το ανθρωπολογικό υλικό από την ανασκαφή του ρωμαϊκού ταφικού κτίσματος αρ. 1. στο νεκροταφείο του Κεραμεικού," *AM* 114, pp. 291–303.

Louggis, T. 1966. "Η εξέλιξη της βυζαντινής πόλης από τον τέταρτο ως τον δωδέκατο αιώνα," *Βυζαντινιακά* 16, pp. 33–67.

Lukacs, J. R. 1989. "Dental Palaeopathology: Methods for Reconstructing Dietary Patterns," in *Reconstruction of Life from the Skeleton,* ed. M. Y. İşcan and K. A. R. Kennedy, New York, pp. 261–286.

Mallegni, F. 1988. "Analisi dei resti scheletrici umani," in *Gortina* I, ed. A. di Vita, Rome, pp. 339–401.

Mango, C. 1980. *Byzantium: The Empire of New Rome,* London.

Mays, S. 1996. "Age-Dependent Cortical Bone Loss in a Medieval Population," *International Journal of Osteoarchaeology* 6, pp. 144–154.

Mays, S., B. Lees, and J. Stevenson. 1998. "Age-Dependent Bone Loss in the Femur in a Medieval Population," *International Journal of Osteoarchaeology* 8, pp. 97–106.

Merenstein, G. B., D. W. Kaplan, and A. A. Rosenberg. 1991. *Silver, Kempe, Bruyn and Fulginiti's Handbook of Pediatrics,* Norwalk, Conn.

Michailidis, M. 1965. "Μακεδονία," *ArchDelt* 20, pp. 475, table 598.

———. 1971. "Μακεδονία-Θράκη," *ArchDelt* 26, pp. 446–447.

Mogle, P., and J. Zias. 1995. "Trephination as a Possible Treatment for Scurvy in a Middle Bronze Age (ca. 2200 B.C.) Skeleton," *International Journal of Osteoarchaeology* 5, pp. 77–81.

Musgrave, J. H. 1976. "Anthropological Assessment," in *An Early Christian Osteotheke at Knossos,* ed. H. W. Catling and D. Smyth, *BSA* 71, pp. 25–47.

Ortner, D. 1984. "Bone Lesions in a Probable Case of Scurvy from Metlatavic, Alaska," *MASCAJ* 3, pp. 79–81.

———. 2003. *Identification of Pathological Conditions in Human Skeletal Remains,* 2nd ed., San Diego.

Ortner, D., W. Butler, J. Cafarella, and L. Miligan. 2001. "Evidence of Probable Scurvy in Subadults from Archaeological Sites in North America," *American Journal of Physical Anthropology* 114, pp. 343–351.

Ortner, D., and M. F. Ericksen. 1997. "Bone Changes in the Human Skull Probably Resulting from Scurvy in Infancy and Childhood," *International Journal of Osteoarchaeology* 7, pp. 212–220.

Ortner, D., E. Kimmerle, and M. Diez. 1999. "Probable Evidence of Scurvy in Subadults from Archaeological Sites in Peru," *American Journal of Physical Anthropology* 108, pp. 321–331.

Papathanasiou, A. 2001. *A Bioarchaeological Analysis of Neolithic Alepotrypa Cave, Greece* (*BAR-IS* 961), Oxford.

Pelekanidis, S. 1960. "Μακεδονία," *ArchDelt* 16, pp. 227–228, table 450.

Petkos, A. 2004. "Αρχαιολογικές έρευνες στην περιοχή αρμοδιότητας της 11ης EBA, στον άξονα της Εγνατίας οδού," in *Αρχαιολογικές Έρευνες και Μεγάλα Δημόσια Έργα, Πρακτικά της Αρχαιολογικής Συνάντησης Εργασίας, Επταπύργιο, Σεπτέμβριος 18–20, 2003,* Thessaloniki, pp. 156–157.

Pitsios, Th. 1985–1986. "Περιπτώσεις πολλαπλών ενταφιασμών του βυζαντινού νεκροταφείου Αβδήρων," *Πελλοπονησιακά* 16, pp. 237–240.

Roberts, C. A. 1987. "Case Report no. 9," *Paleopathology Association Newsletter* 57, pp. 14–15.

Roberts, C. A., and K. Manchester. 1995. *The Archaeology of Disease,* 2nd ed., Ithaca, N.Y.

Roberts, C. A., and J. Wakely. 1992. "Microscopical Findings Associated with the Diagnosis of Osteoporosis in Paleopathology," *International Journal of Osteoarchaeology* 2, pp. 23–30.

Rogers, J. 1982. "Diffuse Idiopathic Hyperostosis in Ancient Populations," in *Proceedings of the IVth European Meeting of the Palaeopathology Association,* Middelburg/Antwerpen, September 15–19, 1982, ed. G. T. Haneveld and W. R. K. Perizonius, Middelburg, pp. 94–103.

Rogers, J., and T. Waldron. 1995. *A Field Guide to Joint Disease in Archaeology,* Chichester.

———. 2001. "DISH and the Monastic Way of Life," *International Journal of Osteoarchaeology* 11, pp. 357–365.

Rogers, J., T. Waldron, P. Dieppe, and I. Watt. 1987. "Arthropathies in Paleopathology: The Basis of Classification According to Most Probable Cause," *JAS* 17, pp. 179–193.

Rogers, J., I. Watt, and P. Dieppe. 1985. "Paleopathology of Spinal Osteophytosis, Vertebral Ankylosis, Ankylosing Spondylitis and Vertebral Hyperostosis," *Annals of the Rheumatic Diseases* 44, pp. 113–120.

Saluja, G. 1988. "The Incidence of *Spina bifida occulta* in a Historic and a Modern London Population," *Journal of Anatomy* 158, pp. 91–93.

Stuart-Macadam, P. 1989. "Nutritional Deficiency Disease: A Survey of Scurvy, Rickets and Iron-Deficiency Anemia," in *Reconstruction of Life from the Skeleton,* ed. M. Y. İşcan and K. A. R. Kennedy, New York, pp. 201–222.

———. 1991. "Anemia in Roman Britain: Poundbury Camp," in *Health in Past Societies: Biocultural Interpretations of Human Skeletal Remains in Archaeological Contexts,* ed. H. Busch and M. Zvelebil, Oxford, pp. 101–113.

Triantaphyllou, S. 2001. *A Bioarchaeological Approach to Prehistoric Cemetery Populations from Central and Western Greek Macedonia* (*BAR-IS* 976), Oxford.

Tritsaroli, P. 2006. "Pratiques funeraires en Grèce Centrale à la periode Byzantine: Analyse à partir des données archaeologiques et biologiques (Tome I & II)" (diss. Musée National d' Histoire Naturelle).

Trotter, M. 1970. "Estimation of Stature from Intact Long Limb Bones," in *Personal Identification in Mass Disasters,* ed. T. D. Stewart, Washington, D.C., pp. 71–83.

Tsigaridas, E., and A. Loverdou-Tsigarida. 1982. "Αρχαιολογικές έρευνες στο Βελβεντό Κοζάνης," *Μακεδονικά* 22, pp. 302–328.

Tsilipakou, A. 1995. "Σέρβια Κοζάνης," *ArchDelt* 50, pp. 609–611, table 184a-b.

———. 1997. "Σέρβια Κοζάνης," *ArchDelt* 52, pp. 814–820.

———. 1998. "Σουρτάρα Γαλανίου Κοζάνης," *ArchDelt* 53, p. 714.

———. Forthcoming. "Σουρτάρα Γαλανίου Κοζάνης," *ArchDelt.*

Ubelaker, D. H. 1989. *Human Skeletal Remains,* Washington, D.C.

Waldron, T. 1985. "DISH at Merton Priory: Evidence for a 'New' Occupational Disease?," *British Medical Journal* 10, pp. 463–514.

Wesolowsky, A. 1973. "The Skeletons of Lerna Hollow," *Hesperia* 42, pp. 340–351.

CHAPTER 9

The World's Largest Infant Cemetery and Its Potential for Studying Growth and Development

by Simon Hillson

This chapter describes a new assemblage of infant remains that at the time of writing is being excavated at Kylindra on Astypalaia, the westernmost island of the Dodecanese group, in Greece. Although most of the children clearly died at birth, or within a few months afterward, a wide range of variation in development is shown. The large number of individuals makes it possible to investigate the pattern of growth in different parts of the body. Using standards derived from modern children, it is possible to estimate the stage of development. From the teeth, the bulk of these ages range from 25 to 50 weeks of development postfertilization (where 37–42 weeks is the usual clinical expectation for normal full-term pregnancy), but some are much smaller and a few are from older children, up to two years of age by modern standards.

Other standards apply to bones of the skull. The basilar part of the occipital bone changes in both size and proportions. Taking this element, the measurements from Kylindra imply that most of the children were in a state of development equivalent to modern full term, but a substantial number were smaller and would be considered preterm by modern standards. Others are considerably larger and some must have survived after birth for a considerable period.

Still other standards apply to long bone development. Lengths of, for example, the femur imply an average state of development equivalent to a modern full-term pregnancy of 40 weeks, but there is again a considerable spread, with the shortest bones implying development equivalent to 24 weeks. It is remarkable that such very small babies were buried in the cemetery—they are very rare finds in archaeology. These preliminary results indicate the potential of this unique assemblage of young children's skeletons. Further research will include a study of the microscopic layers preserved in the tooth enamel, which can be used as a clock against which to measure the development of different parts of the skeleton.

Infant skeletons are a relatively uncommon find on archaeological excavations of cemeteries, perhaps because they were often not buried alongside adults or, due to the delicate nature of the bones, they were not preserved. Classical Greek cemeteries are unusual in that children of all ages were frequently buried, with the very young placed in domestic or trade pots.

These containers have protected the bones, kept them together, and ensured that they will be discovered and recognized if the site is disturbed.

Infant burials are found scattered through most large Classical cemeteries; sometimes they are in separate cemeteries and sometimes they are found with older children and adults in what look like family groups. A good example from the Greek islands is the site of Vroulia on Rhodes.[1] There are, however, more unusual children's burial sites in the Mediterranean region. The largest infant assemblage until recently was the 2nd century B.C. deposit in Well G5:3 in the Athenian Agora,[2] which contained the commingled remains of at least 450 newborn babies. After that, the largest assemblage from a single-site context of any date was the 100 newborn babies found in a Late Roman–Early Byzantine sewer at the site of Ashkelon in Israel.[3] It has been proposed that they might have been the unwanted babies of prostitutes. A large collection of 450 cremation urns was excavated from 1976 to 1979 in the Punic Tophet sanctuary at Carthage.[4] The urns contained both human and animal remains. It was found that most of the human remains represented children between two and twelve months of age, with a substantial proportion representing newborn babies. Opinions are divided on whether these cremations resulted from the sacrifice of children to the goddess Tanit.[5]

EXCAVATION AND RECOVERY OF THE SUBADULT REMAINS

Until recently, little has been known about Astypalaia archaeologically, but the island is known from historical evidence to have been an independent state in Classical times and up until the time of Pliny (1st century A.D.).[6] Its capital was on the site of the present-day small town of Chora, centered on a promontory surmounted by a rocky outcrop that is now occupied by a Venetian citadel.

In 1996, the 22nd Ephorate of Prehistoric and Classical Antiquities carried out the first excavations of two newly discovered large cemeteries associated with the ancient city (Fig. 9.1). The southwest Kylindra cemetery is on the slopes directly below the citadel and the Katsalos cemetery is on a neighboring hill. The Ephorate's archaeologists believe that southwest Kylindra dates mostly to Late Archaic–Early Classical times (ca. 600–400 B.C.), although there are some burials that may be earlier, and a number that date to Roman times (1st century A.D.). The Katsalos burials range from Late Geometric (ca. 750 B.C.) up to the Roman period.

Since 2000, the author and colleagues from the Institute of Archaeology at University College London (UCL) have been collaborating with the 22nd Ephorate in recovering, conserving, recording, and storing the human remains from the two cemeteries. The aim is to establish a research resource, with a study center on Astypalaia, in which to carry out studies on the development of the skeleton and dentition in young children. The Ephorate has been concentrating its efforts on two building plots in the Kylindra site and has now uncovered 2,400 individual burials of infants.

1. Kinch 1914.
2. Little 1998.
3. Smith and Kahila 1992.
4. Stager 1980.
5. Fantar, Stager, and Greene 2000; Schwartz et al. 2002.
6. Pliny the Elder mentions the island in a geographical list as one of the Sporades (Plin. *HN* 4.23).

Figure 9.1. Excavation in progress on the cemetery at southwest Kylindra, just below the town of Chora on the island of Astypalaia. Photo S. Hillson

There are no adult burials on the site, and all the skeletons examined so far represent children under the age of two years. The majority were babies in approximately the state of development expected at birth after a full-term pregnancy. At the time of writing, the UCL team has succeeded in recovering 840 of the skeletons from these burials, and work is proceeding to recover, record, and store the remaining 1,560.

As was common in Classical times, the children were buried in large pots (Figs. 9.1, 9.2). Most are amphoras whose varying forms suggest they came from many parts of the Greek world. Many were clearly well-used vessels, because lead repairs have been found. Somewhat rarer was the use of domestic pots of various kinds. The pots were laid on their sides, in a small pit dug into the hard bedrock. The bodies of the babies had been placed inside each pot through an opening cut in the uppermost side, which was then closed by replacing the piece of pottery that had been cut out. The neck of the pot was blocked with a stone stopper, mortar, or sometimes a piece of pottery. Finally, the pit had been filled in. The whole site was covered with an accumulation of soil, rubbish, and other debris.

In most cases, the opening in the uppermost side of the pot had collapsed, allowing soil and rubble to fall inside on top of the baby's remains. This pot fill usually built up into a large, ball-shaped mass, solidly cemented

Figure 9.2. Kylindra site: child burial in pottery vessel. A door can be seen, cut in the upper side, through which the body was placed in the pot. The neck has been blocked with a stone.
Photo S. Hillson

together by pressure and minerals deposited from the ground water passing through the soil. Fortunately for the UCL team, a pattern of large cracks, starting at the edge of the burial "door" in the side, radiated through the body of the majority of pots and made it possible to expose the skeleton without damage. The 22nd Ephorate archaeologists working on-site have taken great care to lift the whole pot, complete with its heavy soil ball, wrap it, and carry it up the hill. This proved to be no mean task, because the lower side of each pot is cemented into the burial pit and the soil balls are extremely heavy.

In the laboratory, it is the UCL team's job to unwrap the pot and remove the pottery sherds to expose the soil ball. The sherds are repacked and, resting on a padded bench, the soil ball is rolled over to expose its underside. It was discovered at the start of the project that the delicate bones and teeth suffered least damage if they were excavated first, and then the rest of the ball was excavated from underneath (Fig. 9.3). In effect, the fill is excavated in the order of its deposition. Initial experiments in excavating the fill from the top downward, in the more conventional way, showed that the soil was so hard, and so much force was required to remove it that by the time the skeleton at the bottom was reached, it was already damaged.

As a last check, all of the fill removed from the soil ball is sieved through a 2 mm mesh. By trial and error, it was found that this small mesh size was required to recover any tiny elements such as phalanges, middle ear bones, and developing teeth which had been missed during excavation. One problem is that chips of the local rock mimic these small elements, and it has taken considerable practice to work rapidly through sieved residues. These "reverse excavations" of the soil balls are recorded by photographs and drawings. Few have shown a layered sequence of infilling and it is clear that, in the great majority of burials, the soil poured in rapidly after the door in the pot side failed. Sometimes the bones of the

Figure 9.3. Excavation of the ball of soil from inside one of the Kylindra pot burials. Photo S. Hillson

skeleton remain in anatomical position. It is assumed in these cases that the door collapsed, and the pot filled soon after burial and before sufficient decomposition had taken place to remove the ligamentous connections at the joints. In other burials, the bones are scattered through the lower part of the fill, suggesting that sufficient time had passed for decomposition to allow the joints to come apart.

The fill often contains sherds of the burial pot itself, particularly the door which had sometimes landed on the skeleton and damaged it. In some fills there are sherds of other pottery, including some decorated pieces, but only very rarely are there complete vessels of the type normally included as grave goods. This is expected for classical burials, where older children were often accompanied by offerings of small pots, but children younger than the age of weaning were not considered eligible for such treatment.[7]

Much more difficult to manage are burials in which the pot had remained more complete. Instead of containing a ball-shaped mass of fill, they are almost empty, with just a thin layer of clay covering the baby's skeleton. The tiny clay particles must have been washed through the pores and small cracks in the pottery. This clay layer periodically hardened and cracked, so that it looks rather like a dried-up lake bed. The pot remains complete and it is important to keep it so. In such cases, the method evolved is to draw and photograph the pattern of cracks, dividing the clay layer into a number of "slabs." Each of these is labeled, lifted out through the door in the pot side, inverted, and then placed in order on the bench. The skeleton can then be drawn, photographed, and excavated from the underside as before. It is difficult to remove the bones because the clay has the consistency of tough leather, but some of these pots have the best preservation of the skeletons.

The bones are cleaned as far as possible by picking at the soil with wooden sticks or, if the dirt is stubbornly attached, with a scalpel blade. Water is not used for cleaning, because experience has shown that the bones can crack while drying. Instead, drops of acetone are used to soften lumps of soil. Acetone is also applied to bone or tooth surfaces with cotton buds, which when rolled along the bone surface pick up the dirt. Wherever

7. Morris 1992; Garland 1985.

possible, breaks in the bones are not glued. They are only repaired when necessary to help with measurement or when it is judged that the bone fragments need support. A Paraloid B-72 acrylic adhesive is used for repairs. On occasion, it is necessary to consolidate bones to hold them together. Once again, either Paraloid B-72 in acetone or an emulsion of Primal WS24 are used as the consolidants. These are believed to have the best long-term storage characteristics of the currently available consolidants. Teeth are often found isolated in the soil, where they have fallen out of their development chambers inside the jaws (crypts) in which they were forming at the time of death. In addition, they may be preserved in situ inside the jaw. The soil is then carefully cleaned away and the tiny, delicate teeth lifted out. When all the bones and teeth have been cleaned, they are assembled onto standard labeled backgrounds to be photographed as a basic record of what is present. The elements present are recorded on a chart, together with measurements of long bone length, the dimensions of the basilar occipital bone, and the height of the developing tooth crowns. The bones are wrapped in acid-free tissue paper to act as padding and then bagged in labeled Ziploc bags. All the bags have small holes punched in them to keep the air circulating and contain Tyvek labels in case the label on the outside wears away. The bags are stored in a cardboard box with additional acid-free paper as padding.

MORTUARY VARIATION

When the layout of the burials is undisturbed, they typically contain a single child lying on its side with its knees drawn up in "fetal position." The head is toward the neck of the pot and the rump is toward the base of the pot. In about 10% of the burials, the body had been placed in the reverse position with its rump toward the neck of the pot. It is tempting to see the pots as symbolic of the womb, and the normal presentation of a baby at birth is with the head toward the neck of uterus, so that this part emerges first. In about 2%–5% of modern deliveries,[8] the rump is toward the uterus neck—a so-called breech presentation. The 10% of Kylindra burials with rump-to-pot neck is slightly higher, but breech deliveries are difficult to manage, so it might be expected that a death assemblage of babies would include slightly more of them. It is, however, difficult to test the hypothesis that the placement of the burial had some link with the complications of childbirth.

Another question concerns twins. In 1.9% of the burials so far investigated, there were the remains of two children within the same pot. Occasionally, the second individual was represented only by isolated elements that presumably came in with the fill, but in most cases it was clear that the original burial included two children. In undisturbed burials the skeletons can be seen lying side by side. It is possible that two neonatal deaths occurred at a very similar time in two separate births, but the simplest explanation is that these burials represent twins. One part of the project will be to test this explanation, initially through an examination of dental

8. Beischer, Mackay, and Cole 1997.

morphology. Some 1.1%–1.25% of modern births to European women are twins (although in other parts of the world the figure may be as low as 0.6% or as high as 2%),[9] which fits quite well with the proportion of double burials at the Kylindra site.

STUDIES OF DEVELOPMENT

Growth is an important indicator of health and nutrition in living children. It is monitored not only through overall body size and proportions, but also by the size of major bones or wrist bone development observed in X-rays. The teeth provide a useful sequence with which to compare other parts of the body. It is thought that their development is affected less than the skeleton by poor health and nutrition.[10] This can, in any case, be checked by the presence of dental enamel defects, which represent episodes of growth disruption during the formation of the teeth.

The largest direct study of growth in the skeleton was carried out in the 1960s and 1970s by Fazekas and Kósa,[11] using skeletons prepared during autopsies of 138 stillborn babies (and some who died a few hours after birth) from Hungary. They correlated a detailed series of measurements against the overall body length and calculated mean measurements for different stages of gestational age. Sherwood et al.[12] measured bone lengths in X-rays of 72 babies autopsied in Ohio, using the mother's last normal menstrual period to determine the gestational age. They carefully selected those babies who showed no signs of pathological abnormality that might affect the growth rate. Their published results include a series of regression equations to predict age from bone length.

The biggest study of dental development in babies was based on dissections of 787 stillbirths or abortions from throughout the United States. Kraus and colleagues[13] defined stages for crown development in deciduous teeth and gave approximate gestational ages for them. Butler[14] made measurements of developing teeth in the same collections. Deutsch and colleagues[15] measured the size of developing tooth crowns in 100 fetuses collected in Israel. They published their results as regression equations for tooth size based on ages estimated from crown-rump body size. These studies of modern children have established the general timing and sequence of development, but there is little detail on variation in the sequence or in the size and three-dimensional morphology of the different developing elements of the skeleton and dentition.

There has been no direct comparison between skeletal and dental development. For a whole variety of practical and ethical reasons, it is difficult to examine skeletal and dental development as part of the autopsy of a stillborn or aborted baby. Similarly, there are ethical difficulties in taking X-rays of healthy children. Ultrasound methods can be used, but they do not seem to be as reliable as radiographic methods.[16] In any case, there are difficulties in relating measurements taken on exposed "dry" bone and tooth structures to measurements taken on radiographs. To start with, there is the problem of parallax, which makes dimensions in the image slightly different

9. Beischer, Mackay, and Cole 1997.
10. Smith 1991.
11. Fazekas and Kósa 1978.
12. Sherwood et al. 2000.
13. Christensen and Kraus 1965; Kraus and Jordan 1965.
14. Butler 1967a, 1967b, 1968.
15. Deutsch, Goultschin, and Antenby 1981; Deutsch, Pe'er, and Gedalia 1984; Deutsch, Tam, and Stack 1985; Deutsch and Pe'er 1982.
16. Sherwood et al. 2000.

to dimensions in the specimen. In addition, work on the circadian growth rhythm recorded in the layered structure of dental enamel has shown that radiograph-derived standards for the age of particular development stages may be delayed relative to the histological estimates,[17] presumably because radiographs distinguish only those structures sufficiently mineralized to show as contrasts under the particular settings of film and the instrument used.

Archaeological collections can provide an opportunity to study development of the skeleton and dentition directly. The 18th–19th century crypt at Christ Church, Spitalfields, London, contained the remains of 37 children who died within the first year after birth. Their ages and sexes were independently known through a study of coffin plates and parish registers.[18] From a similar date, St. Bride's Lower Churchyard in Farringdon Street, London, included a number of young children whose sexes and ages at death were known. This small number of well-documented specimens has allowed the sequence of development to be verified.[19] Other archaeological collections can, however, still be useful even without the supporting independent evidence of neonatal death or postnatal age at death in early infancy.[20] These studies on modern and historical infants indicate that babies vary considerably in their state of development at birth, due to variation in the length of the gestation period, inherited patterns of growth, or the nutrition and health of the mother.[21] It is known, for example, that experimentally induced fever in pregnant laboratory rats causes defects in the structure of developing teeth in their babies.[22]

In order to investigate variations in infant growth, a large number of skeletons and associated dentition, at a range of different stages of development, is needed. One of the most exciting things about the Astypalaia assemblage is that it will provide a resource for such studies.

DEVELOPMENT OF TEETH IN THE ASTYPALAIA ASSEMBLAGE

By birth, in a normal full-term baby, there are partly formed elements of all the deciduous teeth. Development starts in utero for the teeth at the front of the dentition first, followed by the jaw (Table 9.1). The crowns of the teeth at this stage are just little hollow "caps," but enough detail is seen to be able to identify them (Fig. 9.4). The first incisors have most of their crowns formed at birth, the second incisors are about two-thirds formed, and canine crowns are one-third complete. In the deciduous first molars, the whole basin-like occlusal surface is formed with cusps and a substantial part of the crown sides. The last teeth in the series, the deciduous second molars, have their cusps formed, but these are usually only partially connected by the marginal ridges at the edge of the occlusal "basin."

At Kylindra, while the teeth in a large proportion of the burials are at about this stage of formation, there is a large amount of variation. This could be due to the variation in gestation length. Obstetricians use a "due date of confinement" of 40 weeks (280 days) as a guide to the length of gestation in a full-term pregnancy, but normal or "term" pregnancies may

17. Dean and Beynon 1991; M. C. Dean (pers. comm.).
18. Molleson and Cox 1993.
19. Scheuer, Musgrave, and Evans 1980; Scheuer and Maclaughlin-Black 1994; Liversidge, Dean, and Molleson 1993; Liversidge and Molleson 1999.
20. Tocheri and Molto 2002; Mays 1993; Gowland and Chamberlain 2002.
21. Ulijaszek, Johnston, and Preece 1998.
22. Kreshover and Clough 1953; Kreshover, Clough, and Bear 1953, 1958.

TABLE 9.1. DEVELOPMENT OF DECIDUOUS TOOTH CROWNS

	Development In Utero			*Development after Birth*
Tooth	*First sign at a macroscopic level in utero (weeks after fertilization)*	*First sign of calcification (weeks after fertilization)*	*State of development at birth after full-term pregnancy (40 weeks)*	*Crown complete (age in months after birth)*
Upper 1st incisor	11 weeks	14 weeks	Crown 80% complete	1.5 months
Upper 2nd incisor	11 weeks	16 weeks	Crown 60% complete	2.5 months
Upper canine	11 weeks	17 weeks	Crown 30% complete	9 months
Upper 1st molar*	12.5 weeks	15.5 weeks	Occlusal surface complete	6 months
Upper 2nd molar*	12.5 weeks	19 weeks	Cusps joined into "U" by distal marginal ridge	11 months
Lower 1st incisor	11 weeks	14 weeks	Crown 80% complete	2.5 months
Lower 2nd incisor	11 weeks	16 weeks	Crown 60% complete	3 months
Lower canine	11 weeks	17 weeks	Crown 30% complete	9 months
Lower 1st molar*	12 weeks	15.5 weeks	Occlusal surface complete	5.5 months
Lower 2nd molar*	12.5 weeks	18 weeks	Cusps joined into ring	10 months

* Dentists call these teeth first and second deciduous molars, whereas anthropologists and zoologists would refer to them as third and fourth deciduous premolars. Figures from Kraus and Jordan (1965) and Lunt and Law (1974).

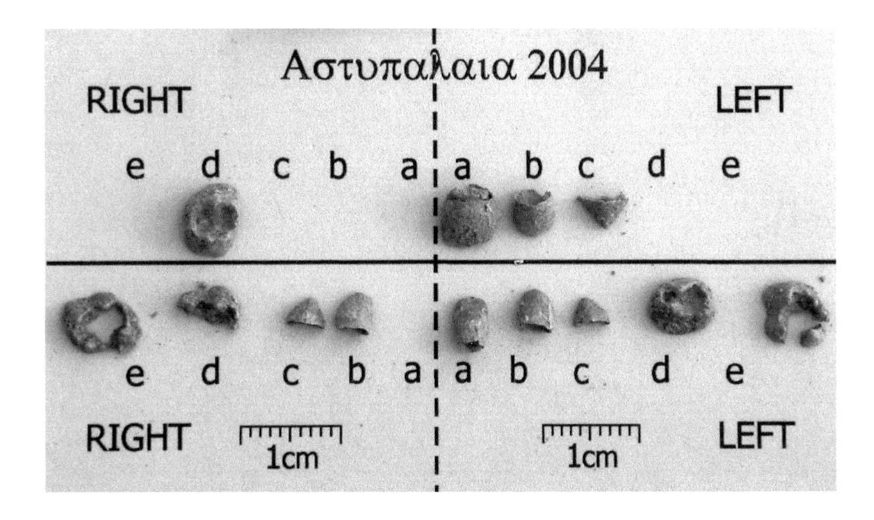

Figure 9.4. Teeth from a baby burial at the Kylindra site. Photo S. Hillson

vary between 37 and 42 weeks (259–293 days).[23] It is also possible that some of the burials at Kylindra could represent preterm births (24–37 weeks) or postterm (>42 weeks), with shorter or longer periods of development. Another possibility is that the variation is due to differences in the timing of tooth formation, such as an earlier or later start of calcification, or different rates of development.[24] The variation in development and its relationship with the timing of birth is therefore an area of potential study.

The most common way in which to record dental development state is a series of codes initially developed by Gleiser and Hunt[25] for radiographic studies of living children. These form part of the recording scheme for the Kylindra site, but they have been found unreliable when being scored by a large number of different people. Instead, it is better to measure the height

23. Beischer, Mackay, and Cole 1997

24. Uliaszek, Johnston, and Preece 1998.

25. Gleiser and Hunt 1955; Moorrees, Fanning, and Hunt 1963a, 1963b.

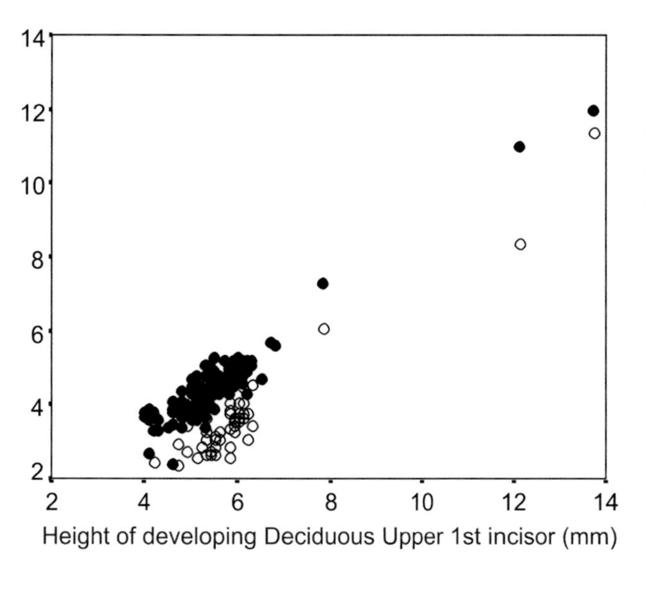

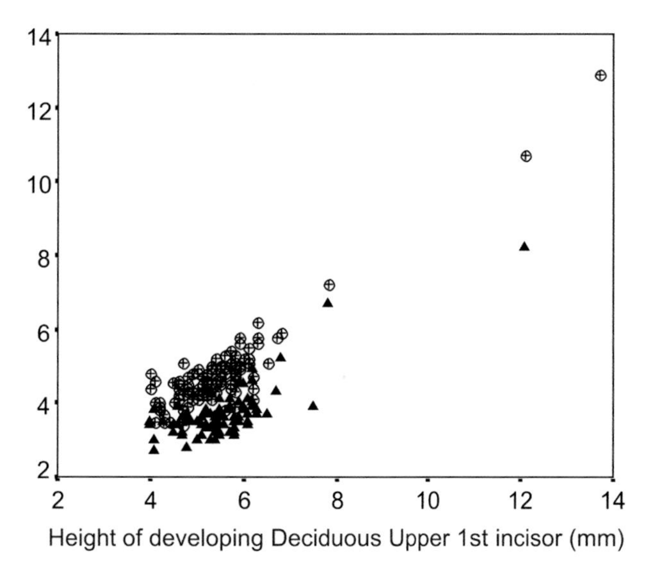

Figure 9.5. Height of developing deciduous upper first incisor plotted against that of other developing teeth

of the developing crown.[26] This approach was originally developed by Deutsch and colleagues[27] for studying teeth dissected from aborted fetuses and has been further developed by Liversidge[28] for studying the development of the somewhat older children buried in Spitalfields. The ages of a number of children were known from coffin plates and parish records, so it was possible to check development against this independently established age. The measurements of different teeth are quite strongly correlated with age at Kylindra (Fig. 9.5, Table 9.2). The measurements can be converted to an estimate of the number of weeks since fertilization using regression formulae derived by Deutsch and colleagues[29] from the fetuses of modern children. If these are plotted for the Kylindra assemblage (Fig. 9.6) the modal estimated age at death is 35 weeks after fertilization and the mean age is 36.7 weeks. This is quite different from the obstetrician's usual

26. The measurements are taken placing one caliper jaw across the open, developing edge of the crown, and the other at the highest part of the occlusal edge or the highest cusp.

27. Deutsch, Pe'er, and Gedalia 1984; Deutsch, Tam, and Stack 1985.

28. Liversidge, Dean, and Molleson 1993; Liversidge 1994.

29. Deutsch, Pe'er, and Gedalia 1984; Deutsch, Tam, and Stack 1985.

TABLE 9.2. CORRELATIONS BETWEEN HEIGHT OF DEVELOPING DECIDUOUS UPPER FIRST INCISOR AND HEIGHTS OF OTHER DEVELOPING TOOTH CROWNS IN THE KYLINDRA ASSEMBLAGE

Deciduous Upper				*Deciduous Lower*				
2nd incisor	*Canine*	*1st molar*	*2nd molar*	*1st incisor*	*2nd incisor*	*Canine*	*1st molar*	*2nd molar*
0.945	0.942	0.806	0.873	0.914	0.858	0.789	0.775	0.83

Note: Pearson's product-moment correlation coefficient. All values are highly significant at $p < 0.01$.

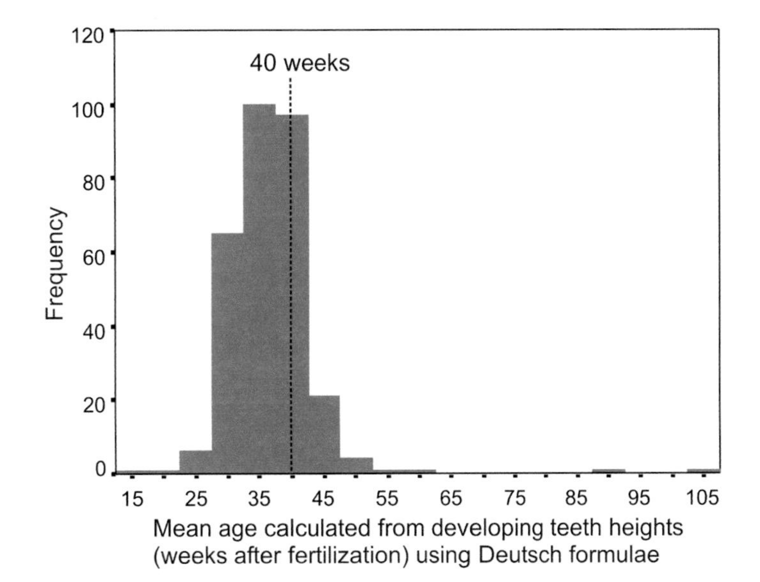

Figure 9.6. Means of ages calculated for different teeth present in each individual. Calculations based on the formulae of Deutsch, Pe'er, and Gedalia 1984.

assumption of 40 weeks for the gestation period, and children born before 37 weeks are usually regarded as preterm.[30] There is also some variation around the mean development state at birth. Ages in the main peak of the graph range from 25 to 50 weeks. The youngest age estimate is 13.4 weeks and the oldest nearly two years. These estimates, based on the development of a group of modern children, can be regarded as only approximations. The youngest ages are so close to those established for first calcification of the deciduous teeth (see Table 9.1) on the basis of studies of modern children that they seem likely to be wrong. A new technique is clearly required to calibrate the tooth development sequence more directly.

An exciting alternative approach to metric dental developmental studies that will be applied to the Kylindra teeth is based on histology. Both the dentine and the enamel that make up the tooth crown are finely layered. Enamel is usually the focus of investigation because being almost entirely mineral, it is better preserved. The main units of enamel structure are bundles of crystals, known as prisms, that are about 4 µm in diameter (1 µm = 0.001 mm). These radiate out from the enamel-dentine junction to the surface of the crown. Along their length they are marked, at roughly 4 µm intervals, by structures called prism cross-striations.[31] From work

30. Beischer, Mackay, and Cole 1997.

31. Boyde 1989; Dean et al. 2001.

with the known age Spitalfields children,[32] it has been confirmed that the cross-striations represent a circadian (approximately 24 hourly) growth rhythm. Counts of cross-striations can therefore be used to determine the time taken for different stages of dental development. The starting point for such counts in deciduous teeth is the so-called neonatal line. All microscope sections of deciduous teeth from children who have survived for more than a few days after birth are clearly marked by this structure, which represents a line of growth disruption at the time of birth. If the neonatal line is not present, then the child must have died at birth or have been stillborn. Cross-striations and measurements of prisms in the enamel formed in utero can be used to determine the age at which different stages of the formation of the teeth were attained, relative to the neonatal line or the enamel being formed at death. This is a very powerful tool for understanding not only the development of the teeth, but also the bones.

GROWTH IN THE SKULLS OF THE ASTYPALAIA ASSEMBLAGE

The bones of the cranial vault in a baby's skull are very thin and in most of the burials they were not well preserved. The most robust part is at the base, including the basilar part of the occipital bone, the body of the sphenoid, and the petrous part of the temporal. The basilar occipital survived in the best condition on Astypalaia, and standard measurements could be taken from it in many skeletons (Fig. 9.7). Ossification of the basilar occipital starts between 11 and 12 weeks after fertilization,[33] so it is possible to find it in the skeletons of babies so premature that their developing teeth were too small to survive, and this gives rise to a larger range of age estimates. There are two measurements of basilar occipital length: the maximum length proposed by Redfield[34] and the sagittal length of Fazekas and Kósa.[35] The Redfield length is slightly longer than the Fazekas and Kósa length, but is more badly affected by postmortem damage. During development, the basilar occipital changes in its outline shape, so the relationship between the measurements changes. The bone starts out longer for both measurements than its width dimension. By approximately 28 weeks after fertilization, the bone width equals the Fazekas and Kósa length, and thereafter exceeds it.[36] By about four months after birth (68 weeks after fertilization), the width exceeds the Redfield length. This can be seen most clearly if the lengths are plotted as ratios of the width (Fig. 9.8). From the Fazekas and Kósa length ratios in the Astypalaia assemblage it can be seen that while most values cluster about those expected at birth, a substantial number are greater than 1, implying that these children were at less than the 28-week stage of development. They must, therefore, have been very premature. From the Redfield length/width ratio, it again is clear that most values cluster around those expected at a normal full-term age of 40 weeks, but a substantial number fall below 1, which implies that they were older than 68 weeks after fertilization and had probably therefore survived for some time after birth. It is striking that very premature babies were buried alongside other, much more developed children.

32. Antoine, Dean, and Hillson 1999.
33. Scheuer and Black 2000.
34. Redfield 1970.
35. Fazekas and Kósa 1978.
36. Scheuer and Maclaughlin-Black 1994.

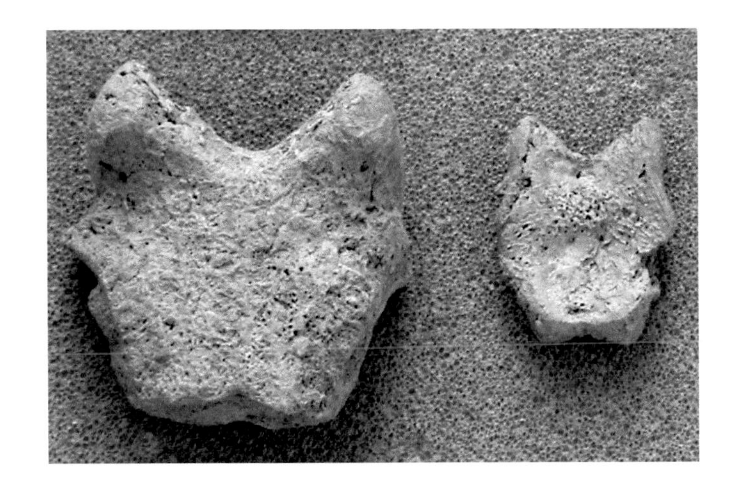

Figure 9.7. Basilar occipital bones, including one of the largest and one of the smallest (from a very premature baby). Scale 3:1. Photo S. Hillson

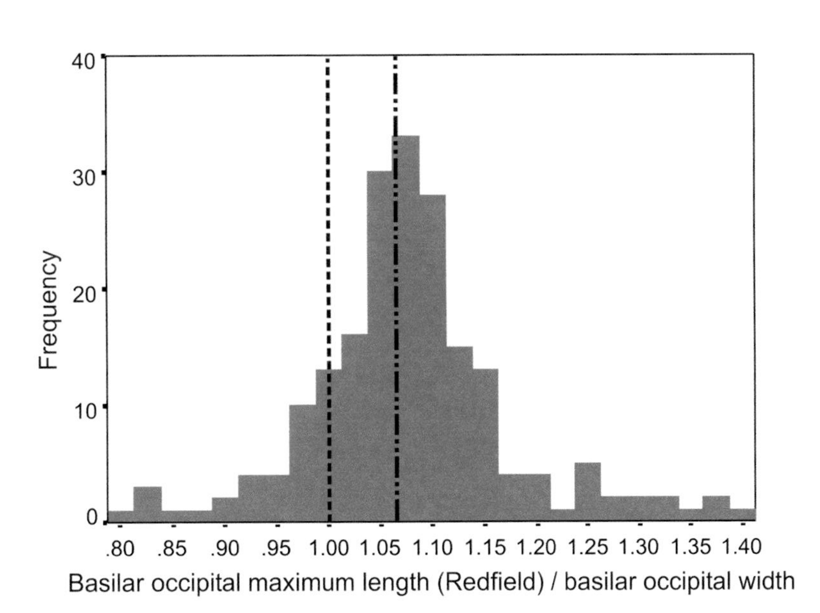

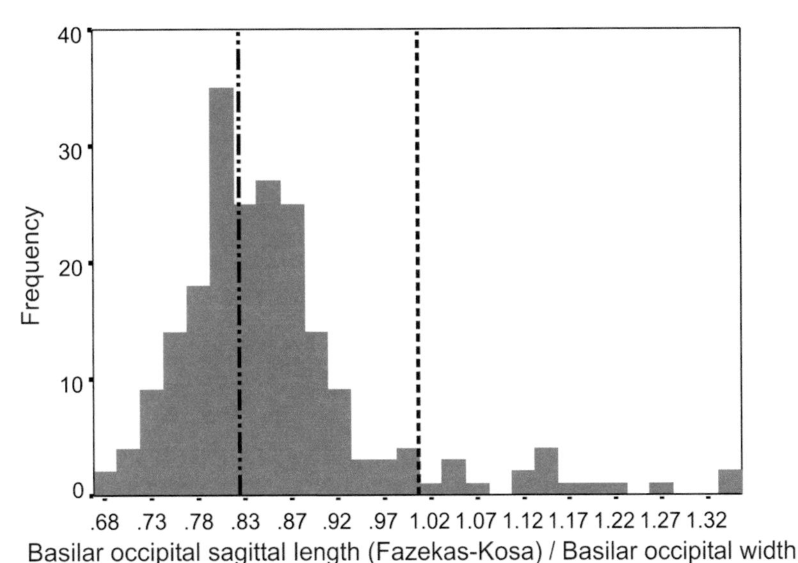

Figure 9.8. Ratio of length to width of basilar occipitals, using both length definitions: *(Top)* Basilar occipital maximum length/basilar occipital width (Redfield 1970). Dashed line shows point at which length equals width, which is around 20 weeks after birth. Ratios less than 1 (width > length) suggest development beyond this point; ratios greater than 1 suggest a developmental age less than 20 weeks after birth. The dotted line shows the approximate ratio expected at birth. *(Bottom)* Basilar occipital sagittal length/basilar occipital width (Fazekas and Kósa 1978). Dashed line shows point at which length equals width, which is around 28 weeks gestational age. Ratios less than 1 (width > length) suggest development beyond this point, and greater than 1, a developmental age less than 28 weeks. The dotted line shows the approximate ratio expected at birth.

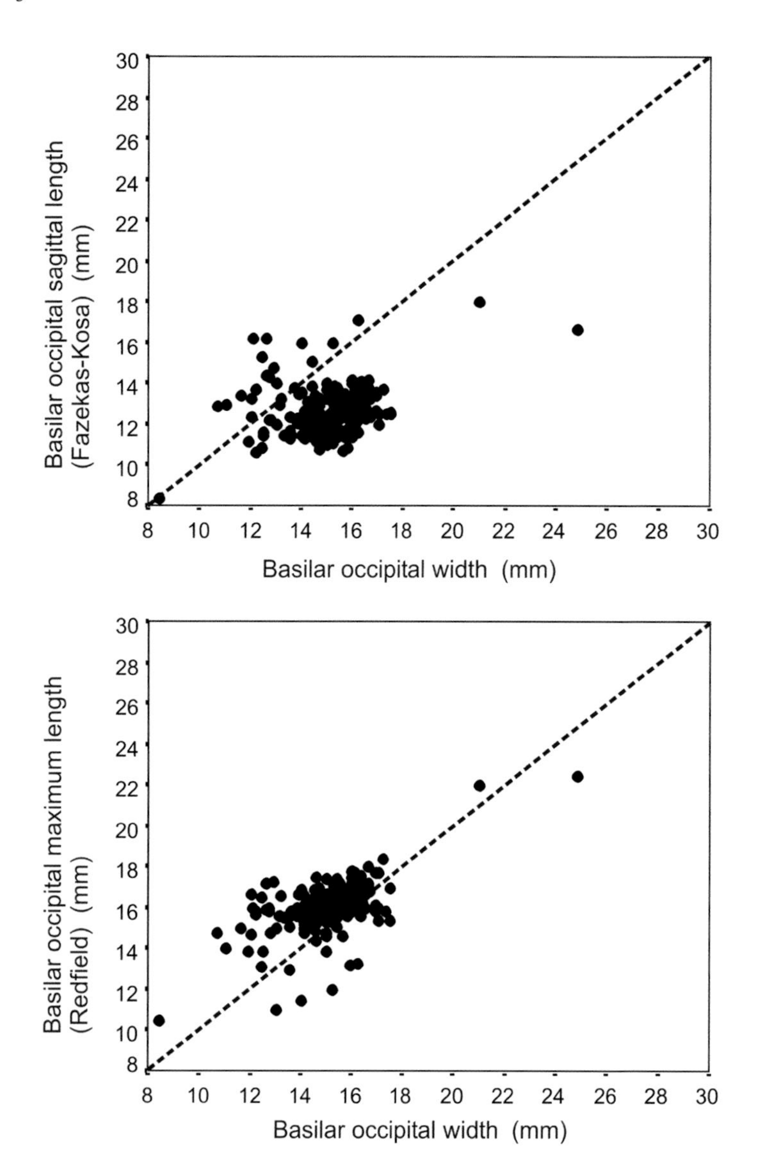

Figure 9.9. Width of basilar occipitals plotted against length: *(Top)* Basilar occipital sagittal length/basilar occipital width (Fazekas and Kósa 1978). *(Bottom)* Basilar occipital maximum length/basilar occipital width (Redfield 1970).

The relationship between the basilar occipital width and lengths at Astypalaia is therefore as expected from other studies, but there is nevertheless a good deal of variation, as can be seen when the measurements are plotted together as scatter diagrams (Fig. 9.9). The possibility of studying growth variation is one of the advantages of the large Astypalaia assemblage.

LONG BONE MEASUREMENTS IN THE ASTYPALAIA ASSEMBLAGE

All long bone measurements show strong correlations with one another (Table 9.3), with the exception of the fibula.[37] It is possible to estimate the developmental age, after fertilization, by using a number of regression formulae.[38] Because different skeletons have different bones preserved, the

37. This is a weak bone subject to distortion, and repairs may also have caused errors in measurement.

38. Scheuer, Musgrave, and Evans 1980; Sherwood et al. 2000.

TABLE 9.3. CORRELATIONS BETWEEN LONG BONE LENGTHS IN THE KYLINDRA ASSEMBLAGE

	Humerus	*Ulna*	*Radius*	*Femur*	*Tibia*
Ulna	0.911				
Radius	0.914	0.831			
Femur	0.955	0.881	0.882		
Tibia	0.951	0.899	0.898	0.963	
Fibula	0.372	*0.289	*0.236	0.463	0.418

Note: Pearson's product-moment correlation coefficient. All values are highly significant at $p < 0.01$, except for those marked "*".

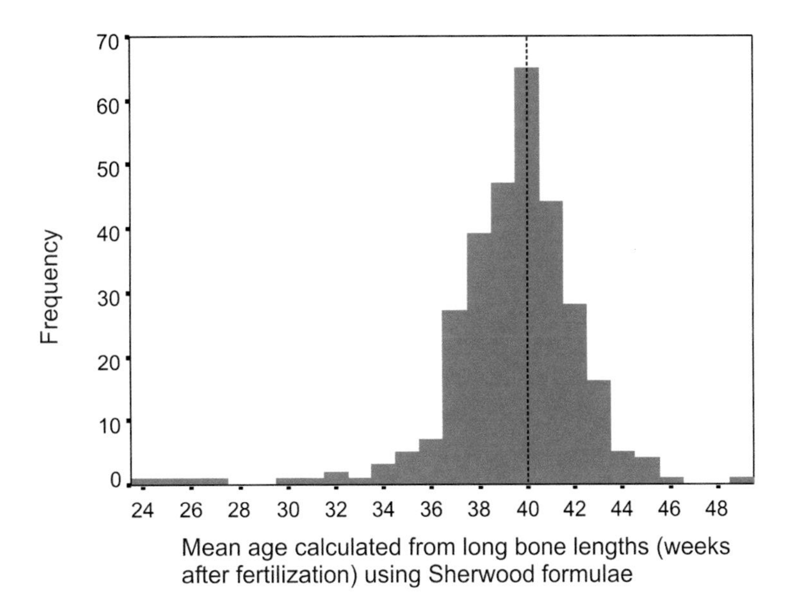

Figure 9.10. Means of ages calculated for different long bones present in each individual. Calculations based on formulae of Sherwood et al. 2000.

number of age estimates included can be maximized by substituting left, right, or the mean of left and right bone measures in the regression. The mean of age estimates for all bones can then be calculated and plotted (Fig. 9.10). From this, the modal age at death for the whole assemblage is 40 weeks, and the mean is 39.4 weeks. This fits rather better than the dental estimate with the expectations of the modern gestation period. The main spread of the peak in the graph is between 34 and 46 weeks after fertilization. In addition, there are some estimates between 24 and 33 weeks. These would count as preterm in modern clinical usage. More of these are visible in estimates based on the occipital bone, but this was a particularly commonly preserved element, especially in the youngest individuals. The two older children seen in the dental age estimates did not have complete long bone shafts and so could not be measured.

If long bones are plotted against one another, occipital bone measurements, and dental measurements, it can be seen that the different parts of the body were following different growth trajectories (Fig. 9.11). Where femur length is plotted against humerus length, basilar occipital width, and upper first central incisor height, it can be seen that the regression

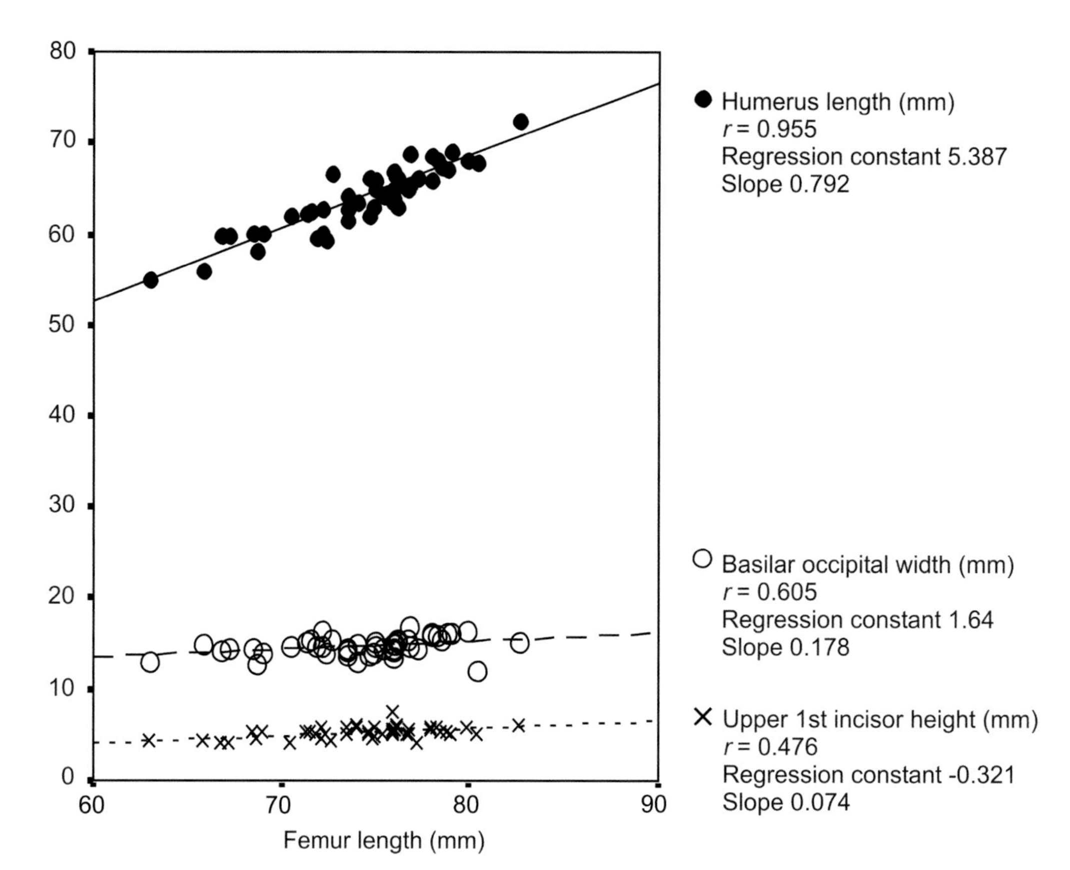

Figure 9.11. Femur length plotted against deciduous upper first incisor height, basilar occipital width, and humerus length.

line slope is considerably greater for the humerus and the slope is smallest for the dental measurement. This suggests that more detailed studies will be able to demonstrate the varying patterns of growth in different parts of the skeleton and dentition.

CONCLUSIONS

The large assemblage of young child burials from the Kylindra cemetery on the island of Astypalaia offers an exceptional opportunity for a detailed study of growth in the dentition and skeleton. In particular, the recovery of this sample makes possible a study of variation in development and a direct comparison of growth trajectories in teeth and bones. These initial studies of dental and skeletal development show considerable variation in the state of development of these babies who must have died either at birth, or within one or two years afterward. A substantial number would be classed as premature by modern standards, and a few are at a very early stage of development. It is possible to discern different patterns of growth for the long bones, the skull, and the dentition. The assemblage will be made available as a resource as part of a center for bioarchaeology research that is being established on the island.

REFERENCES

Antoine, D. M., M. C. Dean, and S. W. Hillson. 1999. "The Periodicity of Incremental Structures in Dental Enamel Based on the Developing Dentition of Post-Medieval Known-Age Children," in *Dental Morphology '98,* ed. J. T. Mayhall and T. Heikinnen, Oulu, pp. 48–55.

Beischer, N. A., E. V. Mackay, and P. E. Cole. 1997. *Obstetrics and the Newborn: An Illustrated Textbook,* London.

Boyde, A. 1989. "Enamel," in *Teeth,* ed. B. K. B. Berkovitz, A. Boyde, R. M. Frank, H. J. Höhling, B. J. Moxham, J. Nalbandian, and C. H. Tonge, New York, pp. 309–473.

Butler, P. M. 1967a. "Comparison of the Development of the Second Deciduous Molar and First Permanent Molar in Man," *Archives of Oral Biology* 12, pp. 1245–1260.

———. 1967b. "Relative Growth within the Human First Upper Permanent Molar during the Prenatal Period," *Archives of Oral Biology* 12, pp. 983–992.

———. 1968. "Growth of the Human Second Lower Deciduous Molar," *Archives of Oral Biology* 13, pp. 671–682.

Christensen, G. J., and B. S. Kraus. 1965. "Initial Calcification of the Human Permanent First Molar," *Journal of Dental Research* 44, pp. 1338–1342.

Dean, M. C., and A. D. Beynon. 1991. "Histological Reconstruction of Crown Formation Times and Initial Root Formation Times in a Modern Human Child," *American Journal of Physical Anthropology* 86, pp. 215–228.

Dean, M. C., M. G. Leakey, D. J. Reid, F. Schrenk, G. Schwartz, C. B. Stringer, and A. Walker. 2001. "Growth Processes in Teeth Distinguish Modern Humans from *Homo Erectus* and Earlier Hominids," *Nature* 414, pp. 628–631.

Deutsch, D., J. Goultschin, and S. Antenby. 1981. "Determination of Human Fetal Age from the Length of Femur, Mandible and Maxillary Incisor," *Growth* 45, pp. 232–238.

Deutsch, D., and E. Pe'er. 1982. "Development of Enamel in Human Fetal Teeth," *Journal of Dental Research* 61, pp. 1543–1551.

Deutsch, D., E. Pe'er, and I. Gedalia. 1984. "Changes in Size, Morphology and Weight of Human Anterior Teeth during the Fetal Period," *Growth* 48, pp. 74–85.

Deutsch, D., O. Tam, and M. V. Stack. 1985. "Postnatal Changes in Size, Morphology and Weight of Developing Postnatal Deciduous Anterior Teeth," *Growth* 49, pp. 202–217.

Fantar, M. H., L. E. Stager, and J. A. Greene. 2000. "An Odyssey Debate: Were Living Children Sacrificed to the Gods in Punic Carthage?," *Archaeology Odyssey* 3, pp. 28–31.

Fazekas, I. G., and F. Kósa. 1978. *Forensic Fetal Osteology,* Budapest.

Garland, R. S. J. 1985. *The Greek Way of Death,* London.

Gleiser, I., and E. E. Hunt. 1955. "The Permanent Mandibular First Molar: Its Calcification, Eruption and Decay," *American Journal of Physical Anthropology* 13, pp. 253–284.

Gowland, R. L., and A. T. Chamberlain. 2002. "A Bayesian Approach to Ageing Perinatal Skeletal Material from Archaeological Sites: Implications for the Evidence for Infanticide in Roman-Britain," *JAS* 29, pp. 677–685.

Kinch, K. F. 1914. *Vroulia,* Berlin.

Kraus, B. S., and R. E. Jordan. 1965. *The Human Dentition before Birth,* Philadelphia.

Kreshover, S. J., and O. W. Clough. 1953. "Prenatal Influences on Tooth Development II: Artificially Induced Fever in Rats," *Journal of Dental Research* 32, pp. 565–572.

Kreshover, S. J., O. W. Clough, and D. M. Bear. 1953. "Prenatal Influences on Tooth Development I: Alloxan Diabetes in Rats," *Journal of Dental Research* 32, pp. 246–261.

———. 1958. "A Study of Prenatal Influences on Tooth Development in Humans," *Journal of the American Dental Association* 56, pp. 230–248.

Little, L. M. 1998. "Babies in Well G5:3: Preliminary Results and Future Analysis" (paper Washington, D.C., 1998).

Liversidge, H. M. 1994. "Accuracy of Age Estimation from Developing Teeth of a Population of Known Age (0 to 5.4 years)," *International Journal of Osteoarchaeology* 4, pp. 37–46.

Liversidge, H. M., M. C. Dean, and T. I. Molleson. 1993. "Increasing Human Tooth Length between Birth and 5.4 Years," *American Journal of Physical Anthropology* 90, pp. 307–313.

Liversidge, H. M., and T. I. Molleson. 1999. "Developing Permanent Tooth Length as an Estimate of Age," *Journal of Forensic Sciences* 44, pp. 917–920.

Lunt, R.C., and D. B. Law. 1974. "A Review of the Chronology of Calcification of Deciduous Teeth," *Journal of the American Dental Association* 89, pp. 599–606.

Mays, S. A. 1993. "Infanticide in Roman Britain," *Antiquity* 67, pp. 883–888.

Molleson, T. I., and M. Cox. 1993. *The People of Spitalfields: The Middling Sort,* York.

Moorrees, C. F. A., E. A. Fanning, and E. E. Hunt. 1963a. "Age Variation of Formation Stages for Ten Permanent Teeth," *Journal of Dental Research* 42, pp. 1490–1502.

———. 1963b. "Formation and Resorption of Three Deciduous Teeth in Children," *American Journal of Physical Anthropology* 21, pp. 205–213.

Morris, I. 1992. *Death-Ritual and Social Structure in Classical Antiquity,* Cambridge.

Redfield, A. 1970. "A New Aid to Aging Immature Skeletons: Development of the Occipital Bone," *American Journal of Physical Anthropology* 33, pp. 207–220.

Scheuer, L., and S. Black. 2000. *Developmental Juvenile Osteology,* San Diego.

Scheuer, L., and S. MacLaughlin-Black. 1994. "Age Estimation from the *Pars Basilaris* of the Fetal and

Juvenile Occipital Bone," *International Journal of Osteoarchaeology* 4, pp. 377–382.
Scheuer, L., J. H. Musgrave, and S. P. Evans. 1980. "The Estimation of Late Fetal and Perinatal Age from Limb Bone Length by Linear and Logarithimic Regression," *Growth* 92, pp. 173–188.
Schwartz, J., F. Houghton, L. Bondioli, and R. Macchiarelli. 2002. "Human Sacrifice at Punic Carthage?," *American Journal of Physical Anthropology* Supplement 34, pp. 137–138.
Sherwood, R. J., R. S. Meindl, H. B. Robinson, and R. L. May. 2000. "Fetal Age: Methods of Estimation and Effects of Pathology," *American Journal of Physical Anthropology* 113, pp. 305–316.
Smith, B. H. 1991. "Standards of Human Tooth Formation and Dental Age Assessment," in *Advances in Dental Anthropology,* ed. M. A. Kelley and C. S. Larsen, New York, pp. 143–168.
Smith, P., and G. Kahila. 1992. "Identification of Infanticide in Archaeological Sites: A Case Study from the Late Roman–Early Byzantine Periods at Ashkelon, Israel," *JAS* 19, pp. 667–675.
Stager, L. E. 1980. "The Rite of Child Sacrifice at Carthage," in *New Light on Ancient Carthage,* ed. J. G. Pedley, Ann Arbor, pp. 1–11.
Tocheri, M. W., and J. E. Molto. 2002. "Aging Fetal and Juvenile Skeletons from Roman Period Egypt Using Basiocciput Osteometrics," *International Journal of Osteoarchaeology* 12, pp. 356–363.
Ulijaszek, S. J., F. E. Johnston, and M. A. Preece. 1998. *The Cambridge Encyclopedia of Human Growth and Development,* Cambridge.

CHAPTER 10

Differential Health among the Mycenaeans of Messenia: Status, Sex, and Dental Health at Pylos

by Lynne A. Schepartz, Sari Miller-Antonio, and Joanne M. A. Murphy

During excavations between 1939 and 1966, Carl Blegen and his team discovered several cemeteries that provide a diachronic view of the burial practices and biology of the Pylians inhabiting the Palace of Nestor and the neighboring area in Messenia during the Late Bronze Age (Figs. 10.1, 10.2). To an extent remarkable for their era, the Blegen team kept detailed records of the tomb excavations and worked to preserve the human skeletal material for biological study. Much of our knowledge of the Late Helladic IIIB Palace and these tombs comes from their detailed publications.[1] J. Lawrence Angel examined part of the skeletal collection in 1957. The results of his analysis, which focused primarily on the aging and sexing of the more complete specimens, were reported with the tomb descriptions. This is the only published information on the Pylos skeletons, aside from their inclusion in Angel's broader study of Mycenaean health and nutrition.[2]

The demographic findings from Angel's study of the Pylos human skeletal sample (Table 10.1) were based upon less than 40% of the total number of excavated burials (estimated to be 140).[3] Even so, because this was the sole source of information on the Pylos population, his results were used in many subsequent studies to document or reaffirm ideas about Mycenaean society, particularly the discussions of burial treatment and status. The role of males as rulers and warriors was emphasized, as was the striking predominance of male burials that Angel determined for several of the Mycenaean cemeteries.[4] For example, Angel reported that the Pylos Grave Circle contained the remains of twenty males and seven females.[5] This unequal sex ratio has been the focus of much Mycenaean scholarship,[6] but is it an accurate picture?

Variation in the human skeletal material from Pylos—whether demographic, morphological, or pathological—cannot be fully understood without considering the burial context. Since the beginning of Mycenaean archaeology, scholars have puzzled over the presence of two or more tomb types at some of the larger sites, such as Mycenae and Pylos.[7] In the early years, it was generally assumed that the architectural wealth of a tomb was indicative of the social status of the people buried in it. Tholos tombs, chamber tombs, grave circles, and cist or pit graves were thought to have

1. *Palace of Nestor* I–III. The tombs are detailed in Vol. III.
2. Bisel and Angel 1985.
3. *Palace of Nestor* III, pp. 79, 107. In addition to the information for Tholos III in *Palace of Nestor* III (N = 16, less one individual included in this analysis), this estimate includes the total count from Angel's field notes on Tholos IV in the Pylos Excavation Archives of the Department of Classics, University of Cincinnati, the Angel archive at the Smithsonian Institution, and revised sample sizes from this study (i.e., 108, plus 15 individuals from Tholos III and 17 individuals from Tholos IV).
4. Bisel and Angel 1985.
5. JLA 1957.
6. Acheson 1999; Cavanagh and Mee 1998; Mee and Cavanagh 1984.
7. Evans 1929, pp. 1–3; Schliemann 1878.

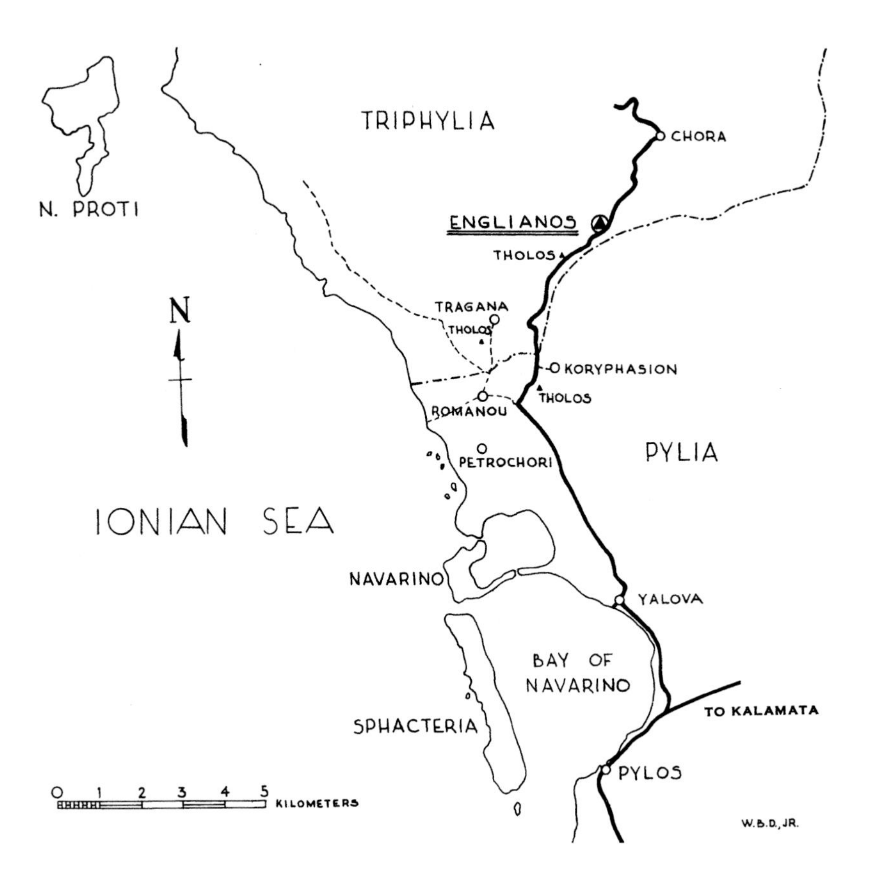

Figure 10.1. Map of the Pylos region with the location of palace structures at Englianos as a focal point.
W. B. Dinsmoor Jr., courtesy Department of Classics, University of Cincinnati

TABLE 10.1. ESTIMATED SAMPLE SIZES FROM PYLOS TOMBS

Tomb	*Blegen*	*Angel*	*This Study*
Grave Circle	21	27	31
Tholos III	16 min	–	1**
Tholos IV	17*	17	–
Tsakalis E-3	2	2	2
Tsakalis E-4	2 min	2	2
Tsakalis E-6	11	3	19
Tsakalis E-8	16	–	16
Tsakalis E-9	2 min	–	9
Kondou K-1	5–6	3	9
Kokkevis K-2	13	–	19
Total	106	54	108

* Blegen based this figure on Angel's estimation, rather than on the excavator's description. Blegen estimates are from the *Palace of Nestor III;* Angel's estimates are based on his field notes in the Palace of Nestor Excavations Archive, University of Cincinnati (JLA 1957); sample sizes for this study were determined using the field notebooks of Lord William Taylour (WDT 1953, 1957, 1958), William P. Donovan (WPD 1956), and Elizabeth Blegen (EPB 1939) in the Palace of Nestor Excavations Archive, University of Cincinnati, in conjunction with studies of the original provenience tags and the skeletal sample.

** Only one specimen was located.

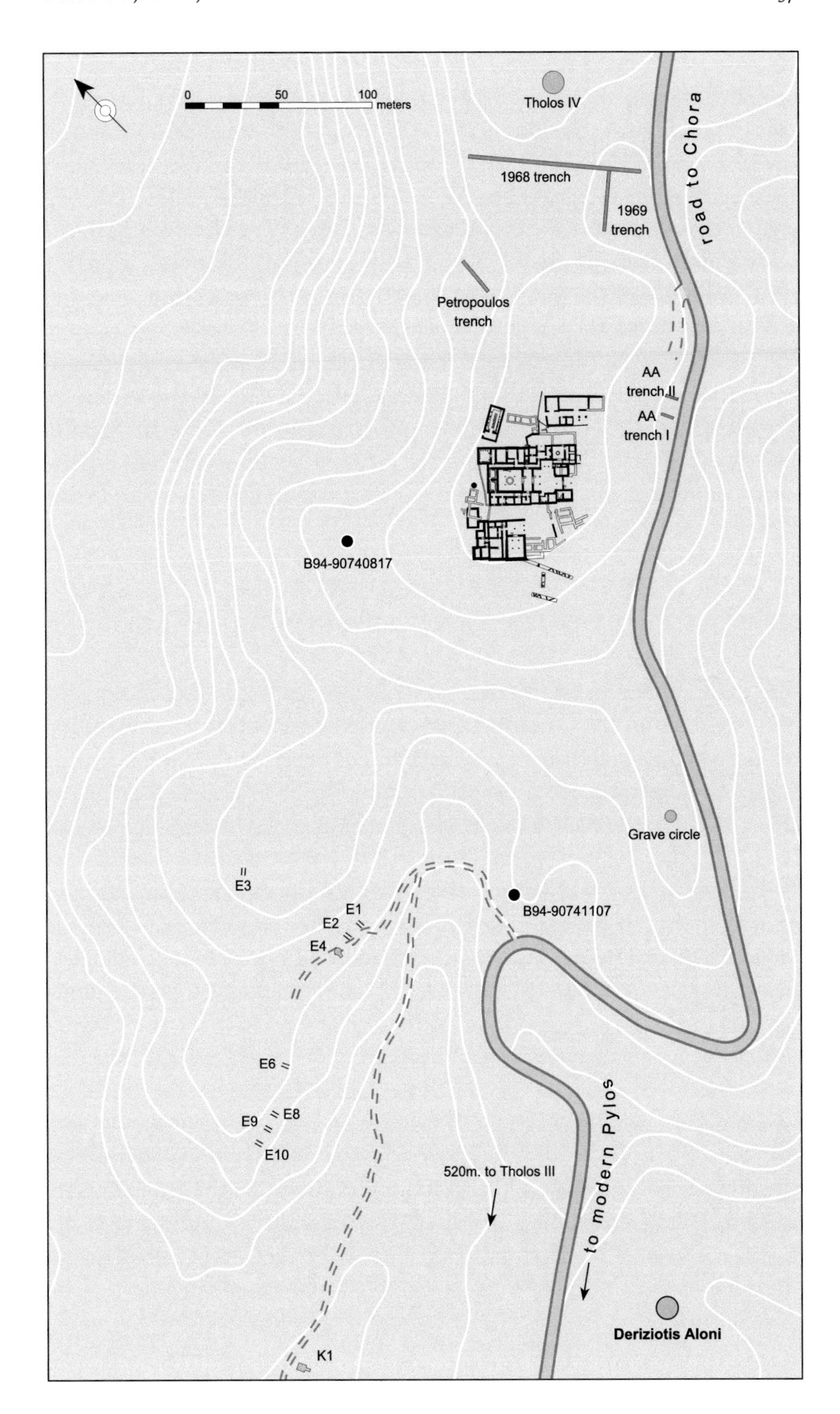

Figure 10.2. Map of the LH IIIB Palace of Nestor and the location of the associated cemeteries and excavation areas. E1–E10 are Tsakalis chamber and pit tombs, K1 is the Kondou chamber tomb. Drawing R. J. Robertson

been used by different echelons of society, with the tholos tombs being the reserve of the monarchy. Over the past century the conviction that there is a simple and direct correlation between wealth and tomb type has waned. Most recently, the tombs at Pylos have been interpreted as ideological tools of the Pylian elites, functioning as expressions of regional power and hegemony.[8]

8. Bennet 1995; Shelmerdine 1997.

Use of cemeteries in the vicinity of the palace structures at Pylos spanned the rise, zenith, and fall of the Pylian palatial system. Therefore, studying the human skeletal material with regard to the diachronic range of the tombs may shed light on the social, political, and economic changes that are thought to have taken place over the life of the polity. As a first step in that direction, this chapter contextualizes the Pylian tombs in the general debate about the correlation between status and tomb type, the ideological role of the tombs, and the relationship between the tombs and Pylian sociopolitical economy by focusing on the evidence for differential life experiences for the people of Pylos. To address this issue, we pursued the following general research questions: What do the samples in the tombs represent? Are they family tombs or special burials for elite warriors? Do individuals from different tomb types differ in health status, life experiences, or basic demography? Are the life experiences different for Pylian females and males?

BURIAL VARIATION AND STATUS AT PYLOS

The Palace of Nestor was the center for a complex economic and hierarchical system. From the Linear B tablets found in the Palace archive, several official positions and levels in the hierarchy can be securely identified: wanax (a ruler, often translated as "king"), other officials or leaders such as the lawagetas (*ra-wa-ke-ta*), basileus (*pa-si-re-u*) or group leader, companions (*e-qe-ta*), officials (*te-re-ta*), and mayors (*ko-re-te*) and vice mayors (*po-ro-ko-re-te*) of 16 major economic districts of the Pylian kingdom. It is also clear from the tablets that some Pylian women had considerable economic power.[9] Recent studies of the palace region by Minnesota Archaeological Research in the Western Peloponnese (MARWP) and the Pylos Regional Archaeological Project (PRAP) show the size of the palace and the adjacent settlement area is between 20–30 hectares.[10] There was a large structure at the site of the palace in Late Helladic (LH) IIIA and another and final structure built over that in LH IIIB.[11] The PRAP survey also demonstrated that during LH III, as the palace grew in size and economic complexity, the number of settlements in the 40 km area investigated around it decreased.[12] Shelmerdine suggests that this change may be indicative of the rise in power of the palace as a center and that people were moving from the outer areas of the provinces toward the focal settlement.[13]

PYLIAN TOMBS AND CHRONOLOGY

Blegen's excavations at the palace and its environs located six burial areas and four different types of tombs of Mycenaean date: the Grave Circle positioned 150 m to the south-southwest of the palace; Tholos III, 1 km southeast of the palace; Tholos IV, 80 m northeast of the palace; the Tsakalis chamber tombs between 210 and 360 m to the west-southwest of the palace; a simple pit (cist) grave among the Tsakalis chamber tombs; the Kondou chamber tomb located 500 m southwest of the palace; and the Kokkevis

9. For references, see Shelmerdine 1997, p. 566.

10. Davis et al. 1997, p. 428. The settlement around the Palace of Nestor spread for 1 km along the top of the Englianos ridge and 200–300 m over the ridge. This is five times the size of nearby Nichoria in LH III A.

11. Killian 1987, p. 209; Bennet 1995, p. 597.

12. Davis et al. 1997.

13. Shelmerdine 1997, p. 553.

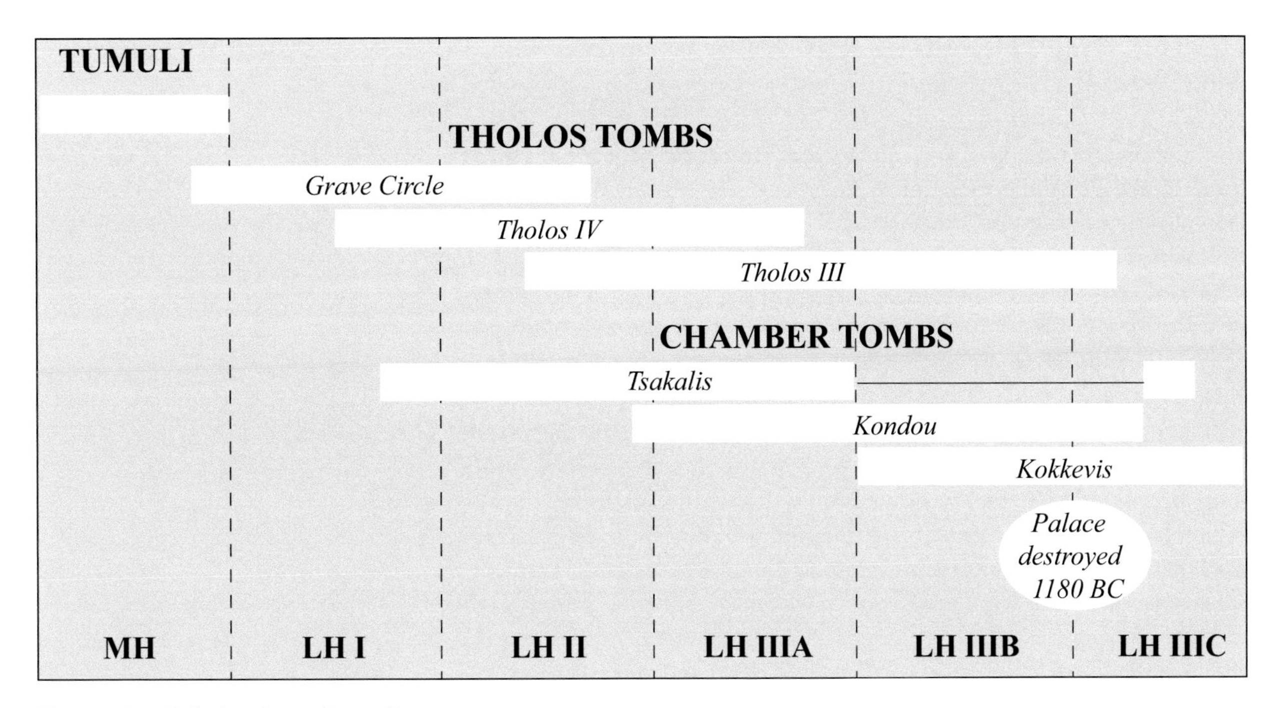

Figure 10.3. Relative chronology of the Pylos cemeteries. This rethinking of the tomb chronology is a result of new research on tomb contents by Murphy.

chamber tomb about 2 km southwest of the palace (Figs. 10.1, 10.2). Each of these tomb types is distinguishable by its unique architectural features. The chamber tombs have an entrance corridor (*dromos*) cut out of the natural rock. This leads into a larger chamber that has also been hollowed out of the hillside. The Grave Circle is a circular stone structure with several funerary and skeletal deposits. Based on similar finds elsewhere in Greece, and the absence of large quantities of stone, Blegen presumed that this circle had little in the way of stone superstructure and that it was simply covered by a pile of earth.[14] Recent studies in the area, and the presence of a field house constructed of large blocks near the Grave Circle, have led to a general consensus among scholars that it was a tholos tomb.[15] The tholoi, which are the most elaborate and massive funereal architecture forms, are built stone tombs. The burial chamber itself resembles an inverted cone with a *dromos* leading into it; these tombs were also covered with mounds of earth. Each type of tomb was usually used for the burial of more than one person and most were used over long periods of time.

The large majority of the tomb artifacts date to LH IIIA, the first palatial period, although several of the cemeteries were in use throughout the Late Bronze Age (Fig. 10.3).[16] A few pots from the chamber tombs were initially dated to LH IIIC,[17] and later redated to LH IIIB.[18] The amount of absolute time that the tombs were in use depends on which chronology one uses: According to the traditional low Aegean chronology, the tombs were in use from MH III–LH IIIB or 1600 B.C. to 1180 B.C., spanning a period of 420 years; according to the high chronology they were in use for 500 years from 1680 B.C. to 1180 B.C.[19] After the fiery destruction of the palace[20] in LH IIIB, there was a dramatic depopulation of the Pylos area, as indicated by the low levels of LH IIIC pottery found during the PRAP survey.[21] Several tombs continued to be used into the LH IIIC period, however.

14. For Blegen's discussion of this issue, see *Palace of Nestor* III, pp. 153–156.

15. The Grave Circle was originally referred to as Tholos V. *Palace of Nestor* III, pp. 71, 153–154; Davis et al. 1997, p. 420; Bennet 1999, p. 11. The Grave Circle is regarded as a tholos in this analysis.

16. *Palace of Nestor* III, pp. 228, 239; Mountjoy 1995, p. 158.

17. *Palace of Nestor* III, p. 228.

18. Mountjoy 1995, p. 158; Bennet 1995; Shelmerdine 1997.

19. See Shelmerdine 1997, table 1. The high chronology is strongly supported by scientific evidence; for an account of the high chronology, see Manning 1995, pp. 217–229. For the scientific reports that support high chronology, see Kuniholm et al. 1996, pp. 780–783. For a summary of the evidence for the low chronology, see Warren and Hankey 1989, pp. 137–169. For a brief and coherent summary of this evidence, see Shelmerdine 1997, p. 540.

20. Bennet 1995.

21. Shelmerdine 1997, p. 581; Davis et al. 1997, pp. 451–453.

CHANGING THEORIES ABOUT MYCENAEAN TOMBS

For the Mycenaeans of Pylos, the architectural form of a tomb and its dimensions, its location relative to the palace and center, its use over an extended time period, and the richness of its contents all played some role in the expression of social aspirations and realities. Theories about the people who used the tombs and the tombs' relationship with society have changed considerably due to two main developments: first, study of the chronology of the tombs; and second, the application of different theoretical frameworks. Initially, scholars assumed that the tholos tombs were used by a hereditary monarchy and that the chamber tombs and cist graves were, respectively, used by people on lower and the lowest social/political/economic strata. This long-standing hypothesis later changed based on Dickinson's chronological study of the construction of the tholos tombs at Mycenae, which showed that several of the tombs were in fact contemporaneous.[22] Darcque's subsequent study clearly showed that they could not be the burial places of a hereditary monarchy.[23] It then became commonly accepted that the tholos tombs were the mortuary architecture of the rich, rather than solely the rulers, while the chamber tombs were used by the lesser elites and the cists by the poorest in the society.[24] The current perspective is that tholos tombs were used by the elite as territorial markers, without assigning any particular political positions to these elites.[25]

The simple correlation of energy expenditure on mortuary architecture and status/wealth was the generally accepted view of Mycenaean tombs until recently. Yet empirical studies by Dickinson and Cavanagh and Mee on the artifacts in Mycenaean tombs revealed that the cist tombs should not be equated with the poorest people and that the contents of some chamber tombs were richer than the tholos tombs.[26] For Pylos, Cavanagh and Mee drew attention to the comparatively rich contents of both the Tsakalis E-6 chamber tomb and the Grave Circle.[27]

Social theories about expressions of status in mortuary treatment, especially those elaborated in the works of Binford, Brown, and Tainter, were adopted by Mycenaean processual archaeologists and gave much credence to the accepted correlation between wealth in death and status in life.[28] More recently, however, the postprocessual movement in archaeology has underscored a greater complexity for mortuary behavior.[29] This perspective, as applied by Voutsakis and Cavanagh and Mee to Mycenaean Greece, suggests that in certain circumstances mortuary behavior does not clearly and simply reflect the status of the recently dead, but can instead create an elevated social position for that person that they may never have held in life.[30]

The tombs of Pylos have been central to the works of scholars reconstructing the ways in which power was created, legitimized, and spread over the kingdom. This debate has mainly focused on the tholos tombs, even though chamber tombs are the most commonly found burial structures dating to the LH period.[31] The earliest Messenian tholos tombs, dating to MH III, were at Koryphasion and Koukounara.[32] By LH I–II, they had been built at several sites: Pylos, Voidokoilia, Tragana, Koukounara,

22. Dickinson 1977, 1982.
23. Darcque 1987, pp. 190–200.
24. Tsountas and Manatt 1987; Dickinson 1983, p. 56. For a summary of the traditional position and reasons why it is erroneous, see Boyd 2002, pp. 11–12.
25. Bennet 1995, p. 596; Mee and Cavanagh 1984; Wright 1987.
26. Dickinson 1983; Cavanagh and Mee 1998, p. 56.
27. Cavanagh and Mee 1998, p. 73.
28. Saxe 1970; Binford 1971; Brown 1981.
29. Hodder 1982; Parker Pearson 1999.
30. Hodder 1982; Cavanagh and Mee 1990, 1998; Mee and Cavanagh 1984.
31. Cavanagh and Mee 1998, pp. 40, 44.
32. Lolos 1989.

Routsi, Nichoria, Peristeria, and possibly Psari.[33] At several additional sites, tholos tombs were built in LH IIIA and used during LH IIIA–LH IIIB: Nichoria, Dara, Mouriatada, Malthi, Ano Kopanaki, and Vigla Kalpaniou.[34] During LH III, however, several of the larger of these tombs were abandoned and smaller ones were built at most of the sites where new tholoi were constructed.

Bennet reasoned that the expansion of Pylos is strongly implied by the fact that most of the tholos tombs at sites in its vicinity go out of use in LH IIIA, "reflecting their effective demotion within the power hierarchy."[35] For example, he suggested that the construction of a new tholos tomb at Nichoria indicated that a new ruling elite had been established—potentially with the support of an external power.[36] Nichoria was apparently expanding in LH IIIA when a *megaron* (throne room) was built. In LH IIIA2, the megaron was destroyed. At the same time, a tholos tomb was built at the edge of the site. Bennet's analysis highlighted the incongruity of a local elite building a tomb when the main sign of hierarchy at Nichoria, the *megaron*, was destroyed. In a further development of Bennet's theory, Shelmerdine argued that the construction of the LH IIIA2 tholos at Nichoria was linked to its assimilation into the Pylian state, with the tomb functioning as a status symbol connecting the local elites at Nichoria with Pylos.[37] Larger tholos tombs continued to be used only at sites with administrative ties to Pylos.[38] Most recently, Cavanagh and Mee suggested that the presence of several tholos tombs in Messenia indicates that there were a number of elite families.[39]

Chamber tomb construction was the other critical component of Mycenaean burial practices in Messenia. The earliest chamber tomb in Messenia, at Volimidia, dates to the MH period.[40] In the Early Mycenaean period, LH I–LH II, the earliest tomb in the Tsakalis cemetery at Pylos, E-8, was constructed, as was a chamber tomb at Volimidia.[41] In LH III, chamber tombs became the most popular type of tomb in all of Greece; at Pylos this growth in popularity is evidenced by the construction of five more in the Tsakalis cemetery. The great increase in chamber tombs in LH III suggests that a greater portion, and perhaps a different segment, of Messenian society were expressing their social position through tomb construction. The chronological development of the chamber tombs and their widespread construction (both large and small) in LH III contrasts with the decreased number of large tholos tombs and the apparent connection between tholos tombs and palatially related sites.

CHARACTERISTICS OF THE PYLOS CEMETERIES

Grave Circle

The burials in the Grave Circle[42] were clearly defined and the only disturbance was from modern agricultural activity.[43] Individuals were placed into four pits of varying sizes. The oldest burials in the cemetery were in pithoi or very large jars; three of these were located in the east side of the

33. Cavanagh and Mee 1998, pp. 44–47.
34. Cavanagh and Mee 1998, pp. 77, 83, fig. 6.2.
35. Bennet 1995, p. 598.
36. Bennet 1995, p. 598.
37. Shelmerdine 1997, pp. 101–102.
38. Shelmerdine 1997, p. 553.
39. Cavanagh and Mee 1998, p. 77.
40. For detailed studies of the chamber tombs, see Cavanagh and Mee 1998; Dickinson 1977; Graziadio 1988.
41. *Palace of Nestor* III, p. 195.
42. Located on the Vayenas property, the Grave Circle was originally referred to as "Vayenas" in the excavation notebooks and in Angel's field notes.
43. *Palace of Nestor* III, p. 148. This lack of grave disturbance contrasts with the heavy damage to the superstructure of the tholos.

circle and one was in the west. The latest burial was laid out in an extended position in a shallow pit (Pit 2) in the center. There were two deposits of disarticulated skeletons: one in the east near the three pithoi and one in the north. The pithos burials may date to late MH or early LH. The objects interred with them included bronze cauldrons, daggers, rapiers, and boars' tusks that may be from a helmet. There were only a few pots associated with the pithos burials or with the extended burial in Pit 2. Several other artifacts were found on or near the body in Pit 2. These included a bronze mirror found near the pelvis, a small knife, a sword/dagger, a juglet, a cylindrical painted *phi* figurine that lay on the chest, and a chert arrowhead that had been placed between the legs. In addition to the above grave goods, there were deposits of bronze objects, obsidian and chert arrowheads, and ceramics in the Grave Circle that are not associated with any skeletal material. Fine quality ornaments of gold, ivory, and silver were recovered throughout the chamber. A large number of pots of LH I–LH IIIA1 date were found in the northern part of the circle, but these also cannot be associated with any individual burials.[44]

Tholos III

Although large quantities of human and animal bones were found in Tholos III (35 baskets were recorded in the field notebook),[45] all of them had been disturbed and moved from their original context. The disarray of the tomb led Blegen to suggest that it had been ransacked in antiquity.[46] There were two pits in the floor of the chamber; both contained disturbed human remains, beads, and broken pottery. Skeletal remains, beads, pottery, and pieces of gold were strewn throughout the *dromos*, the doorway, and the chamber of the tomb.[47]

Tholos IV

The archaeological deposits in Tholos IV[48] were very disturbed and the excavators suspected that it, like Tholos III, had been looted in antiquity. The great quantity of charcoal in the tomb suggested to them that it had been periodically cleansed by lighting fires in the chamber. Human and animal bones were numerous, as were plain-ware sherds, but only four complete pots were recovered.[49] Beads of glass paste, lapis lazuli, amethyst, amber, and gold were abundant, as were worked stone, bronze, and furniture parts. Gold was plentiful, especially gold leaf. In fact, so much gold leaf was attached to the floor in the center of the tomb that the excavators initially thought there had been an intentional gold covering there.[50] There were several pits in the floor; one large semicircular pit was parallel to the northwestern wall of the chamber. Some of the most striking and unique finds of all the cemeteries in Pylos came from this pit. These included a gold pendant in the shape of a figure-of-eight shield, a gold bead seal with the "royal griffin," and an amber spacer-bead, which according to Blegen et al., was very similar to an example found in a Wessex cemetery in England.[51] A stone cist built against the southeast wall of the tomb contained only one human femur, but there were many small finds including a gold signet ring.[52]

44. *Palace of Nestor* III, p. 138.
45. EPB 1939.
46. *Palace of Nestor* III, p. 77.
47. *Palace of Nestor* III, pp. 73–95.
48. Located on Kanakaris land, this tholos is also referred to as the "Kanakaris" tomb in the excavation notebooks and Angel's field notes.
49. The excavator commented that there were pre-Mycenaean sherds.
50. *Palace of Nestor* III, pp. 102–106.
51. *Palace of Nestor* III, p. 105.
52. *Palace of Nestor* III, p. 105.

Chamber Tombs: Tsakalis, Kokkevis, and Kondou

In the Tsakalis cemetery there were five complete chamber tombs (E-3, 4, 6, 8, 9), three unfinished tombs that consisted of only a *dromos* (E-1, 2, 10), one unfinished *dromos* (E-12), and three tombs that were located but not excavated (E-5, 7, 11). One chamber tomb was explored on the land belonging to Kontos (Kondou K-1) and a second Mycenaean tomb was excavated on Kokkevis land (Kokkevis K-2 or γ).[53]

Similar mortuary practices were carried out in all three cemeteries. Individuals were buried in pits or laid out on the floor of the chamber. In some instances (e.g., at Kondou K-1 and Tsakalis E-10), remains of burials were also found in the *dromos*. Several bodies were found laid out in extended position and one (in Tsakalis E-9) was contracted. Most typically, the extended burials appeared to be less disturbed and were described as the latest burials in the tombs.[54] There are also many cases of secondary treatment. Bones were found in jars and scattered over other burials.

The Pylos chamber tombs show far more variability in their contents than the tholos tombs. The most common finds in these tombs were pottery and beads. All tombs contained pots and bronze (mostly knives and swords) and stone artifacts. Tsakalis E-6 stands out as having more numerous and richer grave goods than the other chamber tombs, while Tsakalis E-9 seems to be relatively poorer.[55]

Pit Grave

Tsakalis E-3 was a simple pit that contained four pots and two bodies positioned one on top of the other.[56] The pots lay on top of the upper body; there were no objects with the lower body. Above the pit were several stones associated with an upturned cup and a carnelian bead.[57]

Based on the preceding discussion of Mycenaean burial practices, if the architecture and richness of tombs is a direct measure of access to resources during life, we would predict that individuals from the Grave Circle, Tholos III, Tholos IV, and Tsakalis E-6 at Pylos would have access to the best nutritional resources. Based on our knowledge of Mycenaean food and dietary studies,[58] those individuals should have a higher protein diet (including meat from wild and domestic animals, dairy products, pulses, and potentially marine life) with less dependence on carbohydrate-rich

53. *Palace of Nestor* III, pp. 176–215, 224–237.

54. The extended burials were often located in the central portion of the tomb; their better preservation suggests that they were primary burials and later interments. This interpretation is often supported by the date of the associated vessels, but the effects of tomb robbing and later disturbances in tombs must have played a significant role in the destruction of primary burials.

55. Cavanaugh and Mee 1998.

56. This grave is variably referred to as a pit or cist burial. In *Palace of Nestor* III, p. 177; the skeletons are described as lying in the *dromos*—presumably because E-3 was thought to be an unfinished chamber tomb that was used but never completed. Regardless of its classification to specific burial type, E-3 is notably simpler than the other burials known from Pylos.

57. *Palace of Nestor* III, pp. 176–177.

58. We have information on Mycenaean food resources from diverse sources, including Linear B tablets, frescos, faunal studies, paleobotanical studies, and staple isotope analyses of human bone. See also papers in Wright 2004; Halstead and Barrett 2004; Tzedakis and Martlew 1999; Vaughan and Coulson 2000. See also Chaps. 11, 13, and 14 in this volume.

cariogenic foods. They should therefore show fewer dental pathologies.[59] Conversely, individuals from tombs with simpler architecture and fewer grave goods should show higher levels of dental pathology.

MATERIALS AND METHODS

All of the currently known Pylian skeletal collection was studied by Schepartz and Miller-Antonio in 1998–2003.[60] According to their analysis, the Pylian skeletal sample currently available for study[61] includes 108 individuals from nine tombs (Table 10.1). The Tholos III material that was excavated in 1939,[62] with the exception of one fairly complete cranium, remains unstudied. Unfortunately, the present location of that material, as well as the Tholos IV sample studied by Angel, is unknown.[63] Based on the Blegen team's excavation notes and Angel's estimate of the minimum number of individuals for Tholos IV,[64] the remains of 106 individuals were excavated. If the results of this study are factored into the calculation of the minimum number of individuals (MNI), the total rises to 140.

Age and sex for each skeleton was independently assessed by Miller-Antonio and Schepartz. The entire collection was then reassessed to refine aging and sexing estimates based on a fuller understanding of the observed range of variation in the population. Due to the fragmentary nature of the material, a combination of aging techniques[65] based on pubic symphysis, dental development, and tooth wear[66] were applied. Sex determination was based on pelvic and cranial morphology,[67] with additional metric data from postcrania for some specimens. Specific age estimates were made when possible, but individuals were most frequently assigned to age cohorts.

For the dental analysis, teeth were judged as lost antemortem when the alveolar bone exhibited substantial remodeling to the degree that no root sockets were functional. Caries presence, determined by visual

59. For syntheses of the general relationship between dental pathology and nutrition, see Larsen 1997 and Hillson 1996. For excellent discussions of the biological consequences of dietary variation and social stratification, see Powell 1988; Goodman 1998; Cucina and Tiesler 2003. We focus on dental health in this study as it provides us with the largest sample for analysis. Other indicators of health, such as porotic hyperostosis, were also evaluated. There is limited evidence for these conditions in the Pylos sample, in contrast to the prevalence of dental pathology, as mentioned in the Discussion section.

60. The authors wish to thank Xeni Arapoyianni and Yioryia Hatzi of the 11th Ephorate of Prehistoric and Classical Antiquities, and the staff of the Chora Museum for their help in assisting this research; Shari Stocker and Jack L. Davis, Director of the American School of Classical Studies at Athens, for their invitation to study this material and unflagging support and guidance; the Louise Taft Semple Fund of the Department of Classics, University of Cincinnati, the Institute for Aegean Prehistory for financial support; and Erin Williams for extracting and collating the information from the Blegen excavation notebooks.

61. The entire Pylos skeletal collection examined for this analysis is located in Apotheki 2 of the Chora Museum.

62. EPB 1939.

63. The material from the 1939 excavations of Tholos III was originally transported to the National Museum in Athens as it was recovered prior to the construction of the Chora Museum. The one known cranium, which was not studied by Angel, was found in Apotheki 1 of the Chora Museum. The burials from Tholos IV were studied by Angel in 1957, but they are not catalogued in the Chora Museum or National Museum collections. See postscript of this chapter for updated information on this situation.

64. JLA 1957; EPB 1939; WPD 1956; WDT 1953, 1957, 1958.

65. Buikstra and Ubelaker 1994.

66. Adapted from the techniques of Miles and Molnar, as discussed in Hillson 1996.

67. Buikstra and Ubelaker 1994.

inspection with hand lens magnification,[68] was tallied by individual tooth. Linear enamel hypoplasia, also assessed by visual inspection with hand lens magnification,[69] was denoted as present or absent. The total dental sample includes both teeth present and those missing where antemortem status could be assessed from the condition of the alveolus. Units of analysis included individual teeth and dentitions.

RESULTS

Demography: General Age and Sex Ratios

The notion that Pylian burial customs involved something other than the interment of family groups, or that Pylian society was heavily biased toward male burial, is not supported by the present analysis. As noted above, Angel identified twenty adult males and seven adult females in the Pylos Grave Circle. We found evidence for four more individuals, and a surprisingly different sex ratio: nine males and twelve females, plus nine unsexed adults (Table 10.2). The Kokkevis K-2 chamber tomb is the only burial structure that seems to have had a biased sex ratio of ten males and four females. However, there is a striking underrepresentation of children or subadults under 18 years in all of the tombs; very young children are completely absent.[70] It appears that most Pylian families did not inter their children in the family tomb.

Another important result of this demographic analysis is that the total *N* of the sample (108 burials, plus 32 more individuals if Blegen et al.'s Tholos III and Tholos IV estimates are included),[71] is approximately 45% higher than the excavator's or Angel's estimates, suggesting a much larger proportion of the population was buried in chamber or tholos tombs.

Using very general age cohorts (Table 10.3), age estimates are possible for 60% of the individuals. The most abundant category consists of younger adults (*N* = 24). Using more specific age cohorts for the more complete chamber tomb materials, we find that the majority of the younger adults are in the 25–30 year age cohort; and very few individuals are estimated to be much older than their late 40s. This contrasts with Angel's identification of some quite elderly individuals based on suture closure. This finding also differs from the age distributions for other Late Helladic III skeletal

68. See Rudney, Katz, and Brand 1983. Carious cavities were counted, as were brown stained spots of arrested caries. The latter were only counted when tooth enamel condition and their position, typically adjacent to an affected tooth surface, made an identification of carious decay more probable than postmortem staining from sediments. While these procedures are most practical for field data collection, we recognize their shortcomings and that the true prevalence of caries is probably higher than what is documented here.

69. Buikstra and Ubelaker 1994.

70. It is unlikely that significant numbers of infant and child burials were missed by the excavators, who carefully screened the sediments and recovered numerous small finds, including isolated human teeth.

71. *Palace of Nestor* III, pp. 79, 107.

TABLE 10.2. GENERAL AGE AND SEX DISTRIBUTION

Tomb	*Male*	*Female*	*Adult?*	*Child/SA*
Grave Circle	9	12	9	1
Tholos III	–	1	–	–
Tsakalis	13	12	15	8
Kondou K-1	2	4	1	2
Kokkevis K-2	10	4	1	4
N = 108	34	33	26	15

SA = subadult 13–18 years.

TABLE 10.3. AGE DISTRIBUTION

Tomb	*Child (0–12)*	*Subadult (13–18)*	*Young Adult (19–30)*	*Old Adult (31+)*
Grave Circle	–	1	11	4
Tholos III	–	–	1	–
Tsakalis	7	2	14	5
Kondou K-1	–	2	–	4
Kokkevis K-2	1	3	4	6
N = 65	8	8	30	19

TABLE 10.4. SEX AND AGE DISTRIBUTION

Site	*Young Female (19–30)*	*Old Female (31+)*	*Young Male (19–30)*	*Old Male (31+)*
Grave Circle	4	2	3	2
Tholos III	1	–	–	–
Tsakalis	5	2	7	3
Kondou K-1	–	3	–	1
Kokkevis K-2	2	1	2	5
N = 43	12	8	12	11

samples from the Argolid and the Agora,[72] where Angel also identified many older individuals.[73]

It is possible to provide both sex and general age estimates for approximately 40% of the sample. In Table 10.4 they are presented as younger or older adult females or males. Overall, the proportions of males and females in both adult age cohorts are about equal, although individual tomb groups show some variation, which is not unexpected. For example, there are more younger males in Tsakalis and a few more older ones in Kokkevis.

These basic demographic data suggest that the Pylian burials are not biased in favor of male interment, and that younger males, who might be at greatest risk for death in warfare, are not overrepresented. The adult representations are not different from what we would expect in family tombs. Yet it is important to again emphasize that young children from these social groups were rarely given the formal burial treatment afforded adults. A larger and even more intriguing question, that cannot be answered at this point in time, is why so few Pylian burials—adult or otherwise—have been located.

Dental Health

As might be expected, teeth are the most numerous elements in this sample and for that reason they are the focus of the analysis. There are 68 partial or complete dentitions that can be evaluated. To maximize analytic potential, both individual teeth (*N* = 625) and dentitions (*N* = 68) were evaluated. As we also assessed tooth presence or absence based on alveolar condition,

72. Halstead 1977; based primarily on Angel's data in *Palace of Nestor* III, pp. 79, 107.

73. We suggest that in light of our results, those age estimates should be used with caution.

TABLE 10.5. INDIVIDUAL TEETH

Site	*N*	*N Assessed*	*N AM loss*	*N Caries*	*N Hypoplasia*
Grave Circle/ Tholos III	135	167	8 (4.8%)	5 (3.7%)	6 (4.4%)
Tsakalis	327	341	14 (4.1%)	39 (11.9%)	43 (13.2%)
Kondou K-1	30	88	46 (52.3%)	12 (40%)	1 (3.3%)
Kokkevis K-2	133	172	27 (15.7%)	8 (4.7%)	8 (4.7%)
Total	625	768	95 (12.4%)	64 (8.3%)	58 (7.6%)

N = total number of teeth observed; *N* Assessed = total number of teeth and tooth sockets; AM = antemortem.

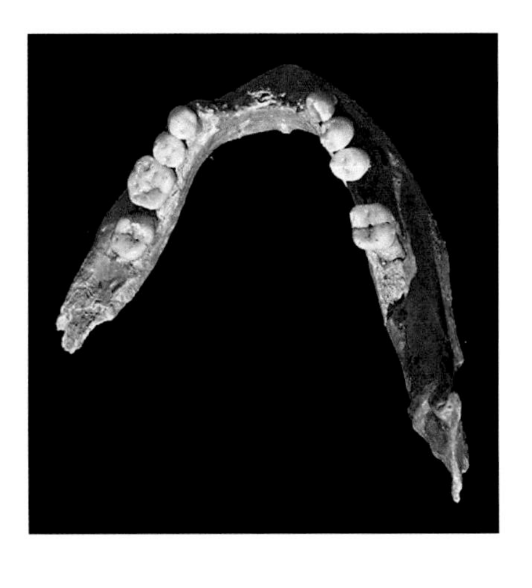

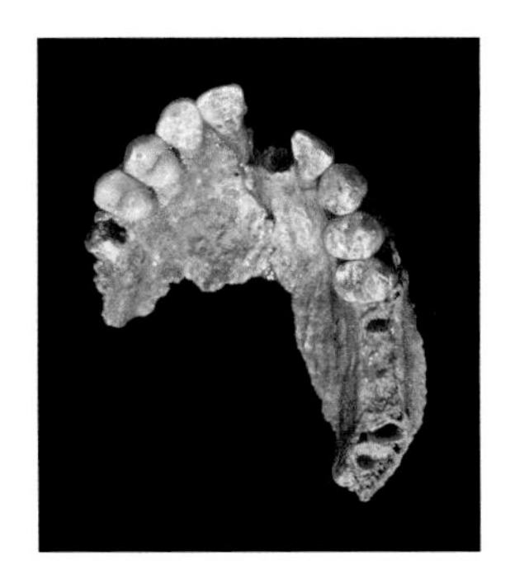

Figure 10.4 *(left)*. Female mandible with slight–moderate occlusal wear, occlusal caries on the LM3 and RM2, and extensive antemortem tooth loss of LI1-C, LM2 (with anteriorly displaced LM3 in response), and RI1-I2, RM1, RM3. Scale 1:2. Courtesy Department of Classics, University of Cinncinnati

Figure 10.5 *(right)*. Maxilla with very slight occlusal wear and extensive remodeling of the left molar sockets. Scale 1:2. Courtesy Department of Classics, University of Cinncinnati

a total of 768 teeth could be evaluated for presence/absence (Table 10.5). Carious lesions were always associated with antemortem tooth loss in individuals over 22 years. Tooth loss was not necessarily associated with heavy tooth wear, as these two examples of young people with major tooth loss illustrate (Figs. 10.4 and 10.5). This suggests that caries infection leading to rapid tooth loss was a serious factor for this population because 12.4% of teeth were lost antemortem. High rates of caries and antemortem tooth also have also been reported for other Mycenaean populations.[74] Linear enamel hypoplasia was only found in mild expressions, and it was fairly uncommon, occurring in only 7.6% of the combined sample (Table 10.5). The rarity of this attribute of "survivors"—those who live through childhood stresses—suggests that some low level of stress was experienced by a small proportion of the individuals studied. From the analysis of the individual teeth, the sample as a whole exhibits moderate levels of all three conditions.

In order to test the relationship between tomb type and these dental indicators of health status, we then compared the frequencies for individual cemeteries. Notably, teeth from the Grave Circle and Tholos III have low levels of antemortem loss, caries, and hypoplasia. Antemortem loss and caries are extremely frequent in the small Kondou subsample (52.3% and 40%, respectively). As caries frequently lead to tooth loss, it is informative to look at the two conditions together. Here, the contrasts between the

74. For example, most recently in the western Peloponnese at Ayia Triada, Tsilivakos et al. (2002) report 17.88% antemortem tooth loss and 7.7% caries.

TABLE 10.6. INDIVIDUAL TEETH: ANTEMORTEM (AM) LOSS AND CARIES COMBINED

Site	*N Assessed*	*N AM and Caries*
Grave Circle/Tholos III	167	13 (7.8%)
Tsakalis	341	53 (15.5%)
Kondou K-1	88	58 (65.9%)
Kokkevis K-2	172	35 (20.3%)
Total	768	159 (20.7%)

N Assessed = total number of teeth and tooth sockets observed; AM = antemortem.

TABLE 10.7. SITE TYPE AND ANTEMORTEM (AM) LOSS/CARIES RELATIONSHIP

	AM/Caries	*No*	*Total*
Grave Circle and Tholos III	13	154	167
Chamber	146	455	601
Total	159	609	

Chi-square = 21.694, $df = 1, p \leq 0.001$.

TABLE 10.8. SITE TYPE AND HYPOPLASIA RELATIONSHIP

	Hypoplasia	*No*	*Total*
Grave Circle and Tholos III	6	129	135
Chamber	52	438	490
Total	58	567	

Chi-square = 4.783, $df = 1, p \leq 0.05$.

Grave Circle and Tholos III and the chamber tombs are even more evident (Table 10.6). The combined level of these two conditions for the entire sample is 20.7%. There is a significant relationship (chi-square = 21.694, $df = 1, p \leq 0.001$) between tomb type (Grave Circle/tholos or chamber) and the frequency of these two conditions (Table 10.7). Similarly, there is a significant relationship (chi-square = 4.783, $df = 1, p \leq 0.05$) between tomb type and hypoplasia occurrence (Table 10.8). In both cases, Grave Circle/tholos teeth have lower levels of pathology.

Most teeth within a dentition are exposed to the same stresses, thus the frequency of dental pathology is highly correlated within a dentition. As a whole, a fairly high proportion of the dentitions from Pylos show some form of dental pathology, but again, the Grave Circle/tholos dentitions appear healthier (Table 10.9).

More importantly, the comparison of dentitions enables us to compare males and females (Table 10.10). The chamber tomb female dentitions have the highest frequencies of antemortem loss and caries (60% and 70%, respectively), while the highest levels of hypoplasia occur in the chamber

TABLE 10.9. DENTAL HEALTH BY DENTITION

Site Type	*N*	*N AM*	*N Caries*	*N Hypoplasia*
Grave Circle/ Tholos III	19	2 (10.5%)	3 (15.8%)	3 (15.8%)
Chamber	49	14 (26.9%)	18 (36.7%)	21 (40.4%)
Total	68	16 (22.5%)	21 (30.9%)	24 (33.8%)

AM = antemortem loss.

TABLE 10.10. DENTAL HEALTH BY DENTITION AND SEX

Site/Sex	*N*	*N AM*	*N Caries*	*N Hypoplasia*
Grave Circle/ Tholos III F	5	1 (20%)	1 (20%)	
Grave Circle/ Tholos III M	4	1 (25%)	1 (25%)	1 (25%)
Chamber F	10	6 (60%)	7 (70%)	4 (40%)
Chamber M	18	4 (22.2%)	9 (50%)	10 (55.6%)
Total	37	12 (32.4%)	18 (48.6%)	15 (40.5%)

AM = antemortem loss.

TABLE 10.11. DENTAL HEALTH BY DENTITION AND TOMB TYPE

Site Type	*N*	*N AM*	*N Caries*	*N Hypoplasia*
Grave Circle/ Tholos III F	5	1 (20%)	1 (20%)	
Grave Circle/ Tholos III M	4	1 (25%)	1 (25%)	1 (25%)
Tsakalis F	6	3 (50%)	5 (83.3%)	4 (66.7%)
Tsakalis M	10	2 (20%)	4 (40%)	8 (80%)
Total	25	7 (28%)	11 (44%)	13 (52%)

AM = antemortem loss.

tomb males (55.6%). If the differences between the Grave Circle/tholos and the Tsakalis cemetery subsample are examined, the distinction between males and females is most notable in the Tsakalis chamber tombs (Table 10.11). Our results indicate that there are significant differences between the dental health of males and females at Pylos and that these differences are correlated with tomb type.

DISCUSSION

The Pylos population is characterized by few indicators of poor skeletal health. For example, there is limited evidence for age-related osteoarthritis and joint surface destruction; this is not unexpected as few Pylians interred in these tombs seem to have lived past 40 years. There are a few healed injuries, but they are not seen in the skeletal elements as expected for Bronze Age combat, such as the forearms, ribs, or crania; nor do they appear with greater frequency in males. Response to some stresses, evident in expanded diploë, bone porosity, and cribra orbitalia, are noted in a small number of females and subadults from both the Grave Circle/tholos and chamber tombs, but these do not seem to have been the result of population-wide stresses. In contrast, the Pylians were characterized by relatively poor dental health and this is especially evident in the chamber tomb burials.

We predicted that if health status was related to resource access (as measured by tomb type and richness), then individuals from chamber tombs should have relatively poorer dental health. Our results support this prediction for all the chamber tombs, even though these tombs vary in the relative "richness" of their contents in terms of the number and quality of grave goods. The notable examples of this variation are the Tsakalis E-3, E-6, and E-9 tombs. E-3 is described above as the simplest burial in the cemetery with a very basic structure (a pit) and four ceramic vessels as grave goods. The two individuals from it are young adults aged 20–25 years with clear signs of dental pathology. Tomb E-9 was also identified as a relatively poor burial,[75] and all of the dentitions that can be evaluated show dental pathology. Thus, both of these tombs provide evidence supporting the suggested general association between tomb type, status, and dental health. In contrast, tomb E-6 might be expected to contain individuals with healthier dentitions because it contained a broader and more luxurious assortment of grave goods. The E-6 tomb is unusual in other ways. It has the greatest number of children preserved ($N = 5$) and the greatest number of interments.[76] The majority of these individuals, including the children, have poor dental health. This is an interesting result given the richness of this tomb, and its deviation from our prediction requires some discussion. The poor dental health of the individuals in this tomb could be the result of greater access to certain nutritional resources, such as more of the high-carbohydrate, cariogenic foods—but not greater access to the animal protein resources that may have characterized the diets of Grave Circle and tholos individuals (perhaps provided during feasting at the palace). At the same time, members of this chamber tomb family also had somewhat greater access to some of the types of grave goods, namely bronze, gold, and ivory, found in the Grave Circle and tholos tombs. Hence, this tomb might reflect interments of a larger family with more personal capital and collective resources, or an attempt to raise the status of the family that built it.[77]

The difference in dental health observed between the Grave Circle/tholos and the chamber tombs may also reflect temporal changes in the resource base of Pylos. Most of the chamber tomb burials are more recent than the Grave Circle, and date to later stages of the palace (Fig. 10.3). The declining dental health of chamber tomb individuals might be indicative of more cariogenic foods in the Pylian diet in general at the height of the palace as population pressure lessened access to varied food resources.[78] This resource change might have had greater impact on females. The conditions underlying the differences between males and females in the chamber tombs may demonstrate early life stress (seen as mild hypoplasia in more vulnerable young males) and then later, greater stresses for females as they encounter reproductive demands on their nutritional resources.

The effect of reproductive stresses on female dentitions is not well understood. Walker, and also Walker and Hewlett[79] documented higher

75. Cavanagh and Mee 1998. The "poverty" of this tomb is reflected in the type, rather than the number, of grave goods. For example, it contained no bronze or gold.

76. The minimum number of individuals suggested for this tomb, 19 burials, is based on the most conservative estimate derived from cranial/dental material, existing associations of elements in the curated collection, and the field notebooks. A higher estimate of 28 individuals is obtained when postcranial elements are added to the calculation.

77. Cavanagh and Mee 1998.

78. A stable isotope study to investigate this possibility was initiated in 2005.

79. Walker 1988; Walker and Hewlett 1990.

levels of dental pathology for females among present-day African foragers and farmers and prehistoric populations of California and discussed the possible causes of these differences. They suggested that factors such as differential use of cariogenic foods, eating patterns (females snacking throughout the day during food preparation), and oral hygiene contribute to poorer dental health. Walker specifically examined the clinical findings on pregnancy and dental condition, citing dental literature that does not provide much evidence for higher caries rates, tooth loss, or periodontal disease in pregnant women even though pregnancy gingivitis is known to occur.[80] However, these modern clinical studies comparing the oral health of pregnant to nonpregnant women are not directly applicable to prehistoric populations for several reasons. Women in prehistory, especially those in food-producing societies, probably began reproducing at younger ages, experienced shorter birth intervals, and thus had higher parity. The potential combined effects of those three reproductive factors on the depletion of maternal nutritional and mineral reserves may have played an important role in the poorer dental health of prehistoric women such as those in the chamber tombs of Pylos. We contend that the old adage "a tooth for every child"—while not describing the dental condition of most women today—potentially held great meaning for the women of Pylos who experienced significant levels of dental decay and antemortem loss while still in their reproductive years.

CONCLUSION

There are several major results of this analysis that have relevance for other studies of skeletal biology in prehistoric Greece. First, the size of the sample, as determined from this analysis, is probably at least 45% greater than what might be determined from published excavation reports and Angel's previous analysis. Yet, the younger sector of the population—infants, children, and subadults—is almost unrepresented. More importantly, the sex ratio is not biased toward males, as had been reported in earlier Mycenaean skeletal biology studies. Younger adults of both sexes are the most frequently represented age cohort, and few individuals lived past 35 years.

Individuals from the Grave Circle and Tholos III appear to have better dental and skeletal health, but this finding might be biased by differentially poorer preservation of the Grave Circle burials and small subsample sizes. In particular, the health of adult females, especially in the chamber tombs that span much of the palace's height, may have been substantially worse than that of males, with higher levels of carious infection, tooth loss, and porotic hyperostosis and related bone changes. This may relate to sex-based dietary differences or additional stresses on young female alveolar health, possibly associated with pregnancy.

This analysis presents the initial stage of research on the health and demography of the Mycenaeans of Pylos. Subsequent work, designed to address the questions of differential health, status, and burial treatment generated by this first study, is in progress. The poorer dental health of chamber tomb burials, and adult females in particular, may reflect dietary

80. Walker 1988. There is no evidence that caries is directly associated with pregnancy, although the other factors that Walker cites could certainly contribute to its higher prevalence in females.

differences. In order to assess this, 50 bone and dental root samples were collected for stable isotope analysis. The sampling strategy was to select representative numbers of males and females from each tomb. In addition, samples from children were included to gain information about dietary variation throughout the life course.[81]

Pylian skeletal biology will be placed into a broader perspective through comparative analyses of other Mycenaean populations. A study of the burials in the Athenian Agora, initiated in 2005, is now highlighting the tremendous variation in health and skeletal robusticity between the populations of Attica and the Peloponnese. The diversity of burial treatments at Pylos is also undergoing intensive reexamination. This includes detailed evaluation of the tomb structures using the Blegen archival materials and field notebooks, and a reexamination of the tomb ceramics; the first systematic analysis of the faunal remains from the burials is also planned.[82] Together, these bioarchaeological approaches to the study of the Pylians and their burials should provide answers to many of the questions this initial study raises.

POSTSCRIPT

This analysis involved only the portion of the Pylos sample that was available up to 2006; subsequently, the burials from Tholos III and Tholos IV were located in the National Archaeological Museum in Athens[83] and another Kokkevis chamber tomb was discovered and excavated by L. Malapani. Schepartz and Miller-Antonio began analysis of this additional material in 2007. Use of the entire Pylian sample may alter the conclusions presented here, but the results of the dietary isotope analysis[84] do support our findings that the life and health of Pylians differed significantly when status and sex are considered.

81. The sampling was conducted by A. Papathanasiou and L. A. Schepartz; the analyses were conducted by M. Richards of the Max Planck Institute.

82. J. Murphy is conducting the new study of the Pylos tombs and their archaeological contents; the faunal analysis will be done by P. Halstead and V. Isaakidou, Sheffield University.

83. We are very grateful to S. Triantaphyllou for her work in the museum collections and for bringing this to our attention.

84. Presented at the 2006 European Paleopathology Association meeting held in Santorini.

REFERENCES

Abbreviations for Archival Sources

EPB = Palace of Nestor Excavations Archives, Elizabeth P. Blegen Notebook, 1939, University of Cincinnati, Cincinnati.

JLA = Palace of Nestor Excavations Archives, J. Lawrence Angel Papers, 1957, University of Cincinnati, Cincinnati.

WDT = Palace of Nestor Excavations Archives, William D. Taylour Notebooks, 1953, 1957, 1958, University of Cincinnati, Cincinnati.

WPD = Palace of Nestor Excavations Archives, William P. Donovan Notebook, 1956, University of Cincinnati, Cincinnati.

Secondary Sources

Acheson, P. E. 1999. "The Role of Force in the Development of Early Mycenaean Polities" in *POLEMOS: Le contexte guerrier en Égée à l'âge du Bronze. Actes de la 7e Rencontre égéenne internationale, Université de Liège* (*Aegaeum* 19), ed. R. Laffineur and W. D. Niemeier, Liège, pp. 97–104.

Bennet, J. 1995. "Space through Time: Diachronic Perspectives in the Spatial Organization of the Pylian State," in *POLITEIA: Society and State in the Aegean Bronze Age* (*Aegaeum* 12), ed. R. Laffineur and W. D. Niemeier, Liège, pp. 587–602.

———. 1999. "Pylos: The Expansion of a Mycenaean Palatial Center," in *Rethinking Mycenaean Palaces: New Interpretations of an Old Idea,* ed. M. L. Galaty and W. A. Parkinson, Los Angeles, pp. 9–18.

Binford, L. R. 1971. "Mortuary Practices: Their Study and Potential," in *Approaches to the Social Dimensions of Mortuary Practices* (Memoirs of the Society for American Archaeology 23), ed. J. A. Brown, Washington, D.C., pp. 6–29.

Bisel, S. C., and J. L. Angel. 1985. "Health and Nutrition in Mycenaean Greece: A Study in Human Skeletal Remains," in *Contributions to Aegean Archaeology: Studies in Honor of William A. McDonald,* ed. N. C. Wilkie and W. D. E. Coulson, Minneapolis, pp. 197–209.

Boyd, M. 2002. *Middle Helladic and Early Mycenaean Mortuary Practices in the Southern and Western Peloponnese* (*BAR-IS* 1009), Oxford.

Brown, J. A., 1981. "Search for Rank in Prehistoric Burials," in *The Archaeology of Death,* ed. R. Chapman, I. A. Kinnes, and K. Randsborg, Cambridge, pp. 25–31.

Buikstra, J. E., and D. H. Ubelaker, eds. 1994. *Standards for Data Collection from Human Skeletal Remains,* Fayetteville, Ark.

Cavanagh, W., and C. Mee. 1990. "The Location of Mycenaean Chamber Tombs in the Argolid," in *Celebrations of Death and Divinity in the Bronze Age Argolid,* ed. R. Hagg and G. Nordquist, Stockholm, pp. 55–64.

———. 1998. *A Private Place: Death in Prehistoric Greece* (*SIMA* 125), Jonsered.

Cucina, A., and V. Tiesler. 2003. "Dental Caries and Antemortem Tooth Loss in the Northern Peten Area, Mexico: A Biocultural Perspective on Social Status Differences among the Classic Maya," *American Journal of Physical Anthropology* 122, pp. 1–10.

Davis, J. L., S. E. Alcock, J. Bennet, Y. G. Lolos, and C. W. Shelmerdine. 1997. "The Pylos Regional Archaeological Project, Part 1: Overview and the Archaeological Survey," *Hesperia* 66, pp. 391–494.

Darcque, P. 1987. "Les tholoi et l'organisation socio-politique du monde mycénien," in *Thanatos: Les coutumes funeraires en Égée a l'age du bronze* (*Aegaeum* 1), ed. R. Laffineur, Liège, pp. 190–200.

Dickinson, O. T. P. K. 1977. *The Origins of Mycenaean Civilization* (*SIMA* 49), Göteborg.

———. 1982. "Parallels and Contrasts in the Bronze Age of the Peloponnese," *OJA* 1, pp. 125–127.

———. 1983. "Cist Graves and Chamber Tombs," *BSA* 78, pp. 55–67.

Evans, A. 1929. *The Shaft Graves and Bee-Hive Tombs of Mycenae and Their Interrelation,* London.

Goodman, A. H. 1998. "The Biological Consequences of Inequality in Antiquity," in *Building a New Biocultural Synthesis,* ed. A. H. Goodman and T. S. Leathermann, Ann Arbor, pp.147–169.

Graziadio, G. 1988. "The Chronology of the Graves of Circle B at Mycenae: A New Hypothesis," *AJA* 92, pp. 343–372.

Halstead, P. 1977. "The Bronze Age Demography of Crete and Greece—A Note," *BSA* 71, pp. 107–111.

Halstead, P., and J. C. Barrett, eds. 2004. *Food, Cuisine and Society in Prehistoric Greece* (Sheffield Studies in Aegean Archaeology 5), Oxford.

Hillson, S. 1996. *Dental Anthropology,* Cambridge.

Hodder, I. R. 1982. "The Identification and Interpretation of Ranking in Prehistory; A Contextual Perspective," in *Ranking, Resource and Exchange: Aspects of the Archaeology of Early European Society,* ed. C. Renfrew and S. Shennan, Cambridge, pp. 150–154.

Killian, K. 1987. "L'architecture des résidences myceniennes; Origine et extension d'une structure du pouvoir politique l'âge du Bronze récent," in *Le système palatial en Orient, en Grèce et à Rome. Actes du Collogue de Strasborg, 19–22 juin, 1985,* ed. E. Levy, Strasborg, pp. 207–213.

Kuniholm, P. I., B. Kromer, S. W. Manning, M. Newton, C. E. Latini, and M. J. Bruce. 1996. "Anatolian Tree Rings and the Absolute Chronology of the Eastern Mediterranean, 2200–718 B.C.," *Nature* 381, pp. 780–783.

Larsen, C. S. 1997. *Bioarchaeology: Interpreting Behavior from the Human Skeleton,* Cambridge.

Lolos, Y. 1989. "The Tholos at Koryphasion: Evidence for the Transition from Middle to Late Helladic in Messenia," in *Transition: Le monde Égée du Bronze Moyen au Bronze Récent* (*Aegaeum* 3), ed. R. Laffineur, Liège, pp. 171–176.

Manning, S. W. 1995. *The Absolute Chronology of the Aegean Early Bronze Age: Archaeology, Radiocarbon and History,* Sheffield.

Mee, C., and W. Cavanagh. 1984. "Mycenaean Tombs as Evidence for Social and Political Organisation," *OJA* 3, pp. 45–64.

Mountjoy, P. 1995. *Mycenaean Pottery: An Introduction,* Oxford.

Palace of Nestor III = C. W. Blegen, M. Rawson, W. Taylour, and W. P. Donovan, *The Palace of Nestor at Pylos in Western Messenia* III: *Acropolis and Lower Town Tholoi, Grave Circle, and Chamber Tombs Discoveries outside the Citadel,* Princeton 1973.

Parker Pearson, M. 1999. *The Archaeology of Death and Burial,* College Station, Tex.

Powell, M. L. 1988. *Status and Health in Prehistory. A Case Study of the Moundville Chiefdom,* Washington, D.C.

Rudney, J., R. V. Katz, and J. W. Brand. 1983. "Interobserver Error Reliability of Methods for Palaeopathological Diagnosis of Dental Caries," *American Journal of Physical Anthropology* 62, pp. 243–248.

Saxe, A. 1970. "Social Dimension of Mortuary Practices" (diss. Univ. of Michigan).

Schliemann, H. 1878. *Mycenae; A Narrative of Researches and Discoveries at Mycenae and Tiryns,* London.

Shelmerdine, C. 1997. "Review of Aegean Prehistory VI: The Palatial Bronze Age of the Southern and Central Greek Mainland," *AJA* 101, pp. 537–585.

Tsilivakos, M. G., S. K. Manolis, O. Vikatou, and M. J. Papagrigorakis. 2002. "Periodontal Disease in the Mycenean (1450–1150 B.C.) Population of Aghia Triada, W. Peloponnese, Greece," *International Journal of Anthropology* 17, pp. 91–100.

Tsountas, C., and J. I. Manatt. 1987. *The Mycenaean Age: A Study of Monuments and Culture of Pre-Homeric Greece,* London.

Tzedakis, Y., and H. Martlew, eds. 1999. *Minoans and Mycenaeans: Flavours of Their Time,* Athens.

Vaughan, S., and W. Coulson, eds. 2000. *Palaeodiet in the Aegean,* Oxford.

Walker, P. L. 1988. "Sex Differences in the Diet and Dental Health of Prehistoric and Modern Hunter-Gatherers," in *Proceedings of the VI European Meeting of the Paleopathology Association, Madrid, September 9–11, 1986,* pp. 261–267.

Walker, P. L., and B. S. Hewlett. 1990. "Dental Health Diet and Social Status among Central African Foragers and Farmers," *American Anthropologist* 92, pp. 383–398.

Warren, P., and V. Hankey. 1989. *Aegean Bronze Age Chronology,* Bristol.

Wright, J. C. 1987. "Death and Power at Mycenae: Changing Symbols in Mortuary Practice," in *Thanatos; Les coutumes funéraires en Égée à l'âge du bronze,* (*Aegaeum* 1), ed. R. Laffineur, Liège, pp. 171–184.

———, ed. 2004. *The Mycenaean Feast,* Princeton.

CHAPTER 11

Regional Differences in the Health Status of the Mycenaean Women of East Lokris

by Carina Iezzi

Archaeological finds in Greece provide a considerable amount of information about Mycenaean civilization, suggesting that the inhabitants of Late Bronze Age Greece had a strict sociopolitical hierarchy and a rigid class structure. They used ostentatious displays of wealth and power, had a martial spirit, and, based on diplomatic records from Egypt and Hittite archives from Anatolia, they were considered an equal power and a force with which to be reckoned.[1]

Yet we know very little about the actual people who helped to build this powerful and intriguing civilization. What we do know about the biology and health of the Mycenaeans comes primarily from the major centers such as Mycenae,[2] and therefore the uppermost levels of society. Much less is known about the health of the majority of the population: the non-elite.[3] This study of health, disease, and physical stress in Late Bronze Age (LBA) East Lokris (Fig. 11.1) provides information about non-elite Mycenaeans in coastal and inland environments.[4]

The first goal of this research is to assess how these health and biomechanical adaptations may have varied in the two different geographic regions, given that each area would have provided unique challenges to populations who lived in them. A second focus is on the health of women. Anthropological studies of warfare indicate that preliterate patrilineal-patrilocal societies tend to be warlike.[5] That Mycenaean society was warlike is documented in the archaeological record. In militaristic societies, males are typically more highly regarded than females. With higher social status comes preferential treatment that is often associated with better health and a longer life span. This type of social structure, in turn, would be predicted to have a negative impact on females. For this reason, the health of the women of Mycenaean-era East Lokris is examined here.

Five paleopathology variables were used to evaluate health status for each region. The results were then subjected to statistical analysis to see if there were any significant differences in their prevalence, and therefore health, between men and women in both regions. The East Lokris data are then interpreted in light of comparative studies and the environmental and subsistence factors specific to each region.

1. Runnels and Murray 2001.
2. Angel 1973, 1945.
3. Although Musgrave and Popham (1991) studied a small LH population from Euboia, and Triantaphyllou (1999) assessed one from Greek Macedonia. These are used as comparative data in the present study.
4. Permission to carry out this project was granted by both Phanouria Dakoronia (Director Emerita of the 14th Ephorate of Prehistoric and Classical Archaeology) and the American School of Classical Studies at Athens (ASCSA). Funding was provided by the J. L. Angel Fellowship and a Research Associateship through the Wiener Laboratory at the ASCSA, the Doreen Spitzer fellowship (ASCSA), and through the State University of New York at Buffalo's Mark Diamond Research Fund and a Dissertation Grant from the College of Arts and Sciences (CAS). Sincere personal thanks to Lynne A. Schepartz; Sherry C. Fox, Director of the Wiener Laboratory; and Sonia Dimakis, Archaeologist, for the 14th Ephorate. Thanks also for editorial comments by three anonymous reviewers.
5. Otterbein 1994.

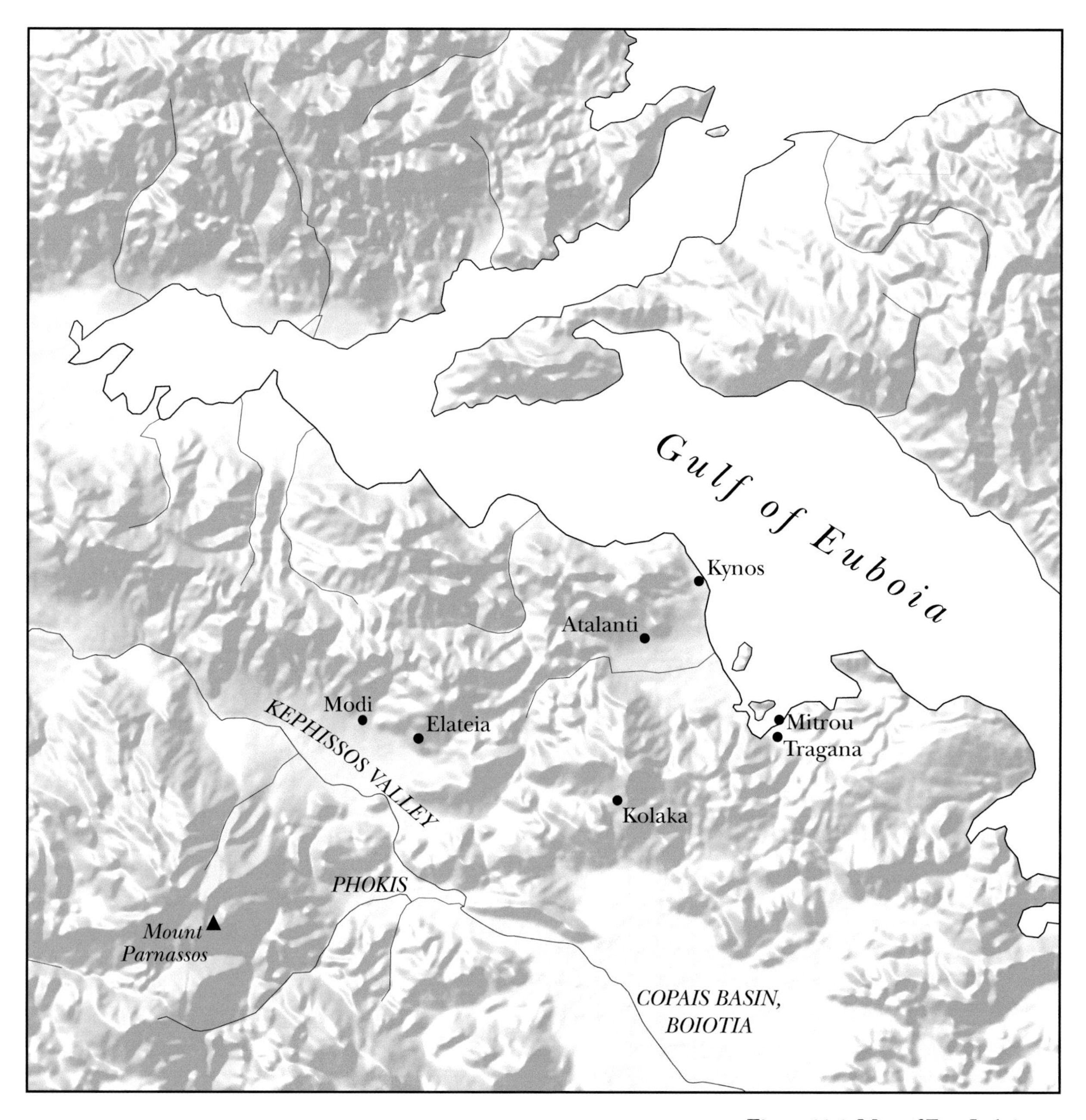

Figure 11.1. Map of East Lokris with coastal (Tragana and Atalanti) and inland (Kolaka and Modi) sites labeled. Mapping Specialists, Ltd.

THE BURIALS

Very little is currently known about the LBA populations in East Lokris. The detailed examination of skeletal material carried out in this project represents the first analysis of a large-scale sample of human remains from this time period and geographic area. All of the burials described here are tentatively dated to the Late Helladic IIIB and IIIC (1325–1100 B.C.). They were recovered from rock-cut chamber tombs by the Greek Archaeological Service during rescue excavations in the 1980s and 1990s subsequent to discovery and damage by modern tomb looters.[6] The vast majority of the skeletons were from secondary commingled burials which

6. A small number of these also contained burials in pithoi and larnakes.

were fragmentary and incomplete due not only to the action of looters, but also to LBA burial customs which included sweeping aside earlier burials to inter later ones. Although the burials had been compromised, the overall quantity of remains provided usable data so that trends in population health could be observed.

Studies of burial customs have shown that tomb type can be a reliable indicator of social status.[7] While tholos tombs seem to have been used by elite groups, or perhaps royalty as evidenced at Mycenae, chamber tombs were likely used by a sub-elite[8]—perhaps what we today might equate with a middle class.[9] The contents of the chamber tombs found in many parts of the Mycenaean world indicate the prosperity of the ordinary population.[10] All of the East Lokrian burials discussed here were derived from chamber tombs and can therefore be viewed as coming from the same general socioeconomic level. In this way, tomb type serves as a general control for potential socioeconomic differences.

MATERIALS AND METHODS

Sites Sampled and the Geography of East Lokris

The topography of central Greece is composed of a sharp juxtaposition of environmental zones created by two parallel mountain chains that divide the coastal areas from the inland. Inland, the presence of these mountain chains creates the lush Kephissos River valley,[11] while the coastal regions provide access to pockets of arable land, as well as sea trade and travel.

The Coastal group used in this study consists of combined data from the chamber tomb cemeteries of Tragana and Atalanti, while the Inland group is the sites of Kolaka and Modi. By combining sites, sample sizes were increased and potential regional health differences could be detected. Within the Coastal group, Tragana is located near the shores of the Gulf of Euboia, while Atalanti is situated somewhat farther from the Gulf, in the fertile Plain of Atalanti. In the Inland group, Kolaka is located on the eastern slopes of Mount Chlomo in an upland plateau at an elevation of approximately 480 m. Modi, the site farthest inland, is located in the foothills of the Kephissos River valley. Because settlements have not been securely associated with most of the cemeteries studied, we must go on the assumption that the tomb inhabitants would, in life, have lived within a few kilometers of each cemetery.[12]

Minimum Number of Individuals (MNI)

Due to the commingled, fragmentary, and incomplete nature of the skeletal materials studied, it was not possible to identify whole individuals in most cases. Therefore the MNI present was determined by identifying all bones and fragments in a given tomb and sorting them by element, side, section, age, and sex. Each of these sets was then assessed for repetitious or analogous areas. The total MNI for East Lokris is 186, with 61% of the sample from the coastal region (Table 11.1).

7. Graziadio 1991.
8. L. V. Watrous (pers. comm.).
9. Dickinson 1994, p. 223.
10. Voutsaki 1995.
11. McInerney 1999, p. 53, Paus. 10.33.4.
12. Kramer-Hajos 2005.

TABLE 11.1. EAST LOKRIS PALEODEMOGRAPHICS

Region	*Total MNI*	*Women*	*Men*	*Unknown*	*Subadults*
East Lokris	186	61	62	20	43
Coastal	113	38	35	15	25
Inland	73	23	27	5	18

Note: The average age of death in the Coastal region was 43 years old (n = 18) for women, and 32 years old (n = 11) for men. In the Inland region, the average age of death was 31 years old (n = 12) for women, and 41 years old (n = 10) for men.

Diagnostic Areas Present (DAP) and Affected (DAA)

Most of the variables (with the exception of osteoporosis) assessed in this study occur only on specific areas of certain bones. Due to the fragmentary and incomplete nature of the skeletal remains, the minimum number of the specific diagnostic areas present (DAP) per variable was counted, rather than the MNI per region. Once the DAP was determined, the minimum number of diagnostic areas affected (DAA) was counted and the percentage affected calculated. Therefore, unless otherwise mentioned, the rates expressed in this study reflect the MNI with a given diagnostic component available for study.

Age Assessment

Age estimates for adults were made using as many methods as possible; these included assessing morphological changes in the sternal end of ribs,[13] pubic symphyses,[14] and the auricular surface of the pelvis.[15] Epiphyseal fusion rates of the following were also used: long bones;[16] iliac crests;[17] ischial and medial clavicular epiphyses;[18] vertebral bodies and spines, sacra, scapulae, sterni, and rib heads.[19] The degree of cranial suture[20] and spheno-occipital synchondrosis[21] fusion were also considered, as were endo- and ectocranial suture closure[22] and hand and wrist ossification.[23]

Average age at death for adults (18 years and older) in each regional group was calculated using only individuals for whom a numeric age range could be assessed using the methods described above. The average of each of these ranges was then determined and the result was then averaged per Coastal and Inland regions (Figs. 11.2, 11.3).

Sex Assessment

Cranial, pelvic, and postcranial morphology[24] were used to estimate sex. Long bone measures were also used for bones that were morphologically impossible to sex. Measurements of the femora[25] and tibiae[26] were compiled, respective histograms generated, distributions assessed, and dividing points for each sex determined.[27] Comparative data from a small collection of modern Greeks[28] as well as ancient Greeks[29] were used to evaluate the accuracy of the dividing points. The results of the morphological and metric data were largely in agreement.

13. İşcan et al. 1984a, 1984b; İşcan and Kennedy 1989.
14. Todd 1920; Suchey, Brooks, and Katz 1988; McKern and Stewart 1957; Gilbert and McKern 1973.
15. Lovejoy et al. 1985.
16. Suchey et al. 1984; Webb and Suchey 1985; McKern and Stewart 1957.
17. McKern and Stewart 1957; Webb and Suchey 1985.
18. Webb and Suchey 1985.
19. McKern and Stewart 1957.
20. Meindl and Lovejoy 1989.
21. McKern and Stewart 1957.
22. Meindl and Lovejoy 1985; Baker 1984.
23. Greulich and Pyle 1959.
24. Bass 1987; Buikstra and Ubelaker 1994; İşcan and Kennedy 1989; İşcan and Derrick 1984; Krogman and İşcan 1986; Phenice 1969.
25. Pearson and Bell 1919; Thieme 1957; Black 1978; Stewart 1979; DiBennardo and Taylor 1982; İşcan and Miller-Shaivitz 1984a.
26. İşcan and Miller-Shaivitz 1984a, 1984b, 1984c.
27. Lower end of histogram were female, upper end male.
28. Lagia et al. 2000.
29. *Lerna* II.

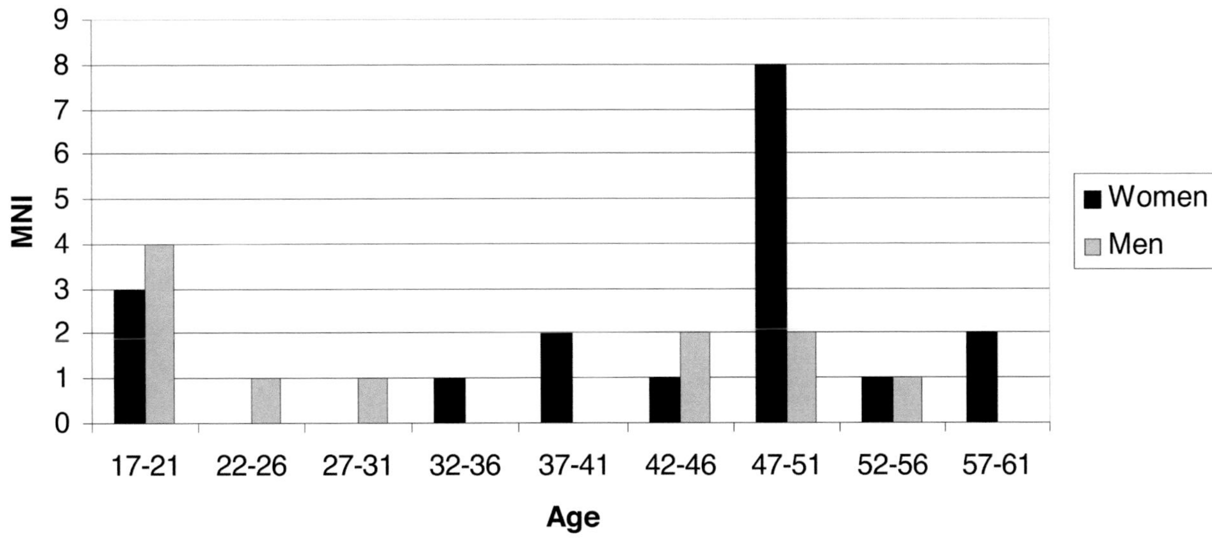

Figure 11.2. Distribution of Coastal adult ages: women versus men

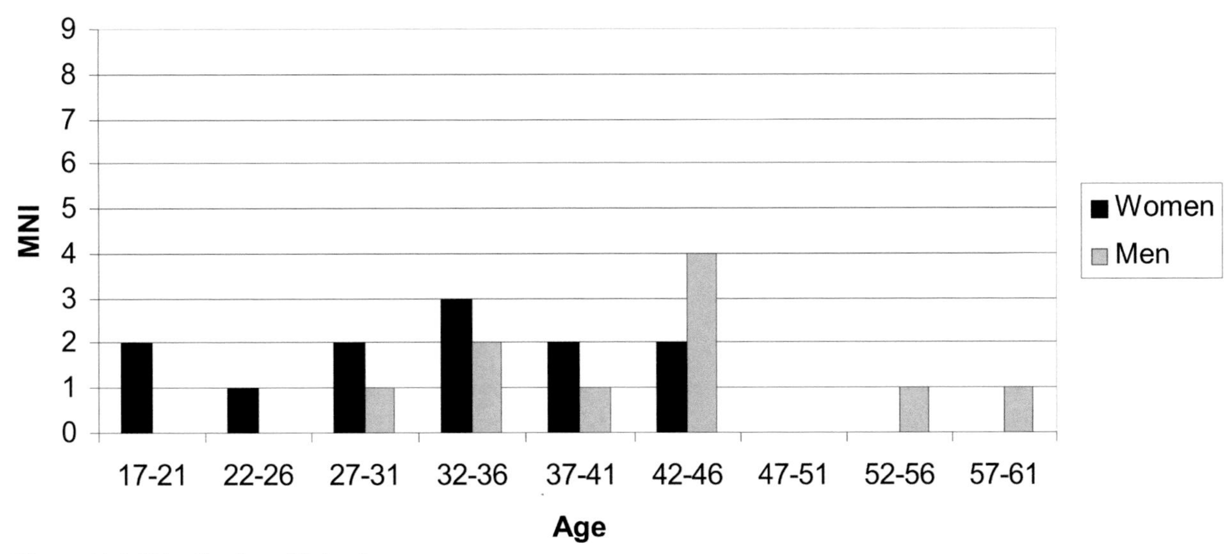

Figure 11.3. Distribution of Inland adult ages: women versus men

Variables

For purposes of this study, the term *health* refers to a combination of five variables that provide information on aspects of biological well-being including disease, nutrition, and biomechanical stressors as they are preserved in the skeletal record. The variables include cribra orbitalia, osteoporosis, osteoarthritis, femoral head/neck conditions, and platymeria. These variables were selected because they met the following criteria: They were detectable in commingled, fragmentary, and incomplete archaeological skeletal material; they reveal different key aspects of health; they were present in large enough numbers to be assessed statistically.

Cribra orbitalia develops during childhood and is expressed as porosity and/or expansion of the bone plates composing the roof of the eye orbit.

In older children and adults, this porosity becomes reduced to small holes or pits if healing and bone remodeling occurs. Generally thought to be the result of an anemia, cribra orbitalia may develop from such conditions as iron deficiency via malnutrition or scurvy, or secondary to another condition such as parasitic infestation, chronic gastrointestinal bleeding, hookworm, or epidemic disease. Alternatively, cribra orbitalia can develop as a result of one of the hereditary anemias such as thalassemia or sickle cell.[30] The DAP for cribra orbitalia were assessed using the minimum number of adult orbit roofs present. The DAA consisted of the minimum number of these orbits affected.

Osteoporosis is the loss of bone mineral density. Morphologically, it manifests as lightweight bones with cortical thinning and sometimes porosity, as observed in radiographic films. Among modern populations, two types of osteoporosis are delineated. Type I is related to the loss of estrogen during menopause,[31] while type II affects both sexes throughout life and is associated with a reduced absorption of calcium, or other factors,[32] such as dietary inadequacy, or secondary to some other condition. Prolonged inactivity or lack of weight-bearing exercise can also be a cause. In women, the likelihood of getting osteoporosis is compounded because they can be affected with both type I and type II. To assess the presence of osteoporosis at East Lokris, all long bones were radiographed. After ruling out taphonomic processes as the source of bone condition, any adult skeletal element that was unusually lightweight and exhibited extremely thin cortical bone accompanied by porosity was considered osteoporotic. Because osteoporosis can affect more than one area of the skeleton, the MNI per geographic region (Coastal or Inland) was used rather than a DAP for this variable. The DAA consisted of the MNI affected. Unsexable adults were included in the Coastal versus Inland comparison.

Osteoarthritis (OA) affects only synovial joints and consists of the loss of one or more areas of articular cartilage followed by osseous changes in subchondral and marginal areas. Frequency of the disease increases considerably with advancing age, and it can also develop from repetitive physical stress or trauma to a joint. Any joint component exhibiting eburnation with or without lipping, pitting, or changes in articular shape was considered osteoarthritic. The DAP were counted based on the minimum number of components of a given joint (e.g., left elbow, right hip) present.

Femoral head and neck conditions such as the cervical fossa of Allen and Poirier's facet are generally thought to be the result of various biomechanical forces affecting the hip joint. The fossa of Allen (or "mountaineer's gait")[33] expresses itself on the anterior portion of the femur at the junction of the femoral head and neck. The condition is said to develop from chronic friction of the tendon of the rectus femoris on the femoral neck. It has also been suggested that the fossa may represent an evolutionary manifestation of bipedal locomotion.[34] Poirier's facet is also present on the anterior femur, just proximal to the area described above. The facet is produced from habitual flexion of the knee and concurrent extension of the hip joint.[35] Generally, researchers agree that these femoral head and neck conditions are produced from biomechanical stressors, possibly related to daily activities such as traversing rough terrain[36] or habitually assuming a squatting posture.[37] The number of femora with at least the proximal

30. Larsen 1997.

31. Ruff 1992, p. 45.

32. Riggs and Melton 1986.

33. Angel 1960, 1964.

34. Kate 1963.

35. Angel 1960, 1964; Kostick 1963; Poirier and Charpy 1911; İşcan and Kennedy 1989, p. 147.

36. Angel 1959, 1960, 1964; Kostick 1963; Meyer 1924; Odgers 1931.

37. Poirier and Charpy 1911; Odgers 1931; Sauser 1936; Angel 1960, 1964; Kostick 1963.

one-fourth present, including the head, were sided and then counted in the DAP. The minimum number of these elements displaying a head/neck condition (no single bone contained more than one type of condition) was counted in the DAA. For the sake of statistical analysis, both conditions were lumped together to increase sample size because both indicate physical stress in the proximal femora.

Platymeria, the anterior-posterior flattening of the upper portion of the femur, is primarily generated by biomechanical stress, though insufficient nutrition,[38] especially in terms of protein and calories,[39] may play a role as well. Platymeria can be described by the meric index (MI) which provides information about the degree of femoral flattening. Anterior-posterior and medial-lateral measurements are taken at the subtrochanteric level of the proximal femoral diaphysis. The MI is then calculated using the following formula:

$$MI = \frac{\textit{subtrochanteric anterior-posterior diameter}}{\textit{subtrochanteric medio-lateral diameter}} \times 100$$

MI values within the 0–84.9 range are said to be platymeric, or flattened. Lower numbers within this range indicate more flattened diaphyses, while higher numbers indicate more rounded ones. By analyzing cross sections of long bones at different points along the diaphysis, the differential effects of mechanical loadings on bone shape can be assessed.[40] It is the shape of the cross section that reflects the type of mechanical loadings placed upon a bone during life. We can therefore use the MI to assess the types of mechanical stressors experienced and, perhaps, the associated activities of affected individuals. The DAP for platymeria were counted using the minimum number of femora with at least the proximal third present. The DAA consisted of the MNI with platymeria.

To test whether health levels in East Lokris were statistically different for regions and sex, a binary logistic regression procedure using SPSS software was performed on the five variables. In this analytical technique a dichotomous dependent variable is regressed on a set of independent variables.[41] This method was used because it was found to be most useful for the type of remains studied (i.e., commingled, fragmentary, and incomplete skeletons) and, when variables are weighted, takes into account small sample sizes. A 95% confidence interval was calculated. A resulting significance (sig.) value of 0.05 or less was deemed statistically significant.

RESULTS

The East Lokris sample has roughly equal numbers of males and females identified for each region (Table 11.1). The average age at death, calculated on the basis of 51 individuals, varies between the regions, with inverse differentials between males and females: 43 years for Coastal females versus 31 years for Inland females and 32 years for Coastal males versus 42 years for Inland males.

Table 11.2 presents the prevalence of the five health indicators. The comparison of the Coastal and Inland adult samples reveals some interesting differences and similarities. Both groups have similar levels of OA,

38. Gillman 1874; Hoyme and Bass 1962.

39. Bisel and Angel 1985; Larsen 1981; Parsons 1914; Holtby 1917–1918.

40. Ruff 1992.

41. Aldrich and Nelson 1984; Fox 1991; Hosmer and Lemeshow 1989; McCullagh and Nelder 1989; Agresti and Finlay 1986.

TABLE 11.2. RESULTS OF STATISTICAL ANALYSIS OF VARIABLES

	Coastal			*Inland*		
	All Adults	*Women*	*Men*	*All Adults*	*Women*	*Men*
Cribra orbitalia	22% (*n* = 9) *8%	32% (*n* = 6)	18% (*n* = 3)	67% (*n* = 14) Sig. = 0.00 *19%	50% (*n* = 4)	29% (*n* = 2)
Osteoarthritis	31% (*n* = 13)	33% (*n* = 5)	50% (*n* = 6)	35% (*n* = 18)	31% (*n* = 4)	50% (*n* = 9)
Osteoporosis	9% (*n* = 10)	11% (*n* = 4)	0%	14% (*n* = 10)	9% (*n* = 2)	0% 0%
Femoral head/neck	16% (*n* = 8)	9% (*n* = 2)	16% (*n* = 3)	18% (*n* = 6)	0% 0%	31% (*n* = 4)
Platymeria and meric index (MI)	87% (*n* = 48) MI = 81.12	88% (*n* = 21) MI = 80.80	81% (*n* = 17) MI = 81.94	96% (*n* = 26) Sig. = 0.00 MI = 76.06	100% (*n* = 11) MI = 74.55	100% (*n* = 10) MI = 76.44

* Using regional group total adult minimum number of individuals (MNI) to calculate percentage of individuals affected; *n* is total sample size.

osteoporosis, and femoral head/neck features. The only major differences are in the prevalence of cribra orbitalia (67% Inland and 22% Coastal) and platymeria (87% Coastal and 96% Inland), which are both statistically significant at the 0.00 level. Both the Coastal and Inland groups had similar rates of OA, but patterning of the condition throughout the body varied. The Coastal group experienced a higher rate of upper body OA while the Inlanders had roughly equal rates affecting the upper and lower body. Further, vertebral OA was much more prevalent Inland, as were Schmorl's nodes, which develop through a combination of advancing age and physical stress.

The comparisons of women and men show clear patterns that are also distinctive to regions. Cribra orbitalia is highest in women and achieves its greatest frequency in the Inland sample. Osteoporosis is also more common in women, as predicted from its etiology, and no males were found to have osteoporosis. At first glance, Coastal women appear to have had a somewhat higher prevalence of osteoporosis than Inland women. This may have been partially due to the advanced age (postmenopausal) of an unusually large number of women in the Coastal group (Figs. 11.2, 11.3). However, there was more than twice the number of unsexable individuals afflicted with osteoporosis in the Inland group. Because no identifiable males exhibited the condition in either region, it is possible that the unsexable individuals are primarily, if not all, women. This situation is particularly relevant for the Inland women, who also had the higher rate of cribra orbitalia—a condition that can be associated with osteoporosis. If the unsexable individuals affected by osteoporosis are counted as women, a much higher rate is seen in the Inland group.

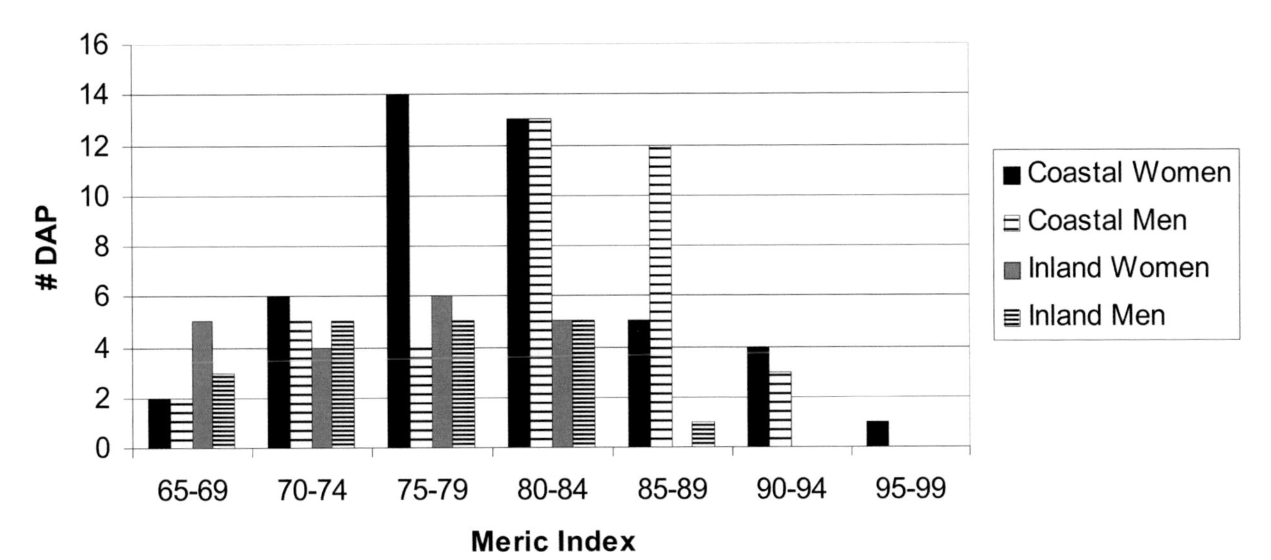

Figure 11.4. Coastal and Inland, women and men: meric index distribution

When the sexes were compared per regional group, it was found that rates of OA were much lower for women than for men. Within these lower rates, however, distribution patterns varied even more. Coastal women were less affected by upper body stressors, though they were about equally affected as their men by lower body stressors. Conversely, Inland women were more affected by upper body stressors and much less affected by lower body stressors.

Overall, rates of the combined femoral head/neck conditions were similar in both regional groups. Between the sexes, women had a significantly lower rate than men. In the Coastal region, women had almost half the rate of their men, and they exhibited only Allen's fossa. Inland, no women exhibited either of the conditions, though their men had the highest rate of all. However, sample sizes in the present study were small, so results may vary in future studies.

Platymeria was the single most prevalent condition affecting East Lokrians. The Inland group had a significantly higher frequency of platymeria and a lower average MI (flatter bones) than did the Coastal group (Fig. 11.4). These data provide further information about how the two regions may have differed. Women of the Coastal region had a moderately higher rate of platymeria compared to their men, but both sexes shared very similar average MIs. Inland, all measurable female and male femora exhibited platymeria, and also shared similar average MIs. However, these were much lower than those of the Coastal group.

Inland, the significantly higher prevalence of platymeria and, more importantly, much lower MIs indicate that the people there were subject to more severe lower body biomechanical stressors. Conversely, in the Coastal group, while platymeria rates were high, MIs were also high, thus indicating rounder femoral diaphyses.

In East Lokris as a whole, women and men had similar average MIs. This indicates that as a population, both sexes were subject to about the same intensity of lower body stressors. Both sexes in the Coastal group had almost identical MIs. This indicates that they were subject to similar levels of lower body physical stress, though the women did have a somewhat higher rate of platymeria overall.

DISCUSSION

Diet and Disease Impacts on Health

According to osteological studies throughout prehistory, the acquisition of sufficient nutrients and calories was undoubtedly an issue; according to Bisel and Angel,[42] the Mycenaean diet would have contained mostly carbohydrates (grains and cereals) and been deficient in calories, protein, and iron.[43] This has been confirmed by subsequent isotope analyses, which have also documented a reliance on terrestrial resources.[44] Based on dental, isotopic,[45] and archaeological[46] data from East Lokris, the populations conform to this general pattern although there were subsistence differences between the regions. The Coastal group was more involved in agriculture, while the Inland group was more dependent upon a mixed economy.

Diets heavily dependent upon grain, such as those of agriculturalists, contain large amounts of phytate which inhibits iron absorption and can lead to anemia. The presence of cribra orbitalia in the Coastal group could be due to this cause, but the much higher levels inland may require a different explanation. Thalassemia may have been a factor underlying cribra orbitalia at Modi due to the marshy environment along the shores of the Kephissos River, but it was not likely the cause at Kolaka, where the prevalence was the highest. Here, topographic conditions included a high elevation and dry rocky environment that would not be welcoming to *Plasmodium falciparum*-carrying mosquitoes. Overall, it seems more likely that the high rate of cribra orbitalia in the Inland group, as a whole, was due to some other parasitic infection such as hookworm, which can result from poor hygiene, overcrowding, and living in close proximity to animals. While some cases of thalassemia may have occurred in East Lokris, it was probably not a primary cause of poor health there. Across the mainland, anemia declined throughout the Bronze Age due to Mycenaean dam-building, swamp-draining projects,[47] and improved farming methods.[48]

The differences in the health of men and women in East Lokris also need explanation. In terms of nutrition in a patriarchal society such as that of the Mycenaeans, males would typically receive the lion's share of food, or at least more high-quality food, including more protein; if not always, then at least in times of drought or food shortage. This preference, when given to male children over female children, would hinder female physiological development including the attainment of peak bone mass. If this is not attained by early adulthood, it never will be.[49] Therefore, females may begin life with a nutritional deficit that would affect their health for the rest of their lives and lead to a stronger tendency toward developing osteoporosis at earlier ages.[50]

The availability of calcium in foods is important, but equally important is how well it is absorbed by the body. Phytic acid, a stored form of phosphorus in many plant tissues, not only binds iron but also calcium,

42. Bisel and Angel 1985.
43. A. Papathanasiou (pers. comm.); Triantaphyllou 1999.
44. Triantaphyllou 1999; Papathanasiou 1999.
45. Iezzi 2005.
46. Dakoronia 1993.
47. Dickinson 1994, p. 162.
48. *Lerna* III.
49. Grauer and Stuart-Macadam 1998, p. 32.
50. Grauer and Stuart-Macadam 1998, p. 32.

thereby making it unavailable to the body. Calcium absorption is further affected by the proportion of fat and protein in the diet, vitamin D levels, presence of disease, and the health of the individual.[51] In this way, a heavy dependence on grain consumption could lead not only to cribra orbitalia but to osteoporosis as well. Thus, dietary differences could have been one of the causes of osteoporosis, especially among Coastal women.

Larsen[52] points out that in some early agricultural societies, women may have had less adequate diets than men. This could then place them at a higher risk of disease because their immune systems would not be fully equipped. Pregnancy and lactation can lead to osteoporosis when the diet is deficient and/or when lactation is continued for several years. Osteoporosis in the Coastal women could have also been partially due to prolonged breastfeeding. At the other end of the spectrum, insufficient postmenopausal estrogen may have been a contributing cause in the females of that group, given the unusually high number of mature women.

It is possible that Inland women, if insufficiently nourished and subject to higher pathogen loads, may not have been able to breastfeed their children for an optimal length of time. These children, specifically females, would not have the full immunological benefits of breast milk and would therefore not be as protected from parasitic infection. Thus, the cycle would continue. Further support for this is shown in the higher rate of subadults with an earlier average age at death inland.[53]

In addition to these dietary and pathogenic vectors that are generally considered, East Lokris women may have been exposed to higher pathogen loads from having more of a role in preparing the dead for burial. Vermeule[54] describes how, in ancient times, women were responsible for the preparation of the corpse. At nearby Tanagra, painted larnakes depict what appears to be ritual lamenting by women mourners.[55] Together, these data indicate that women played an important role in the funerary process, perhaps as early as Mycenaean times.

Comparison with Other Populations

Bisel and Angel[56] found that the average percentage of LBA Mycenaeans with evidence of cribra orbitalia/porotic hyperostosis was 9% (n = 215). Using adjusted East Lokris data, the Coastal group was on par with Bisel and Angel's findings, while the Inland group had more than twice the norm. A LBA inland population from Greek Macedonia studied by Triantaphyllou[57] was found to have 20.0% (n = 3) of adults with cribra orbitalia; this is similar to the level for coastal East Lokris, but again the high prevalence inland stands out. The condition was attributed to anemia and all of the individuals affected were women. It has been observed in a number of bioarchaeological studies that in certain settings, women exhibit higher rates of acquired anemia than do men.[58] McGeorge suggested that this condition is related to the physiological stress and nutrient losses involved with menstruation, pregnancy, and lactation.[59] Alternatively, if the cribra orbitalia was caused by thalassemia or sickle cell anemia during childhood, its presence could indicate that females were subject to metabolic stress throughout their childhood years.[60]

51. Grauer and Stuart-Macadam 1998, p. 32.
52. Larsen 1987.
53. Iezzi 2005.
54. Vermeule 1979, pp. 13–14.
55. Dickinson 1994, p. 229.
56. Bisel and Angel 1985.
57. Triantaphyllou 2001.
58. Larsen 1997, p. 39; Stuart-Macadam 1998; Triantaphyllou 1999, p. 189.
59. McGeorge 1988.
60. Triantaphyllou 1999, p. 189.

McGeorge[61] also notes that rates of osteoporosis in Late Minoan times seem to have been quite high compared to modern times. She attributes this high rate, as well as premature death, to the nutritional drains of pregnancy and lactation coupled with insufficient dietary intake, unsanitary living conditions, poor hygiene, and infections. Harrison et al.[62] point out that nutrition and environment affect age at menarche, with suboptimal nutrition and stressful environmental factors causing delays in its onset and affecting nutrient stores and growth in the body. Wells[63] concluded that in prehistory, chronic malnutrition in women was the likely cause of their truncated life span compared to men; this view was later accepted by Hamilton[64] and Zivanovic.[65] It seems quite feasible that lifelong, chronic, suboptimal nutrition affected the East Lokrian women, most particularly those of the Inland group.

Physical Activity

Larsen[66] found that among agriculturalists there is a reduction in arthritis associated with increased sedentism. Bisel and Angel[67] compiled data on LBA Mycenaean skeletal remains and determined that 63% (n = 79) were affected by vertebral OA. While there was considerable variation in vertebral OA rates within East Lokris, neither region had a rate approaching this Mycenaean comparative sample. Based solely on this general aspect of OA it seems that the East Lokrians had a relatively less physically demanding life compared to the Mycenaean norm. However, the vertebral data indicate that the Inland group was more involved in the flexion and lateral bending involved with strenuous activities, perhaps such as heavy lifting,[68] hauling,[69] shoveling,[70] and generalized physical stress.[71]

Overall, Coastal and Inland women had similar rates of OA, but this was distributed differently throughout the body. Based on the lower rate of upper body OA among Coastal women, it seems that they were less involved in agricultural duties than their men were. Among the men, OA of the shoulder and arm would most likely have been related to activities such as preparing fields, planting, and harvesting. It is likely that the women were responsible for the less physically demanding task of tending to herds, grinding grain, and some craft-making activities. Archaeological data in support of an agricultural lifestyle in the Coastal group comes from the nearby Late Bronze Age port of Kynos. Here, one or more complexes of storerooms and workshops were present and several of these had clay bins for storing agricultural products. Pot manufacture and bronze working seem to have been carried out here, based on evidence including part of the floor of a pottery kiln, sherds from misfired vessels, large pieces of burned kiln walls, and a bronze-melting kiln.[72]

Angel[73] found that Allen's fossa ("reaction area") was quite frequently present in ancient Greek skeletal material, and he attributed its development to the vigorous traversing of hilly or mountainous terrains, as in rapidly descending steep slopes.[74]

In Greek prehistoric skeletal collections, females tend to exhibit much lower rates (39%) of Allen's fossa than men do (58%), but higher rates (44%) of Poirier's facet (25% among men).[75] Poirier's facet has been

61. McGeorge 1988.
62. Harrison et al. 1977, p. 350.
63. Wells 1975, pp. 1246–1247.
64. Hamilton 1982, p. 145.
65. Zivanovic 1982, pp. 94–95.
66. Larsen 1982.
67. Bisel and Angel 1985.
68. Merbs 1983; Stirland 2000.
69. Stirland 2000.
70. Lane 1887.
71. Kelley 1982.
72. Dakoronia 1993.
73. Angel 1960, 1964.
74. Angel 1959, 1960, 1964.
75. Angel 1964, p. 133.

observed in individuals who are thought to have habitually assumed a squatting posture.[76] The topography of East Lokris's inland region includes more mountainous or hilly terrain and the people there would have had to interact with this on an intimate level. Reflective of this is a high rate of both femoral head/neck conditions among Inland men, but none among the women there. These data further indicate that there was a division of labor and activity by sex in both geographic groups. Coastal and Inland men appear to have been most involved in activities that included lower body stressors, such as regularly traversing rugged terrain, and also squatting or sitting for long periods of time, as some craftsmen do. Interestingly, women in neither group seem to have been involved in activities that necessitated sitting or squatting for prolonged periods of time. Both East Lokris groups are comparable to Angel's data in that, overall, women had a lower rate of Allen's fossa than their men and, therefore, seem to have been less involved than them in traversing rugged terrain on a regular basis.

The average MI for the Inland group (MI = 76.1) is on par with Bisel and Angel's[77] compilation of Mycenaean metrics (an average MI of 76.8 [n = 98]) and, therefore, normal for the period. However, the average MI for the Coastal group was higher, and therefore their bones were rounder than either the Inland group or the Mycenaean norm. The Coastal group had relatively fewer individuals with platymeria and those who did have it had it to a less extreme degree. This means that the other measurable femora in the Coastal group were not platymeric and, in fact, had quite round diaphyses. Therefore, the Coastal group was subject to different types of lower body biomechanical stressors; ones that were not as physically strenuous as those encountered Inland. Rounder femoral cross sections have been associated with populations that are involved in heavy labor but with a restricted travel range, and increased sedentism with accompanying decreased dietary quality. These are the hallmarks of an agricultural subsistence base. Ruff[78] found that there was a reduction in mechanical loading of the femur, and probably the lower limb in general, among agriculturalists.

Larsen's[79] review of ethnographic evidence found that hunter-gatherers were subject to more biomechanical stress because they regularly had to traverse their environment searching for food. Populations subsisting via this means should have relatively flatter long bones due to constant locomotion. Conversely, agriculturalists should have rounder bones because they are more sedentary. Ethnographic studies further show that subsistence tasks that entail long-distance travel, such as among hunter-gatherers, are usually assigned to males, but among agriculturalists males do more sedentary work.[80]

Based on this evidence, it appears that the Inland group had more of a reliance on hunting and gathering, or a mixed economy with pastoralism that required more mobility than did the Coastal group.

CONCLUSIONS

The natural features of the land in central Greece had an important effect on prehistoric cultures by influencing the size and location of settlements

76. Poirier and Charpy 1911; Odgers 1931; Sauser 1936; Angel 1960, 1964; Kostick 1963.
77. Bisel and Angel 1985.
78. Ruff 1992.
79. Larsen 1981.
80. Ruff 1992, pp. 48–49; Ruff 1987; Murdock and Provost 1973.

and the form of subsistence, determining local and regional boundaries,[81] and generating the tendency for local centers to arise.[82] These same factors also affected the health and biology of the Mycenaean people of East Lokris, with more negative impacts on the health of women.

Osteological analysis of skeletal samples from two distinct regions of East Lokris indicates that the Mycenaeans living nearer the coast generally had better health. This is reflected in lower rates of metabolic disorders and fewer biomechanical stressors compared to inland populations. Levels of cribra orbitalia, osteoporosis, and indicators of physical stress were all more prevalent inland.

The more rugged inland environment, and possibly somewhat different social conditions, negatively impacted the health of those populations. Dental and isotopic evidence from past studies indicate that while both the Coastal and Inland groups depended upon some degree of cultivation, the Coastal population was more involved in agriculture. The evidence for the inland regions suggests a more mixed economy. These subsistence differences, which would have been largely dictated by environment, were important factors affecting the health of each group.

The women of East Lokris suffered from higher rates of metabolic disorders and nutritional issues than their men did. Fortunately, they also seem to have experienced lower rates of biomechanical stress. But this may have partially been the result of a reduced work capacity due to poor health. The higher rates of cribra orbitalia and osteoporosis among women, especially those of the Inland group, suggest that there may have been a division of labor by sex, which may have placed women at a higher risk of exposure to pathogens from working in less hygienic settings. Examples of such conditions might include the responsibility of women to tend flocks or other domesticated animals, or preparing the dead for burial, a task sometimes assigned specifically to women. A lifetime of nutritional deficiency, possibly related to living in a patriarchal society with unequal access to high-quality foods, coupled with unhygienic working conditions could have led to the higher levels of cribra orbitalia observed among women.

The equally high rates of platymeria and low MIs for both sexes from inland sites indicate that the more rugged terrain there was the probable cause of both these adaptations and a mixed-economy diet. Lower rates and severity of platymeria are found in the Coastal group, indicating that they had rounder femoral diaphyses; this characteristic has been associated with the activity patterns of agricultural populations.[83] Both men and women were similarly affected, suggesting that there must have been a division of labor within the Coastal population that had similar impacts on both sexes.

Compared to their Coastal contemporaries, women of the Inland group were involved in occupations that entailed more upper body stressors, as evidenced by their higher rates of upper body OA. Perhaps they spent much of their time tending to small gardens and producing the secondary products from livestock while men may have engaged in hunting, as is often described for traditional societies.[84] The higher rates of lower body OA in Coastal women was likely related to different agricultural duties and possibly the extended life span of these populations.

81. McInerney 1999, p. 76.

82. Hope Simpson and Dickinson 1979, p. 235.

83. Ruff 1992.

84. Hudson 1976.

There seems to have been more of a clear difference of labor/activity between the sexes in the Inland group than in the Coastal group. Further, the Inland women closely adhere to the Mycenaean norms for age at death, while the Coastal women are unusual in their extended life spans.

Overall, the regional health differences found within East Lokris, especially among the women, indicate that both environmental and somewhat different behavioral modes were the source. However, with the exception of cribra orbitalia and platymeria rates, the differences between the Coastal and Inland groups were not necessarily great, although they were consistently different.

The distance of East Lokris from the major Mycenaean centers seems to have provided cushioning from many of the negative health and social effects of the LH III destructions that ravaged much of the Mycenaean world. The present study adds to the small but growing body of information about the Late Bronze Age in central Greece—an intriguing area and one which undoubtedly holds many secrets patiently awaiting discovery. Further in-depth study of the physical anthropology and archaeology of this region will enlighten us about the lives and culture of its populations and their role in larger Mycenaean interactions.

REFERENCES

Agresti, A., and B. Finlay. 1986. *Statistical Methods for the Social Sciences,* 2nd ed., San Francisco.

Aldrich, J. H., and F. D. Nelson. 1984. *Linear Probability, Logit, and Probit Models,* Beverly Hills.

Angel, J. L. 1945. "Skeletal Material from Attica," *Hesperia* 14, pp. 279–363.

———. 1959. "Femoral Neck Markings and Human Gait," *Anatomical Record* 133(2), p. 244.

———. 1960. "Human Gait, Hip Joint, and Evolution," *American Journal of Physical Anthropology* 18(4), p. 361.

———. 1964. "The Reaction Area of the Femoral Neck," *Clinical Orthopaedics* 32, pp. 130–142.

———. 1973. "Human Skeletons from Grave Circles at Mycenae," in *Circle B at Mycenae,* ed. G. Mylonas, Athens, pp. 379–397.

Baker, R. K. 1984. "The Relationship of Cranial Suture Closure and Age Analyzed in a Modern Multi-Racial Sample of Males and Females" (diss. California State Univ., Fullerton).

Bass, W. M. 1987. *Human Osteology: A Laboratory and Field Manual,* 3rd ed., Columbia, Mo.

Bisel, S. C., and J. L. Angel. 1985. "Health and Nutrition in Mycenaean Greece: A Study in Human Skeletal Remains," in *Contributions to Aegean Archaeology: Studies in Honor of W. A. McDonald,* ed. N. C. Wilkie and W. D. E. Coulson, Minneapolis, pp. 197–209.

Black, T. 1978. "A New Method for Assessing the Sex of Fragmentary Skeletal Remains: Femoral Shaft Circumference," *American Journal of Physical Anthropology* 48, pp. 227–231.

Buikstra, J. E., and D. H. Ubelaker, eds. 1994. *Standards for Data Collection from Human Skeletal Remains,* Fayetteville, Ark.

Dakoronia, P. 1993. "Homeric Towns in East Lokris: Problems of Identification," *Hesperia* 62, pp. 115–127.

DiBennardo, R., and J. V. Taylor. 1982. "Classification and Misclassification in Sexing the Black Femur by Discriminant Function Analysis," *American Journal of Physical Anthropology* 58, pp. 145–151.

Dickinson, O. T. P. K. 1994. *The Aegean Bronze Age,* Cambridge.

Fox, J. 1991. *Regression Diagnostics,* Newbury Park.

Gilbert, B. M., and T. W. McKern. 1973. "A Method for Aging the Female *Os Pubis,*" *American Journal of Physical Anthropology* 38, pp. 31–38.

Gillman, H. 1874. "The Mound-builders and Platycnemism in Michigan," *Smithsonian Annual Report,* 1873, pp. 364–390.

Grauer, A. L., and P. Stuart-Macadam. 1998. *Sex and Gender in Paleopathological Perspective,* Cambridge.

Graziadio, G. 1991. "The Process of Social Stratification at Mycenae in the Shaft Grave Period," *AJA* 95, pp. 403–440.

Greulich, W. M., and S. I. Pyle. 1959. *Radiographic Atlas of Skeletal Development of the Hand and Wrist,* Stanford.

Hamilton, M. E. 1982. "Sexual Dimorphism in Skeletal Samples," in *Sexual Dimorphism in Homo sapiens: A Question of Size,* ed. R. L. Hall, New York, pp. 107–163.

Harrison, G. A., J. S. Weiner, J. M. Tanner, and N. A. Barnicot. 1977. *Human Biology,* Oxford.

Holtby, J. R. D. 1917–1918. "Some Indices and Measurements of the Modern Femur," *Journal of Anatomy* 52, pp. 363–382.

Hope Simpson, R., and O. T. P. K. Dickinson. 1979. *A Gazetteer of Aegean Civilisation in the Bronze Age* 1: *The Mainland and Islands,* Göteborg.

Hosmer, D. W., and S. Lemeshow. 1989. *Applied Logistic Regression,* New York.

Hoyme, L. E., and W. M. Bass. 1962. "Human Skeletal Remains from the Tollifero (Ha6) and Clarksville (Mc14) Sites, John H. Kerr Resevoir Basin, Virginia," *Bureau of American Ethnology Bulletin* 182, pp. 329–400.

Hudson, C. 1976. *The Southeastern Indians,* Knoxville, Tenn.

Iezzi, C. A. 2005. "Regional Differences in the Health Status of Late Bronze Age Mycenaean Populations from East Lokris, Greece" (diss. State Univ. of New York, Buffalo).

İşcan, M. Y., and K. Derrick. 1984. "Determination of Sex from the Sacroiliac Joint: A Visual Assessment Technique," *Florida Scientist* 47, pp. 94–98.

İşcan, M. Y., and K. A. R. Kennedy. 1989. *Reconstruction of Life from the Skeleton,* New York.

İşcan, M. Y., and P. Miller-Shaivitz. 1984a. "Determination of Sex from the Femur in Blacks and Whites," *Collegium Anthropologicum* 8, pp. 169–177.

———. 1984b. "Determination of Sex from the Tibia," *American Journal of Physical Anthropology* 64, pp. 53–58.

———. 1984c. "Discriminant Function Sexing of the Tibia," *Journal of Forensic Sciences* 29, pp. 1087–1093.

Kate, B. R. 1963. "The Incidence and Cause of Cervical Fossa in Indian Femora," *Journal of the Anatomical Society of India* 12(2), p. 69.

Kelley, M. A. 1982. "Intervertebral Osteochondrosis in Ancient and Modern Populations," *American Journal of Physical Anthropology* 59, pp. 271–279.

Kostick, E. L. 1963. "Facets and Imprints on the Upper and Lower Extremities of Femora from a Western Nigerian Population," *Journal of Anatomy* 97, pp. 393–402.

Kramer-Hajos, M. T. 2005. "Mycenaean East Lokris," *Archaeological Institute of America 106th Annual Meeting Abstracts, January 6–9, 2005,* Boston, pp. 107–108 (abstract).

Krogman, W. M., and M. Y. İşcan. 1986. *The Human Skeleton in Forensic Medicine,* 2nd ed., Springfield, Ill.

Lagia, A., K. Moraitis, C. Eliopoulos, and S. K. Manolis. 2000. "Sex Determination and Age Estimation from the Human Skeleton: Application and Evaluation of Modern Methods for Greek Populations," *Proceedings 22th [sic] Panhellenic Meeting of HSBS, Skiathos Island, Greece, May 25–28, 2000* (abstract).

Lane, W. A. 1887. "A Remarkable Example of the Manner in which Pressure-Changes in the Skeleton May Reveal the Labour-History of the Individual," *Journal of Anatomical Physiology* 21(3), pp. 385–406.

Larsen, C. S. 1981. "Functional Implications of Postglacial Size Reduction on the Prehistoric Georgia

Coast, U.S.A," *Journal of Human Evolution* 10(5), pp. 489–502.

———. 1982. "The Anthropology of St. Catherine's Island 3: Prehistoric Human Biological Adaptation," *Anthropological Papers of the American Museum of Natural History* 57, pp. 159–270.

———. 1987. "Bioarchaeological Interpretations of Subsistence Economy and Behavior from Human Skeletal Remains," *Advances in Archaeological Method and Theory* 10, pp. 339–445.

———. 1997. *Bioarchaeology: Interpretating Behavior from the Human Skeleton*, Cambridge.

Lerna II = J. L. Angel, *The People*, Princeton 1971.

Lovejoy, C.O., R. S. Meindl, T. R. Pryzbeck, and R. P. Mensforth. 1985. "Chronological Metamorphosis of the Auricular Surface of the Ilium: A New Method for the Determination of Adult Skeletal Age at Death," *American Journal of Physical Anthropology* 68, pp. 15–28.

McCullagh, P., and J. A. Nelder. 1989. *Generalized Linear Models*, London.

McGeorge, P. J. P. 1988. "Health and Diet in Minoan Times," in *New Aspects of Archaeological Science in Greece* (Fitch Laboratory Occasional Paper 3), ed. R. E. Jones and H. W. Catling, Athens, pp. 47–54.

McInerney, J. 1999. *The Folds of Parnassos: Land and Ethnicity in Ancient Phokis*, Austin, Tex.

McKern, T. W., and T. D. Stewart. 1957. *Skeletal Age Changes in Young American Males: Analysis from the Standpoint of Age Identification* (Environmental Protection Research Division Technical Report EP-45), Natick, Mass.

Meindl, R. S., and C. O. Lovejoy. 1985. "Ectocranial Suture Closure: A Revised Method for the Determination of Skeletal Age at Death Based on the Lateral Sutures," *American Journal of Physical Anthropology* 68, pp. 57–66.

Merbs, C. 1983. *Patterns of Activity-Induced Pathology in A Canadian Inuit Population* (National Museum of Man Mercury Series, Archaeological Survey of Canada 119), Ottawa.

Meyer, A. W. 1924. "The 'Cervical Fossa' of Allen," *American Journal of Physical Anthropology* 7, pp. 257–269.

Murdock, G. P., and C. Provost. 1973. "Factors in the Division of Labor by Sex: A Cross-Cultural Analysis," *Ethnology* 12, pp. 203–225.

Musgrave, J. H., and M. Popham 1991. "The Late Helladic IIIC Intramural Burials at Lefkandi, Euboea," *BSA* 86, pp. 273–296.

Odgers, P. N. B. 1931. "Two Details about the Neck of the Femur: 1. The Eminentia; 2. The Empreinte," *Journal of Anatomy* 65, pp. 352–362.

Otterbein, K., ed. 1994. *Feuding and Warfare: Selected Works of Keith F. Otterbein*, Philadelphia.

Papathanasiou, A. 1999. "A Bioarchaeological Analysis of Health, Subsistence, and Funerary Behavior in the Eastern Mediterranean Basin: A Case Study from Alepotrypa Cave, Greece" (diss. Univ. of Iowa).

Parsons, F. G. 1914. "The Characteristics of the English Thigh Bone," *Journal of Anatomy and Physiology* 48, pp. 238–267.

Pearson, K., and J. Bell. 1919. *A Study of the Long Bones of the English Skeleton*, Cambridge.

Phenice, T. W. 1969. "A Newly Developed Visual Method of Sexing the *Os Pubis*," *American Journal of Physical Anthropology* 30, pp. 297–302.

Poirier, P., and A. Charpy. 1911. *Traite d'anatomie humaine* 1, Paris.

Riggs, B. L., and L. J. Melton 1986. "Involutional Osteoporosis," *New England Journal of Medicine* 314, pp. 1676–1686.

Ruff, C. 1987. "Sexual Dimorphism in Human Lower Limb Bone Structure: Relationship to Subsistence Strategy and Sexual Division of Labor," *Journal of Human Evolution* 16, pp. 391–416.

———. 1992. "Biomechanical Analyses of Archaeological Human Skeletal Samples," in *The Skeletal Biology of Past Peoples: Advances in Research Methods*, ed. S. Saunders and M. A. Katzenberg, New York, pp. 37–58.

Runnels, C., and P. Murray. 2001. *Greece Before History: An Archaeological Companion and Guide*, Stanford.

Sauser, G. 1936. "Eminentia Colli Femoris Dorsalis (ein neuer osteologischer Befund am menschlichen Schenkhals)," *Zeitschrift für Rassenkunde* 3, pp. 286–291.

Stewart, T. D. 1979. *Essentials of Forensic Anthropology: Especially as Developed in the United States*, Springfield, Ill.

Stirland, A. J. 2000. *Raising the Dead: The Skeleton Crew of Henry VIII's Great Ship, the* Mary Rose, Chichester.

Stuart-Macadam, P. 1998. "Iron Deficiency Anemia: Exploring the Difference," in *Sex and Gender in Paleopathological Perspective*, ed. A. L. Grauer and P. Stuart-Macadam, Cambridge, pp. 45–63.

Suchey, J. M., S. T. Brooks, and D. Katz. 1988. *Instructions for Use of the Suchey-Brooks System for Age Determination of the Female* Os Pubis, Bellvue, Colo.

Suchey, J. M., P. A. Owing, D. V. Wisely, and T. T. Noguchi. 1984. "Skeletal Aging on Unidentified Persons," in *Human Identification: Case Studies in Forensic Anthropology*, ed. T. A. Rathburn and J. E. Buikstra, Springfield, Ill., pp. 278–297.

Thieme, F. P. 1957. "Sex in Negro Skeletons," *Journal of Forensic Medicine* 4, pp. 72–81.

Todd, T. W. 1920. "Age Changes in the Pubic Bone: I. The Male White Pubis," *American Journal of Physical Anthropology* 3, pp. 285–334.

Triantaphyllou, S. 1999. "A Bioarchaeological Approach to Prehistoric Cemetery Populations from Western and Central Greek Macedonia" (diss. Univ. of Sheffield).

———. 2001. *A Bioarchaeological Approach to Prehistoric Cemetery Populations from Central and Western Greek Macedonia* (*BAR-IS* 976), Oxford.

Vermeule, E. 1979. *Aspects of Death in Early Greek Art and Poetry*, Los Angeles.

Voutsaki, S. 1995. "Social and Political Processes in the Mycenaean Argolid: The Evidence from the Mortuary Practices," in *POLITEIA: Society and State in the Aegean Bronze Age* (*Aegaeum* 12), ed. R. Laffineur and W. D. Niemeier, Liège, pp. 55–66.

Webb, P. A. O., and J. M. Suchey. 1985. "Epiphyseal Union of the Anterior Iliac Crest and Medial Clavicle in a Modern Multiracial Sample of American Males and Females," *American Journal of Physical Anthropology* 68, pp. 457–466.

Wells, C. 1975. "Ancient Obstetrical Hazards and Female Mortality," *Bulletin of the New York Academy of Medicine* 51, pp. 1235–1249.

Zivanovic, S. 1982. *Ancient Diseases,* New York.

CHAPTER 12

Anthropological Research on a Byzantine Population from Korytiani, West Greece

by Christina Papageorgopoulou and Nikolaos I. Xirotiris

The Byzantine period, in what is today considered to be the geographical area of Greece, was characterized by extended movements of population and several political upheavals. Our main information for this period comes from the few existing written records and sporadic archaeological excavations.[1]

The historical sources allow us to piece together the basic political and military history of the whole Byzantine Empire and provide considerable information about isolated events, but they are significantly biased in their focus on rulers and dynasties and they do not provide much information about the economy, society, ordinary people, or daily life.[2] In comparison, the archaeological evidence is abundant and informative, but it has not been used as fully as it ought to be. In addition, because the interest of earlier excavators was on the Classical or prehistoric periods, the Byzantine remains from these archaeological sites were frequently ignored or, even worse, completely destroyed in the rush to uncover earlier levels.[3] For these reasons, we have practically no information on the anthropological structure of the various populations of Byzantine Greece, and a very limited understanding of their everyday life, health conditions, demographic movements, and general biological existence.

The written sources for this period are particularly poor for the region of Thesprotia (Epirus), the focus of this chapter. During the 10th and 11th centuries A.D., Thesprotia and the neighboring region of Epirus were part of the *Thema*[4] of Nikopoleos[5] while from the 13th century A.D. they belonged to the *Despotate* of Epirus.[6] These regions, located on the western border of the Byzantine Empire, had experienced since the 4th century A.D. invasions from neighboring populations (such as the Normans, Franks, Venetians, and Genovese)[7] who tried to conquer parts of the Empire. Especially during the 10th and 11th centuries A.D., the region was very poor with a low socioeconomic level and no known cultural centers. It was not until the 14th century A.D. that Thesprotia flourished again.[8]

A salvage excavation of a Byzantine cemetery in Korytiani, a village 25 km east of Igoumenitsa (Epirus), enables us to study, with new methodologies, many aspects of the everyday life of a medieval rural population during a period from the end of the 10th to the beginning of the 11th century A.D.[9] One hundred and seventy-six pit graves, covered with schist

1. Gregory 2005, pp. 13–20.
2. Gregory 2005, p. 14; Savvidis 1986, pp. 15–16.
3. Gregory 2005, p. 17.
4. After the 7th or the 8th century, the Byzantine Empire was divided into provinces called *Themata*.
5. Magdalino 2002, p. 178
6. Savvidis 1986, p. 82.
7. Savvidis 1986, p. 14.
8. C. Pitsakis (pers. comm.).
9. The recovery of this sample was excellent due to the exceptional cooperation with the 8th Ephorate of Prehistoric and Classical Antiquities, and especially with the excavator of the site, G. Kordatzaki; the supervisor of the excavation program, G. Riginos; and the anthropologists K. Kouvaris, E. Milka, and C. Papageorgopoulou. All photographs were taken by C. Papageorgopoulou.

slabs, were excavated on the summit, the west slope, and at the foot of the site, which sits on a hill. All the graves were orientated east–west and they represent mostly primary burials of single individuals. Bodies were placed in an extended position with the upper limbs folded against the chest, the abdominal area, or the pelvis. Most of the graves contained artifacts consisting of bronze objects, ceramic and glass pots, and one coin.

MATERIALS AND METHODS

Two hundred and two relatively well-preserved skeletons were recovered from the cemetery. The estimation of age at death for the children and the juvenile individuals was made according to Scheuer and Black,[10] while the adults were aged using the pubic symphysis[11] and the auricular surface of the os coxae,[12] the cranial sutures,[13] and, although only as a general indicator, the attrition of teeth.[14] Determination of sex was restricted to the adult individuals. The morphology of the os coxae was the main sex indicator; the skull, the femur, and other available bones were used to refine the assessments.[15]

In order to reconstruct the biological profile of this Byzantine period population, we studied demography, metric and nonmetric traits, the paleopathological lesions, and conducted chemical analyses of bones.[16] In the present publication, we will focus on the study of diet and health, as determined from dental and bone pathological lesions and chemical analyses. We attempt to reconstruct the nutritional deficiencies, the dietary habits, and the health status of this rural Byzantine period population.

Dental Pathological Lesions

The permanent teeth and jaws (*N* = 2,340) were examined for caries,[17] calculus,[18] periapical lesions,[19] alveolar bone loss,[20] dental wear, antemortem

10. Scheuer and Black 2000 (teeth development after Ubelaker 1978, p. 161; clavicle, scapulae, pelvis, long bone, epiphyseal fusion after Scheuer and Black, 2000, pp. 251, 269, 285, 295, 372, 373, 306, 392, 413, 424).

11. Brooks and Suchey 1990, pp. 227–238.

12. Meindl and Lovejoy 1989, p. 137.

13. Meindl and Lovejoy 1985, pp. 57–66.

14. Miles 1962, p. 881; Brothwell 1981.

15. Buikstra and Ubelaker 1994, pp. 18–20.

16. The chemical analysis was conducted in the chemical laboratory of the Greek Institute of Geology and Mineral Exploration in Xanthi (IGME).

17. The exact location and the intensity (degree) of each lesion was recorded using a six-scale system: (1) dark stain on enamel surface; (2) dark stain on enamel surface with slight destruction of the enamel's surface; (3) minimal destruction of enamel, small pits, or small fissures (not to be confused with normal anatomical pits on the enamel surface); (4) large carious lesion, but with less than two-thirds of the tooth crown destroyed; (5) large carious lesion, with almost two-thirds of the tooth crown destroyed, pulp exposure; (6) complete destruction of the tooth crown (for detailed description of the scale system and the protocols used for all pathological conditions, see Papageorgopoulou 2002).

18. A three-scale system was used for scoring the intensity of the calculus after Brothwell 1981.

19. Radicular cysts and periapical granulomas macroscopically observable in the interna and externa lamina of the alveolus crest.

20. The presence of alveolar loss (horizontal/vertical) and intensity (distance between the cementoenamel junction and the alveolus crest) was recorded. Degrees modified after Brothwell 1981: (1) beginning of alveolar bone resorption, (2) one-quarter root exposure, (3) one-third root exposure, (4) one-half root exposure, (5) complete root exposure.

tooth loss, linear enamel hypoplasia,[21] and congenital anomalies of the maxillofacial region.[22] The intensity (degree),[23] the exact location, and, in some cases, the pattern and the direction of wear were also recorded.[24] Prevalence of oral pathologies was examined by sex and age and comparisons were made between mandibular and maxillary teeth, and among different tooth classes (incisors, canines, premolars, and molars). SPSS 8.0 was used for the statistical analysis and chi-square tests were performed between all comparisons in order to test whether the differences are statistically significant.

Bone Pathological Lesions

The postcranial and cranial bones ($N = 4{,}756$) were evaluated for degenerative diseases of the joints, infectious disease, fractures, metabolic diseases, neoplastic conditions, and congenital anomalies.[25]

For the evaluation of degenerative joint disease of the vertebral column, a comparative analysis of the pattern and severity of articular surface osteoarthritis of the vertebrae and osteophytosis of the vertebral bodies was performed.[26] Each vertebra was examined for lipping on the four parts of the body (upper left, upper right, lower left, lower right), and for lipping, porosity, and eburnation on each apophyseal surface (superior left, superior right, inferior left, inferior right). The joint surfaces on the ribs and the costal facets were not examined due to incompleteness. Schmorl's nodes were also recorded. Over 1,200 vertebrae (379 cervical, 555 thoracic, and 308 lumbar) were studied. The analysis was made separately for the adult, older adult, and juvenile individuals. The adult and older adult individuals were grouped by sex.

The sample of long bones examined for degenerative disease of the joints totaled 1,317. On each bone, the presence of osteoarthritis was recorded separately for the proximal and the distal ends.[27] Scapulae and patellae were also examined in order to record the percentage of osteoarthritis among the different peripheral joints of each skeleton. The percentage was counted separately for each bone, for each joint, and for each individual.

21. Hypoplasia degrees: (1) the horizontal lines are slightly present, (2) the horizontal lines are well formed, (3) intense narrow grooves are present on the crown.

22. Identification of teeth pathologies after Hillson 1996; Brothwell 1981; Alt, Rosing, and Teschler-Nicola 1998.

23. Wear degree modified after Murphy 1959: (1) minimal signs of wear on the anterior teeth; (2) wear facets observable on both anterior and posterior teeth; (3) small dentine patches observable in all teeth; (4) extensive dentine patches; (5) appearance of secondary dentine; (6) extended loss of dentine and enamel, pulp exposure; (7) crown obliterated, root functioning in occlusal surface.

24. Pattern and direction of wear after Molnar 1971.

25. Identification of the skeletal pathologies mainly after Ortner and Putschar 1981; Mann and Murphy 1990; Roberts and Manchester 1995; Aufderheide and Rodríguez-Martín 1998.

26. Degrees for recording articular surface osteoarthritis and osteophytosis of the intervertebral joints: (1) barely discernible (osteophytes and/or lipping), pinpoint size porosity (articulate surface); (2) elevated ring (osteophytes), sharp ring (lipping), larger pits (articulate surface porosity); (3) curved spicules (osteophytes and/or lipping), eburnation (articulate surface). Ankylosis was scored separately.

27. Degrees of osteoarthritis: (1) pinpoints pits, slight lipping, barely discernible osteophytes and/or periarticular resorption; (2) larger pits, lipping forms a sharp ridge, well-formed osteophytes, clearly present periarticular resorption, eburnation (only polish); (3) large osteophytes, eburnation polish with grooves, severe erosion.

All individuals were examined for infectious diseases (specific and nonspecific), and for fractures or traumatic lesions.[28] Every bone was first examined macroscopically, and then bones exhibiting pathological lesions were examined radiologically.

Trace Element Analysis

The samples for chemical analysis were taken from the best preserved skeletons, a total of 116 individuals from both sexes and all age groups. Additionally, 12 samples of soil from inside the cemetery and 3 samples from near the cemetery were extracted and analyzed for comparative purposes.

Twenty three major and minor elements[29] were analyzed using atomic absorption spectroscopy (Perkin-Elmer Analyst 100 and Perking-Elmer M2100). The selection of these elements was based on the current literature on trace element analysis.[30] For some of the selected elements there is no extensive literature on their relevance to studies of skeletal material and dietary reconstruction, but we include them here in order to present a more complete database for the concentration of various elements in skeletal remains.

Bone samples were taken with a stainless steel trephining drill from the middle shaft of the right femur. The superficial periosteal and endosteal surfaces were abraded. Each bone sample was cleaned ultrasonically with double-distilled water, heated to 105°C for four hours and then homogenized in an agate mortar in 200 gauge mesh. The homogenized powder was transferred to porcelain crucibles and then heated to 550°C for 24 hours to remove the organic portion.

In order to measure the atomic absorption, it was necessary to prepare the samples in liquid form. The pretreatment procedure depends on the sample but certain specifications should always be followed.[31] The chosen method should give a fairly dilute and stable standing solution, a concentration of the element of interest to a level that is convenient for its determination, and a solution that does not contain large quantities of substances leading to chemical interference in the flame. Taking into consideration the above factors, we used four different methods for measuring the 23 elements (Table 12.1). Similar methods were used for the preparation of the soil samples (Table 12.2).[32]

To minimize the measurement error in the atomic absorption analysis, extreme care was taken with regard to the precise preparation of the control samples, the adjustment of the instrument to known calibration points, the stability of electronic systems, and the stability of the flame.

28. A standard descriptive protocol was used in order to acquire complete information on infectious disease: detailed description of the lesion—side affected, state of healing, comparison with other affected bones of the same individual. For traumas—the type of trauma, cause, healing procedure, and complications after Lovell 1997.

29. Barium (Ba), Cadmium (Cd), Caesium (Cs), Chromium (Cr), Cobalt (Co), Copper (Cu), Iron (Fe), Lead (Pb), Lithium (Li), Manganese (Mn), Molybdenium (Mo), Nickel (Ni), Rubidium (Rb), Silicon (Si), Strontium (Sr), Titanium (Ti), Vanadium (V), Zinc (Zn), Calcium (Ca), Magnesium (Mg), Phosphorus (P), Potassium (K), Sodium (Na).

30. Sandford and Weaver 2000, pp. 329–350.

31. Amos et al. 1975.

32. Cesium in the soil samples was not analyzed due to technical problems in the atomic absorption spectrometry.

TABLE 12.1. CHEMICAL PRETREATMENT OF BONE SAMPLES

Method Used	*Elements*
1. HNO3 for low elemental concentrations	Ba, Sr, Cr, Ti, V, Mo, Cd, Co, Ni, Mn, Pb, Li, Rb, Cs
2. Fussion mixture	Ca, Si, Mg, Fe, Cu, Zn
3. HCI	K, Na
4. Indirect method of P by measuring Mo	P

TABLE 12.2. CHEMICAL PRETREATMENT OF SOIL SAMPLES

Method Used	*Elements*
1. Geochemical	Pb, Cd, Co, Ni, V, Li, Rb, Mo, Sr, Cr
2. Fussion mixture	Si, Ca, Mg, Ba, Fe, Mn, Zn, Cu, Ti
3. HCI	K, Na
4. Chromatometric	P

In order to assess whether the elemental concentrations follow a normal distribution, a nonparametric test (the Kolmogorov-Smirnov one-sample test) was performed. For the elements that showed significant deviations from the normal distribution, nonparametric tests (i.e., Kruskal-Wallis) were employed for the comparisons between age and sex groups, while for the elements with normal distributions, a one-way analysis of variance (ANOVA) was performed.

THE BIOLOGICAL PROFILE OF THE KORYTIANI POPULATION

Age and Sex Estimation

It was possible to determine with certainty the sex for 40 males and 44 females, while 47 adult skeletons remain unsexed. Estimation of age at death revealed that 27.7% of the individuals were infants aged between 0–7 years, 7.4% were juveniles aged 15–24 years, 58.9% were adults aged 25–60 years, and the remaining 6.0% were older adults, aged over 60 years. The age classes and raw numbers are shown in Table 12.3.

TABLE 12.3. AGE DISTRIBUTION

Age Groups	*N of Individuals*
Infants I (0–7 years)	56
Infants II (8–14 years)	0
Juveniles (15–24 years)	15
Adults (25–60 years)	119
Older Adults	12

Caries

Of the 2,340 permanent teeth examined, 251 (10.7%) exhibited carious lesions. Maxillary teeth were more affected than mandibular teeth (5.9% vs. 4.8% respectively). The molars were the most affected (5.9%), followed by the premolars (2.8%), the incisors (1.1%), and the canines (0.9%). Molars

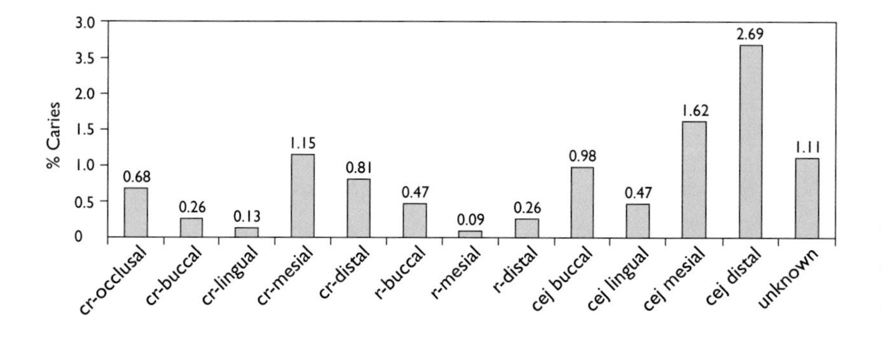

Figure 12.1. Caries location (cr = crown, r = root, cej = cemento-enamel junction)

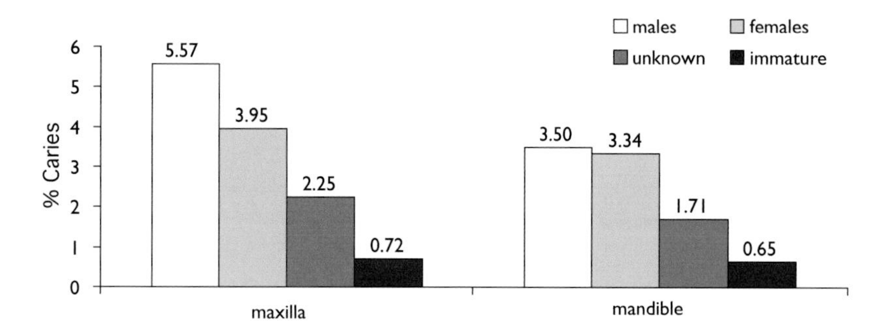

Figure 12.2. Caries by jaw and by sex

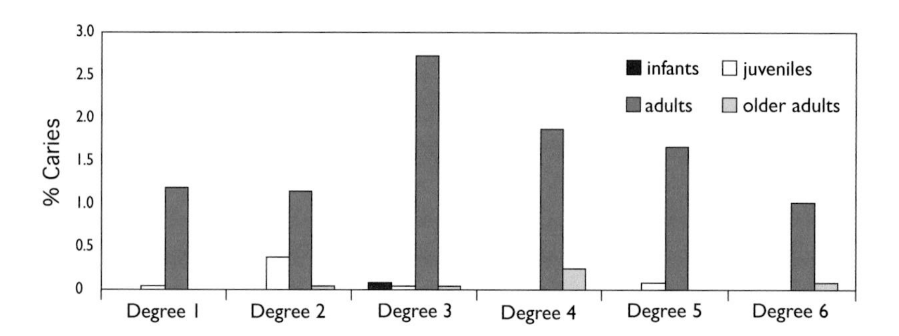

Figure 12.3. Caries degree by age groups

exhibit higher percentages of severe caries in comparison to the other tooth types; this is common in all human populations,[33] irrespective of their era or geographical region.

The area of the tooth most affected by caries was the distal and mesial cementoenamel junction (CEJ), followed by the crown (Fig. 12.1). In general, for modern populations the crown is the most affected area while in prehistoric and historic populations higher percentages of caries are observed on the CEJ.[34] Caries at the root surface and CEJ develop with the gingival recession that is related to long-standing plaque deposits. According to Tóth,[35] attrition and low sugar consumption can also contribute to this condition. The high percentage of caries on the CEJ can probably be attributed to the high incidence of alveolar bone resorption and attrition at Korytiani, as described below.

Males are affected more by caries (4.5%) than females (3.6%). Their maxillary teeth have a higher percentage of caries than their mandibular teeth, while females show almost the same distribution between the maxillae and the mandible (Fig. 12.2). With respect to the severity of caries, both males and females suffer from severe lesions.

33. Hillson 1996, p. 280.
34. Hillson 1996, p. 282.
35. Tóth 1970.

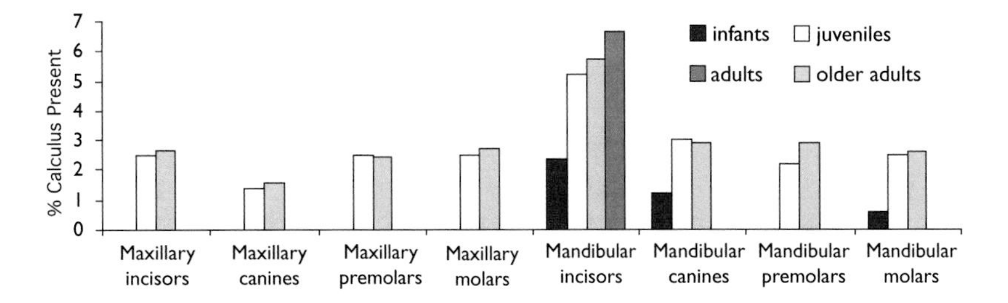

Figure 12.4. Calculus deposits on maxillary and mandibular teeth by age

The caries rate is related to age; the older adults (22.22%) are the most affected group, followed by the adults (12.83%), the juveniles (3.55%), and the infants I (1.2%). Molars are the first teeth affected by caries. Most of the juveniles had lost their first molar antemortem, and subadults exhibit caries only on the permanent first molar. The adults and older adults suffer from carious lesions in all tooth types. The intensity of caries increases with respect to age; infants I and juveniles suffer from slight and medium carious lesions, while adults and older adults have carious lesions of all sizes (Fig. 12.3). Infants I show carious lesions only on the crown of permanent molars. Juveniles display carious lesions mostly on the crown, and there are few cases of caries on the CEJ. Adults and older adults exhibit caries on all tooth surfaces, but the most common are lesions at the CEJ.

Dental Calculus

Medium-sized calculus deposits, which in most cases cover all tooth surfaces, occur on 21.5% of the examined teeth. All tooth types are affected, with the incisors showing the highest frequencies (7.82%) and the heaviest deposits of calculus. Canines have the lowest frequency of calculus deposits at 4.15%. Mandibular teeth exhibit more calculus than maxillary teeth. Most populations, historic and contemporary, display a higher frequency of calculus on the lingual surface of the lower incisors and on the buccal surface of the upper molars probably because those teeth are closest to the ducts of the salivary glands, which play one of the most important roles in the formation of calculus.[36]

Males (9.87%) are more affected by calculus deposits than females (4.96%). Both males and females show higher percentages of calculus on anterior than on posterior teeth, and especially on the mandibular incisors. In general, the pattern of calculus accumulation on males and females does not differ substantially.

All age groups exhibit calculus. The adults (23.55%) and the juveniles (21.58%) are the most affected groups, followed by the older adults (6.6%) and the infants I (4.19%). It is obvious that calculus formation in this population starts from an early age. Juveniles and adults exhibit calculus deposits on all tooth types (Fig. 12.4), while infants I exhibit calculus formation on the mandibular anterior teeth and the molars. Juveniles and adults exhibit the highest frequency of calculus on mandibular canines and incisors. Thus, it can be assumed that the mandibular anterior teeth are the first teeth of the dentition affected by calculus in this population. This is consistent with the typical pattern of calculus throughout the dentition in contemporary clinical studies.[37]

36. Hillson 1996, pp. 255–256.
37. Hillson 1996, pp. 255–256.

Periodontal Disease

Of the 2,340 permanent teeth and alveoli evaluated, 36.71% show alveolar bone loss. The most affected region is around the molars (12.22%), followed by the premolars (10.64%), incisors (8.46%), and canines (5.38%). No statistically significant differences were found between the mandibles and maxillae.

On both the mandible and maxilla, males (13.21%) exhibit higher percentages of alveolar resorption than females (11.5%). Females do not show differences between the mandible and the maxillae, while males show a higher percentage of alveolar resorption on the mandible. In addition, females do not exhibit differences between anterior and posterior teeth, while males show slightly higher percentages of alveolar resorption for the alveoli of the anterior teeth (about 2% more).

All individuals, except infants I, exhibit alveolar bone resorption. Adults (39.04%) and older adults (100%) show expected levels, while juveniles have a relatively high percentage (34.4%) of alveolar bone loss. Juvenile periodontitis is a rare form of periodontal disease. Besides the common factors that lead to alveolar bone loss, infectious disease, hormonal disturbances, and inadequate nutrition can produce or contribute to juvenile periodontal disease. In addition, a monocrop, refined carbohydrate-based diet is conducive to alveolar resorption,[38] whereas a varied diet with adequate trace elements and minerals seems to prevent it.[39]

Periapical Lesions

Only 34 (1.08%) of the 3,138 observed alveoli show periapical lesions. Males (0.61%) exhibit a higher frequency of lesions in all tooth types than females (0.29%), who, in addition, do not show abscesses on the canines. For both females and males, the posterior teeth are more affected than the anterior teeth. Infants I and juveniles do not show periapical abscesses. Adults exhibit a higher percentage of abscesses (0.89%) in comparison to older adults (0.19%). Considering the lack of severe attrition in this population (see below), we attribute the periapical lesions to severe caries.

Dental Wear

Overall, the population of Korytiani exhibits a medium dental wear rate. Specifically, although more than half of the teeth exhibit dental attrition, this is restricted to the enamel, or has led to slight exposure of the dentine. Only 19 of the 2,340 teeth displayed pulp exposure due to severe wear. Incisors (17.65%) and molars (17.65%) exhibit the highest percentage and most severe wear (Fig. 12.5), followed by premolars (15.81%) and canines (10.26%). Mandibular teeth (34.2%) exhibit higher frequencies of wear than maxillary teeth (27.5%). In most populations the maxillary teeth wear to a greater degree than do the mandibular teeth.[40] The reverse situation is also observed in some populations, including Korytiani. An explanation for these contrasts is still unknown, but it may be possible that face form and the timing of premolar eruption influence the wear differential between the upper and lower teeth.[41]

38. Leigh 1925, pp. 179–199; Davies, Picton, and Alexander 1969, pp. 74–77; Hillson 1979, pp. 147–162.
39. y Edynak 1989, pp. 17–36.
40. Molnar 1971, pp. 175–190.
41. Molnar and McKee 1988, pp. 125–136.

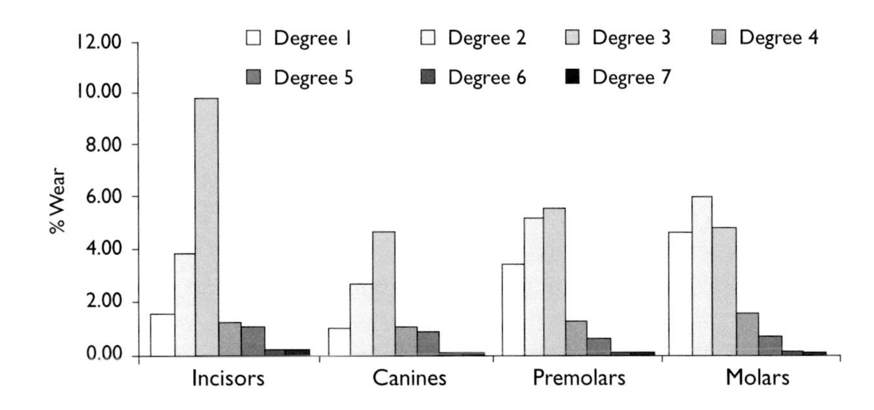

Figure 12.5. Degree of wear by tooth type

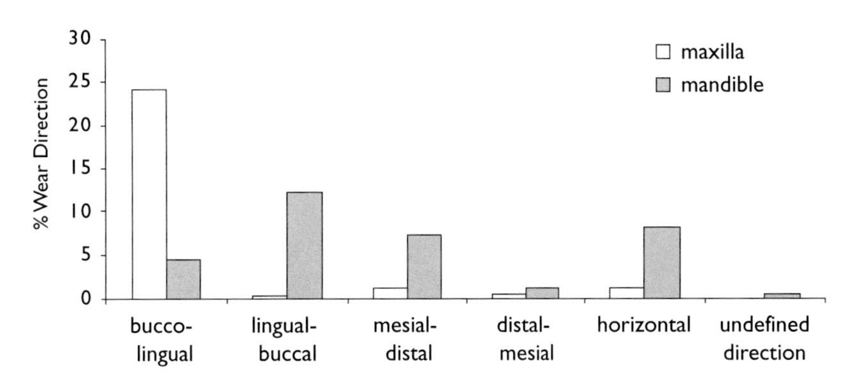

Figure 12.6. Wear direction by jaw

For the Korytiani individuals, the wear direction and the form of the occlusal surface wear, both in the maxillary and mandibular teeth, follow the normal patterns: maxillary teeth wear in a bucco-lingual direction, while mandibular teeth wear in lingual-buccal and mesio-distal directions (Fig. 12.6). The most common pattern of wear on the occlusal surface was flat, with less frequent partially concave and total concave patterns. However, there were numerous exceptions to this expected pattern, such as the bucco-lingual direction of mandibular tooth wear, the lingual-buccal wear of the maxillary teeth, and the notched and rounded form of the occlusal surface. These are extremely uncommon patterns, and they may indicate malocclusion or paramasticatory use of the dentition.

Comparison of the frequencies of wear between males and females shows that males (24.7%) exhibit more worn teeth and a higher percentage of slight and medium wear while females exhibit fewer worn teeth (18.3%), but higher percentages of severe wear. In addition, males exhibit more wear on the posterior teeth than the anterior teeth. Females do not exhibit notable differences in anterior and posterior tooth wear. This pattern of accelerated tooth wear and differences between anterior and posterior tooth wear among females is recorded for many aboriginal societies.[42] The proposed explanation is the use of teeth for different nonmasticatory functions, and sex-based differences in diet.

On the maxillae, both males and females experience wear on their teeth in the common bucco-lingual direction, and on the mandible the most common wear is in the lingual-buccal direction. Both males and females exhibit flat occlusal surfaces. Females do not show the rounded form of wear and display lower percentages of the notched and concave forms than males.

42. Molnar 1971, pp. 175–190; Molnar, McKee, and Molnar 1983a, pp. 51–65; Molnar, McKee, and Molnar 1983b, pp. 562–565; Richards 1984, pp. 5–13.

Older adults (80%) exhibit higher percentages of wear, followed by adults (71.62%), juveniles (48.63%), and infants I (3.59%). Infants I show slight wear, juveniles slight and medium, while adults and older adults show a range from slight to medium and severe wear. Although the population in general does not exhibit severe degrees of wear, the permanent teeth of the infants I and juveniles have dentine exposure. This indicates that the process of wear starts at an early age but proceeds at a slow rate.

The notched and rounded wear forms appear only on the adults and the older adults. Infants I show only the flat wear pattern, while juveniles have flat wear, but also a low percentage of concave and partially concave wear.

Linear Enamel Hypoplasia

Of the 2,340 permanent teeth studied, 8.16%[43] exhibit hypoplastic defects. The most affected teeth were the premolars (2.69%), followed by the incisors (2.52%), and the canines (2.18%). Mandibular teeth (4.36%) are more affected by hypoplasia than maxillary teeth (3.80%).

When examining the prevalence of hypoplasia by sex, we find that males (4.67%) exhibit a higher percentage of hypoplasia than females (1.62%) on the mandibular and maxillary teeth, and on all tooth types. Seventeen (9.6%) individuals exhibit hypoplastic enamel defects. These include three juveniles and 14 adults. The mean number of hypoplastic teeth per individual is 11.24. It must be noted that each tooth with hypoplasia displayed at least three hypoplastic disruptions of the enamel. The presence of hypoplasias on more than one tooth and the number of disruptions per tooth indicate that those individuals were chronically stressed throughout childhood.

Paleopathological research[44] on archaeological populations shows that 20% to 80% of individuals in agricultural groups exhibit enamel hypoplasia, and that hypoplasia gradually increases in frequency from hunter-gatherer to intensively agricultural populations. In comparison with these data, the percentage of hypoplasias in our sample is low and may indicate a population with a diverse nutritional base of good quality. Nevertheless, linear enamel hypoplasia is present in some individuals, and although the specific cause cannot be determined, the mere existence of the hypoplastic defects indicates a stress of sufficient magnitude to disrupt the normal growth process from birth to about 10 years of age.[45] Multiple factors may have caused these defects, such as nutritional stress, infectious diseases, trauma, genetic causes or deformities, systemic diseases, or developmental disturbances.[46] The higher frequency in males suggests that they were more susceptible to the stress that caused the growth disruption in comparison to females.

Congenital Abnormalities: Mesiodens

Two cases of mesiodens (supernumerary tooth located in the midline of the maxillae) were recorded in the Korytiani sample: one in a male (burial 78) and one in a female individual (burial XXXIII). The female's mesiodens (Fig. 12.7:a) is situated vertical to the median palatine suture, exactly below the incisive foramen. In the male individual (Fig. 12.7:b) the tooth is situated in the floor of the nasal cavity 4 mm behind, and parallel

43. Seventeen of 177 individuals.

44. Goodman and Armelagos 1989, pp. 271–305.

45. Goodman and Rose 1991, pp. 279–293.

46. Goodman and Rose 1991, pp. 279–293.

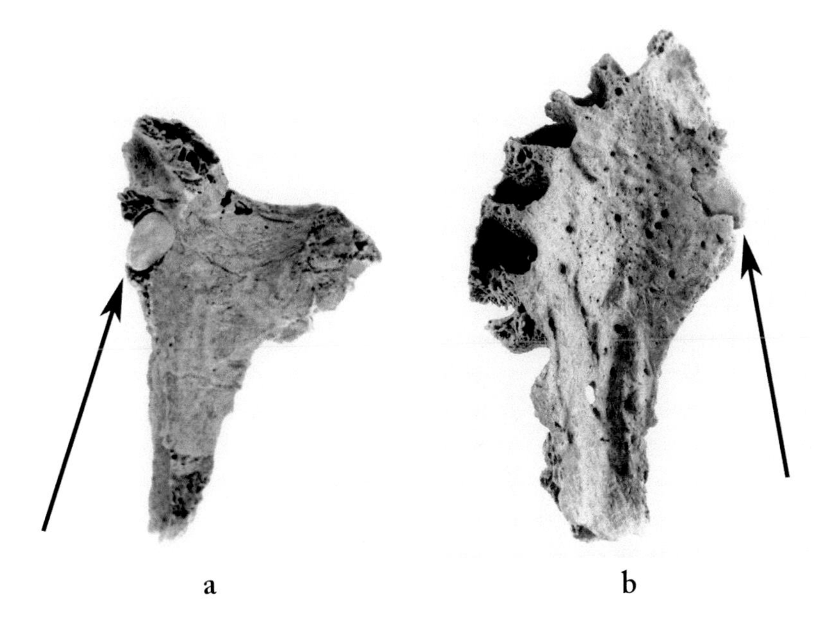

Figure 12.7. Mesiodens (indicated by arrows): (a) female adult (burial XXXIII); (b) male adult (burial 78)

to, the incisive foramen. The enamel of both teeth is extremely thin and hypomineralized. Both crowns have a cylindrical shape and the root is almost absent.[47] The eruption and position of the neighboring teeth has not been affected. Only the roots of the left central and lateral incisors show minor resorption.

Degenerative Joint Disease of the Spine

Of the 809 vertebral bodies of adult individuals examined, 11.6% show osteophytosis. The lumbar are the most affected vertebrae (19.4%), followed by the thoracics (13.9%) and the cervicals (11.9%). The vertebral bodies exhibit bilateral symmetry in the formation of osteophytosis. Between the upper and the lower vertebral surfaces, slight differences were recorded for the thoracic segment (Figs. 12.8, 12.9).

A total of 846 adult vertebrae were examined for osteoarthritis. Of these, 209 (24.7%) exhibit osteoarthritic changes on the articular surfaces. The thoracic vertebrae (28.7%) are the most affected, followed by the lumbar (28.0%) and the cervical (16.1%). Differences were recorded between the left and the right articular surfaces of all vertebrae and between the superior and the inferior articular surfaces (Figs. 12.10, 12.11).

Osteoarthritis is more common than osteophytosis in the adult subsample. A small proportion (6.4%) of all vertebrae was affected both by osteophytosis and osteoarthritis, while 18.9% were affected by only osteoarthritis and 8.4% by only osteophytosis. On the basis of this evidence, it is suggested that osteophytosis can appear independently of osteoarthritis and vice versa. It cannot be assumed which condition appears earlier, because there are vertebrae affected by only osteophytosis and by only osteoarthritis, and the presence of one condition does not prohibit the other.

As might be expected, 36.3% of the older adult vertebrae examined exhibit osteophytosis. Like adults, older adult individuals show a higher percentage of osteophytosis on the lumbar (40.6%), followed by the thoracic (38.5%), and cervical vertebrae (26.3%). Both superior and inferior surfaces

47. It was possible to measure the mesiodens of the female individual. Crown mesio-distal = 4.5 mm, crown bucco-lingual = 5.5 mm, root bucco-lingual = 4.5 mm, root mesio-distal = 3.5 mm, total crown length = 6 mm, total root length = 3 mm.

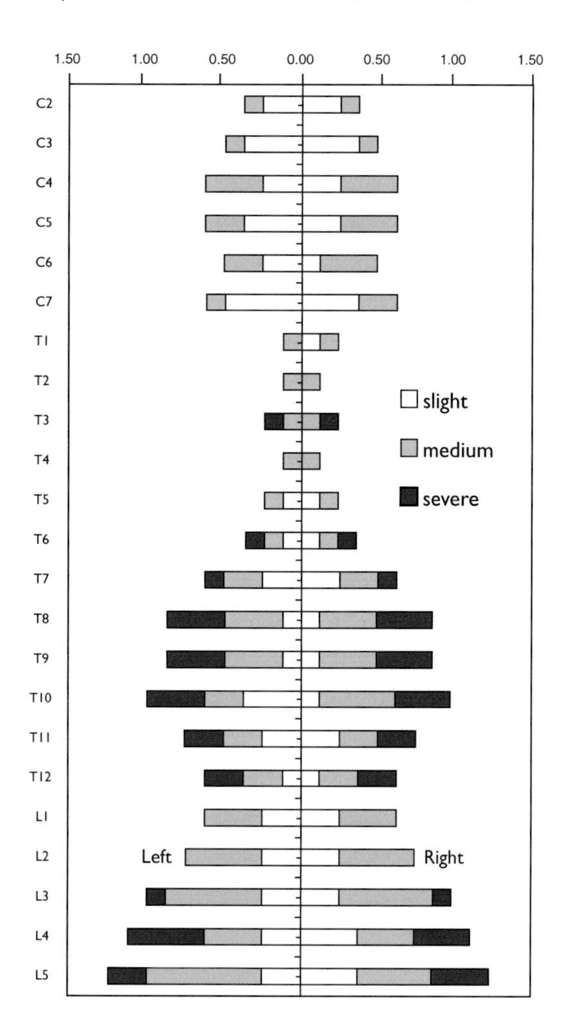

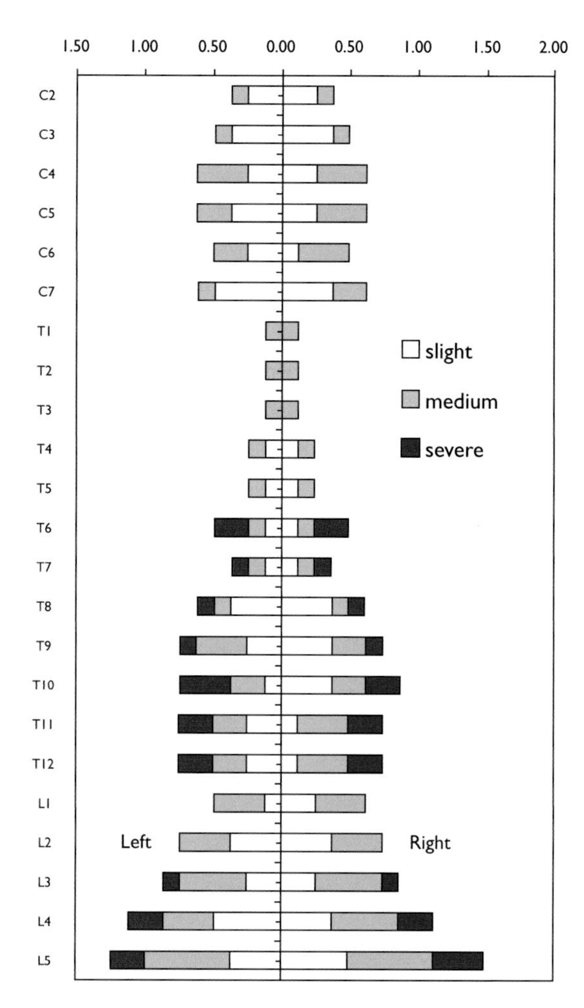

Figure 12.8 ***(left).*** **Degree of osteophytosis on the left and right sides of the superior vertebral bodies for adult individuals**

Figure 12.9 ***(right).*** **Degree of osteophytosis on the left and right sides of the inferior vertebral bodies for adult individuals**

of the lumbar vertebral bodies exhibit bilateral asymmetry. The left sides of the lumbar vertebrae display more severe osteophytosis than the right sides.

Older adults also have a higher frequency (53.3%) of osteoarthritic changes of the articular surface. The lumbar (69.2%) are the most common vertebrae affected, followed by the thoracic (58.0%) and cervical vertebrae (31.8%). Osteoarthritis in the older adults displays asymmetry between the left and the right sides, and the differences recorded are not in the presence, but rather in the different degrees, of osteoarthritis on the same vertebrae. It must be noted, though, that the sample of older adults is small and the above differences were not statistically significant.

Males (19.4% of adults, 42.0% of older adults) exhibit a slightly higher percentage of osteophytosis and osteoarthritis than females (17.9% of adults, 9.4% of older adults), and the distribution among the different vertebral types is not the same. Females display a progressive increase in both osteophytosis and osteoarthritis from the first thoracic to the fifth lumbar vertebra, while males show a low percentage of osteoarthritis on the lower lumbar vertebrae and higher percentages on the upper lumbar and upper thoracic vertebrae and high percentages of osteophytosis on the mid-cervical, the lower thoracic, and the lower lumbar vertebrae.

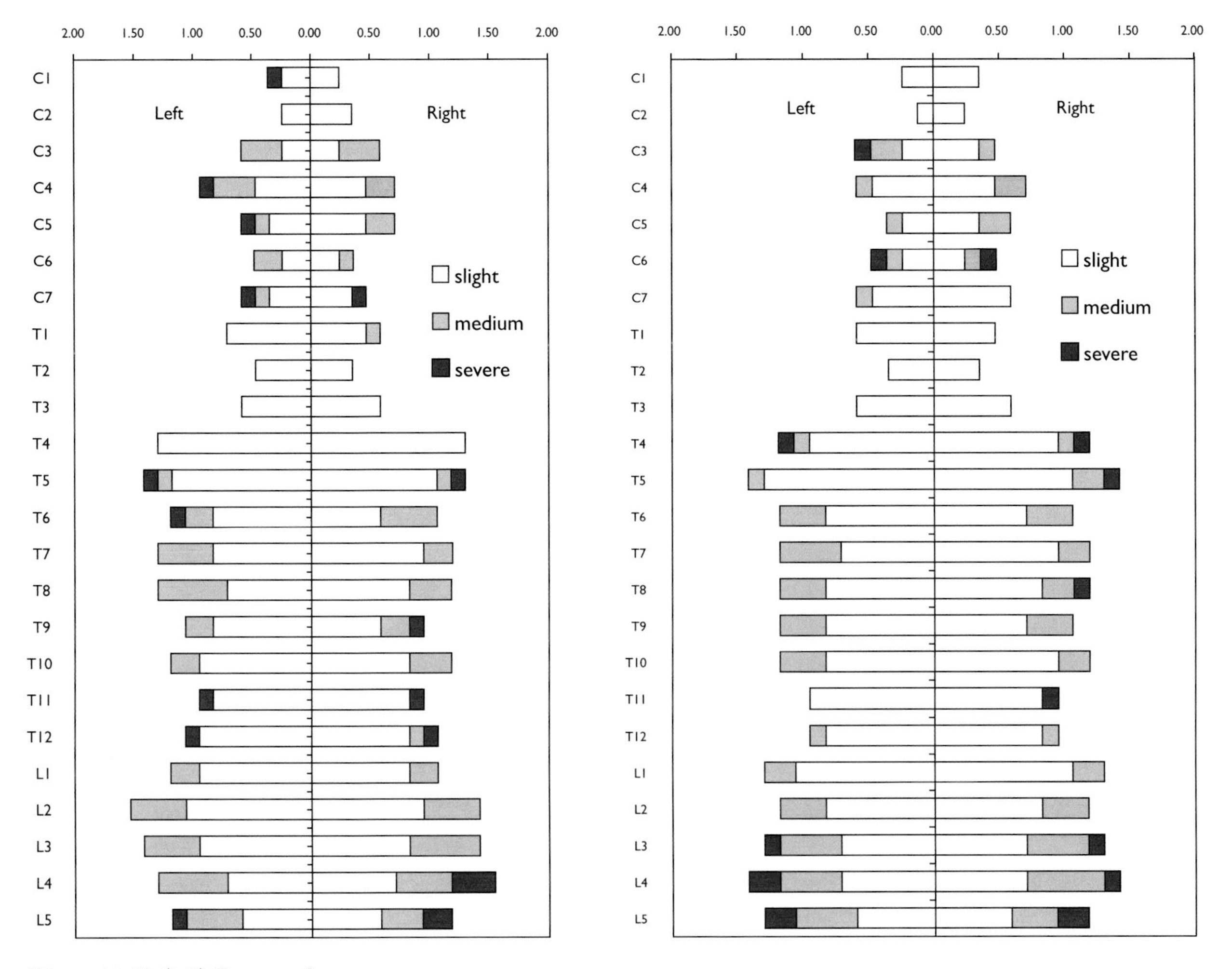

Figure 12.10 *(left)*. Degree of osteoarthritis on the superior articular surface of vertebrae for adult individuals

Figure 12.11 *(right)*. Degree of osteoarthritis on the inferior articular surface of vertebrae for adult individuals

The articulation of the odontoid process of the second cervical vertebra with the first cervical vertebra (*dens fovea*) was examined separately. Out of a sample of 116 vertebrae, 18.6% exhibited osteoarthritic lesions. Males (11.1%) are more often affected than females (2.9%), and no differences were observed in the frequencies between adults (7.5%) and older adults (7.7%).

Eighty three vertebrae exhibit at least one Schmorl's node lesion on the superior and/or inferior vertebral surface. The lesions are more common on the lower thoracic vertebrae (11.5%), while the lumbar (5.2%) and cervical (0.8%) vertebrae have lower frequencies. Males (10.3%) are more affected than females (0.6%). With respect to age, older adult individuals do not exhibit Schmorl's nodes, while adults (7.8%) and juveniles (7.0%) exhibit almost equal frequencies. Schmorl's depressions on subadults are relatively rare and may be attributed to traumas such as severe falls or heavy lifting. In a probably agricultural population, such as Korytiani, evidence of physical stress is not uncommon even from an early age.

Degenerative Joint Diseases of Long Bones

One-third (32.9%) of the individuals show osteoarthritic changes in at least one long bone. The femur is the most affected bone, followed by the

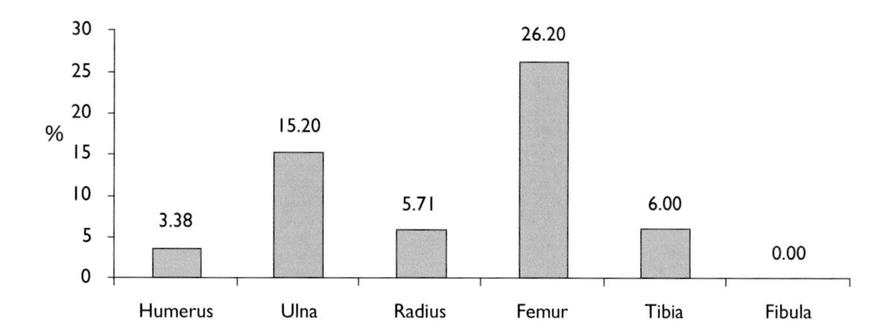

Figure 12.12. Osteoarthritis on long bones

Figure 12.13. Distal end of right femur showing osteoarthritic changes: marginal lipping, porosity (inferior view)

ulna, the tibia, the radius, and the humerus (Fig. 12.12). The knee joint (Fig. 12.13) is the most affected joint, followed by the right elbow, the hip, and the shoulder. All long bones besides the humerus exhibit slight, medium, and severe osteoarthritic changes but severe osteoarthritis is not very common. The upper limbs show higher frequencies of osteoarthritic lesions on the right side, while on the lower limbs no such pattern is observed (Fig. 12.14).

The osteoarthritic changes in this population are affected by age and physical stress. Although older adults (20.2%) exhibit the greatest amount of osteoarthritis, the presence of juveniles (Fig. 12.15) with arthritic changes indicates that specific physical activities or biological factors may have predisposed them to osteoarthritis. The appearance of higher percentages of severe osteoarthritis for adult rather than older adult individuals further supports the above hypothesis and probably reflects specific activities for a certain period of time.

Males (14.5%) are more affected by osteoarthritis of the long bones than are females (11.4%), but female individuals exhibit more cases of severe osteoarthritis. Males show severe osteoarthritic changes only on the lower limbs while females have them on both the lower and the upper limbs (Fig. 12.16). Males and females also exhibit differences in the distribution

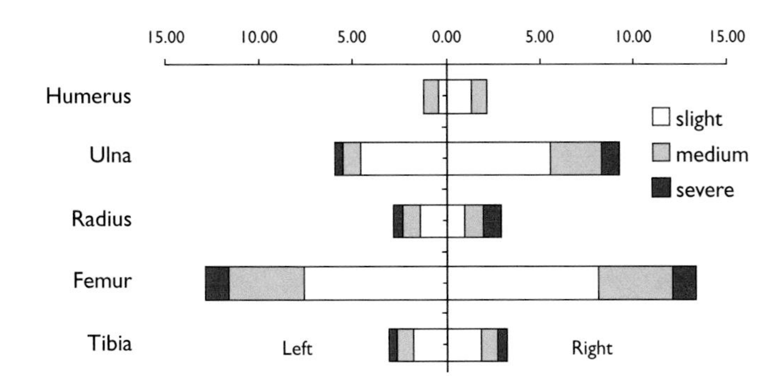

Figure 12.14. Degree of osteoarthritis on left and right long bones

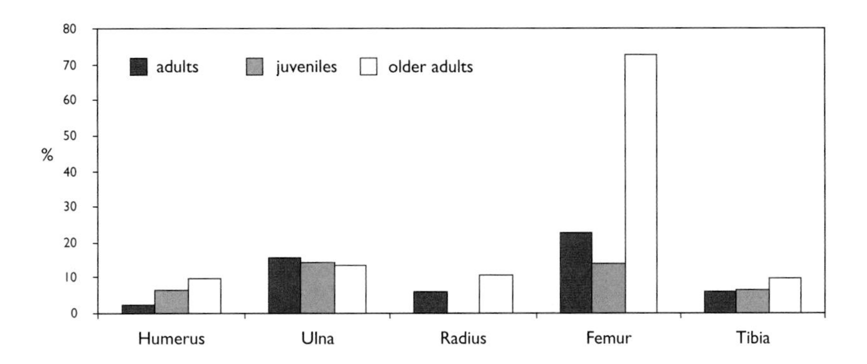

Figure 12.15. Osteoarthritis on long bones by age

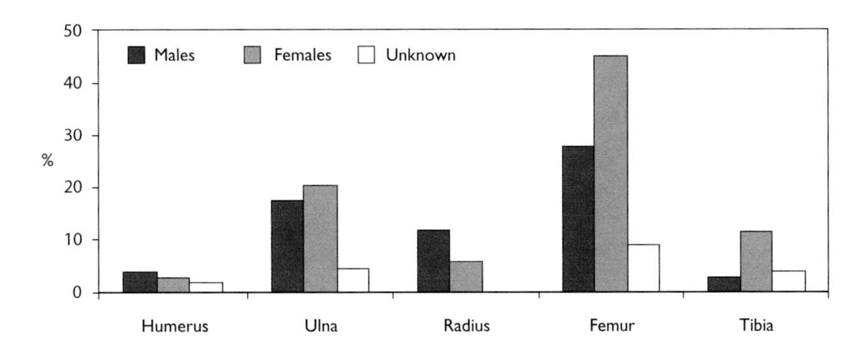

Figure 12.16. Osteoarthritis on long bones by sex

of osteoarthritis among the various peripheral joints. The most affected joint among long bones is the knee for males and the elbow for females. This pattern of sexual dimorphism was also observed for spinal arthritis and it may be attributed to the involvement of males and females in a division of physical labor that places different stresses on the skeleton—a pattern that persists even today in the rural populations of Greece.[48]

Infectious Diseases: Periostitis

Although no cases of osteomyelitis were observed at Korytiani, periostitic lesions (Fig. 12.17) were very common and affected 6.6% of the lower extremities in the sample. The most affected bone is the tibia (10.3%), followed by the fibula (6.6%) and the femur (3.3%). The upper extremities were not affected in this sample.

48. Efstratoglou 1998, pp. 27–54.

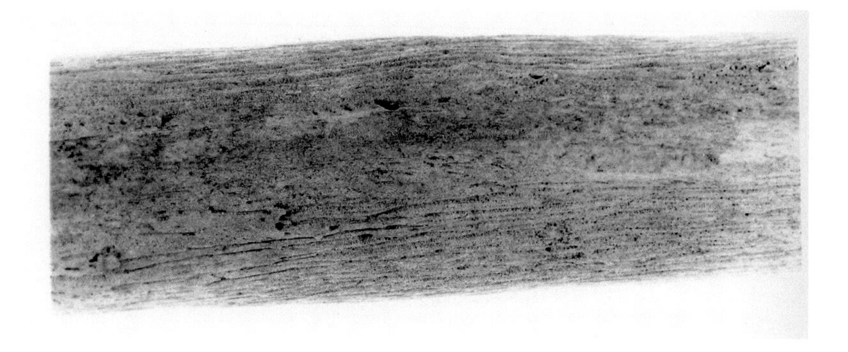

Figure 12.17. Periostitic lesion on tibia (detail)

Males (7.7%) exhibit a higher percentage of periostitis than females (4.3%). Juveniles (24.9%) are more affected than adults (19.9%) and older adults (15.1%). The relatively high incidence of periosteal changes for juveniles is noteworthy. Their periostitis has the form of woven bone, characteristic of unhealed periostitis; for the adults and older adults the skeletal tissue was incorporated into the normal cortical bone, and the surface of the bone in most cases had the smooth, undulating and inflated characteristics of the healed form. X-rays were made of all cases of periostitis. They show striations, nodes, new bone formation, and cortical thickening. No intramedullary inflammation (i.e., on the inside of the bone) was observed.

At present, the only diagnosis that can be made is primary periostitis, but the exact etiology (e.g., trauma of soft tissues, inflammation, or disease syndrome) is unknown. It could be assumed, for example, that such periosteal changes in the long bones of the lower extremities, and especially tibiae, is the result of treponematosis, but since no superficial cavities or intramedullary inflammations were recorded,[49] this diagnosis was excluded. Pyogenic osteomyelitis was excluded as well, because no drainage canals or intramedullary inflammations were noticed.[50]

Fractures

Four adult individuals, including three males and one of unknown sex, have fractured bones. The fractures occurred on three fibulae (one example of which is shown in Fig. 12.18) and one ulna. All three fibula fractures were located on the upper third of the right diaphysis, close to the proximal end.

From the four fractures, three (burials 9, 43, LII) were oblique and closed, and one (burial LX) was an oblique or transverse fracture. That it was probably open is indicated by the presence of active periostitis (infection of the bone) near the fracture area. An interesting case is the adult individual of unknown sex who exhibits an ununited fracture on the middle diaphysis of the left ulna. The distal end is missing (probably postmortem) and no signs of any pathological lesions were visible on the left radius.

The fibular fractures were in general well treated, aligned, and healed. No signs of angulation, malposition, and fusion of articular surfaces, or severe infections were recorded. Considering that a lower limb fracture needs immobility for successful healing, it can be assumed that all three individuals were immobilized for some time. Pathologists and paleo-

49. Elting and Starna 1984, pp. 267–275.

50. Ortner and Putschar 1981.

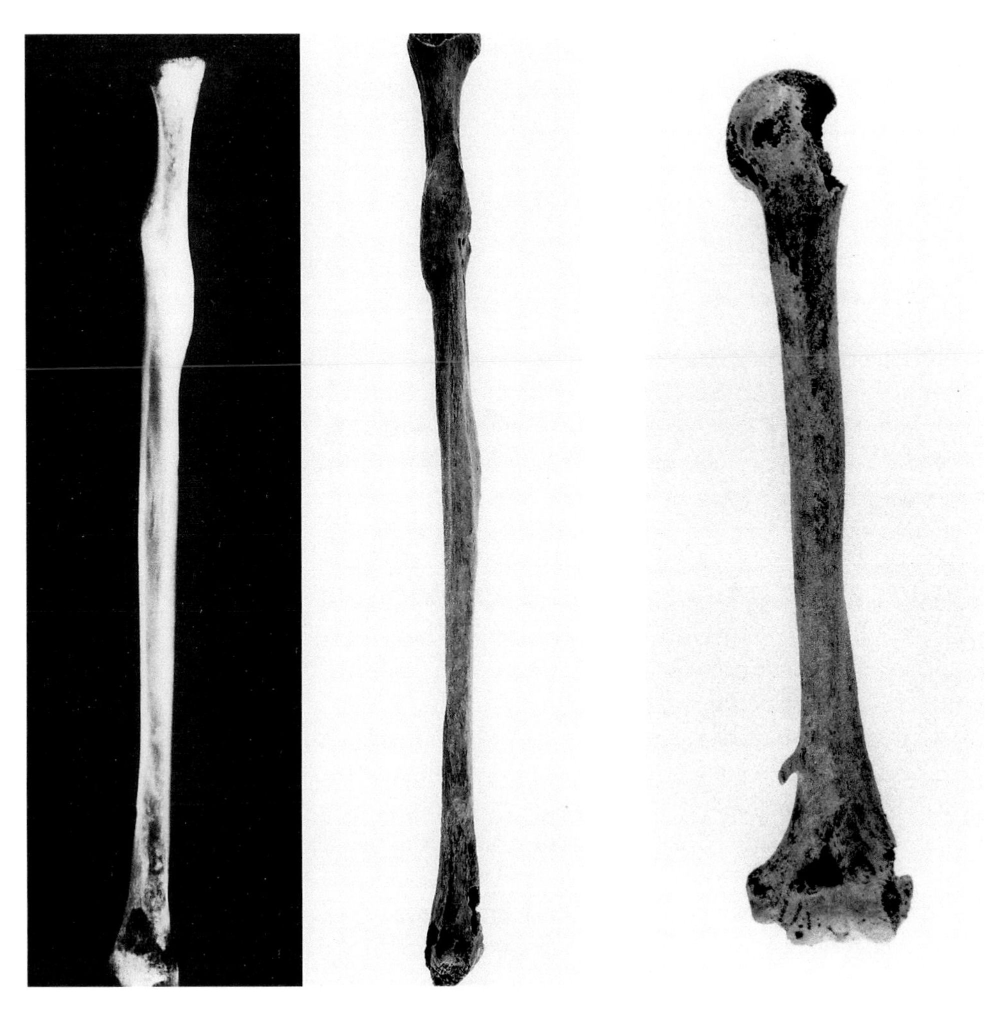

Figure 12.18. Fibular fracture, adult male (burial 43A). X-ray *(left)* and photograph *(center)*.

Figure 12.19 *(right)*. Supra condyloid process of humerus, adult male (burial 14)

pathologists suggest that there are three basic conditions for the successful healing of a fracture: repositioning and realignment of the broken ends, immobilization of the broken bone, and maintenance of circulation and muscle tone through careful exercise of the affected limb. It is probable that the necessary medical care was being offered to these individuals, and we can assume that the necessary medical knowledge was probably present in this medieval population.

Exostosis

The supracondyloid process or supratrochlear spur[51] is a small, roughly triangular and hook-shaped exostosis projecting 5–7 cm above the medial epicondyle of the humerus. Three cases were recorded: one on the left humerus of a male adult from burial 14 (Fig. 12.19), one on the right humerus of another male adult, and one on the right humerus of an adult individual of unknown sex. All three exostoses are situated on the medial epicondyle ridge 5 mm above the medial epicondyle. Only the individual in burial 14 preserved the complete projection, seen as a hook of 10 mm in length and 3 mm in width.

51. Mann and Murphy 1990.

A similar exostosis was recorded for the right scapula of a male adult. It is located 9.5 cm below the inferior glenoid tubercle and has a triangular form with a hook that is turned upward facing the glenoid cavity. The length of this exostosis is 9 mm, the width is 8 mm, and the breadth is 4 mm.

X-rays were taken in all the above cases, confirming that none of the exostosis examples was related to a pathological condition.

Neoplastic Conditions

A possible case of osteochondroma was recorded for the right radius of an adult male individual from burial 48B (Fig. 12.20). Osteochondroma is a form of benign bone tumor. It may occur on any bone that develops by endochondral ossification, and the most common location is the proximal or distal end of long bones, at or near an epiphyseal line. In this specific case, the bony growth is situated between the groove for the *extensor indicis* and *extensor digitorum* muscles and the dorsal tubercle exactly at the point of the groove for the *extensor pollicis longus* on the distal radius. The dorsal tubercle and the groove for the *extensor carpi radialis longus* are not clearly visible due to the abnormal bony growth. The right ulna of the same individual does not exhibit any pathological changes.

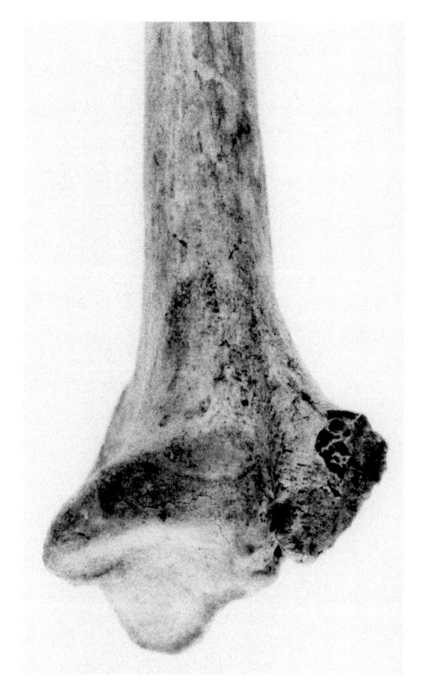

Figure 12.20. Abnormal bone growth on the distal right radius (dorsal aspect), adult male (burial 48B)

One case of osteoid osteoma was recorded on the left humerus of a male adult (burial XLVIA). The bone exhibits a cortical swelling at the point of the deltoid tuberosity that may be described as an oval bony growth with a length of 55 mm and a width of 12 mm. The structure of this newly formed bone is no different from the surrounding bone and the surrounding bone tissue does not show any changes such as cortical thickening or lytic lesions. The X-rays display a cortical thickening only at the point of the bony growth; the medullary cavity is normal, and no trauma or infection is observed in the rest of the bone.

Congenital Abnormalities

Spondylolysis is the separation of the neural arch of lumbar vertebrae that typically occurs at the *pars interarticularis* between the superior and inferior articular facets. The etiology of spondylolysis is not yet clear. It has been described either as a congenital malformation, or the result of repeated microtrauma to the *pars interarticularis*,[52] or as the result of both genetic and mechanical factors.[53] Recent findings have related the condition to erect posture and bipedal locomotion. Although the congenital nature of the defect is not excluded, it is most commonly attributed to fatigue fracturing and certain activities.[54]

We found three male individuals with spondylolysis of the fifth lumbar vertebra; the separation is in all cases bilateral observed on the *pars interarticularis*, one of the most common sites. No other signs of pathological condition have been recorded. Although it is suggested that the separation of the *pars interarticularis* is more likely a traumatic separation, which appears to be a common consequence of habitual physical stress,[55] initiated by an acute overload event that causes microfractures and exacerbated by repeated stress leading to nonunion of the microfractures, we hesitate to

52. Stewart 1953, pp. 937–950.
53. Bridges 1989, pp. 321–329.
54. Merbs 1996, pp. 201–228.
55. Merbs 1989, pp. 161–189; Merbs 1995, pp. 2328–2334; Merbs 1996, pp. 201–228.

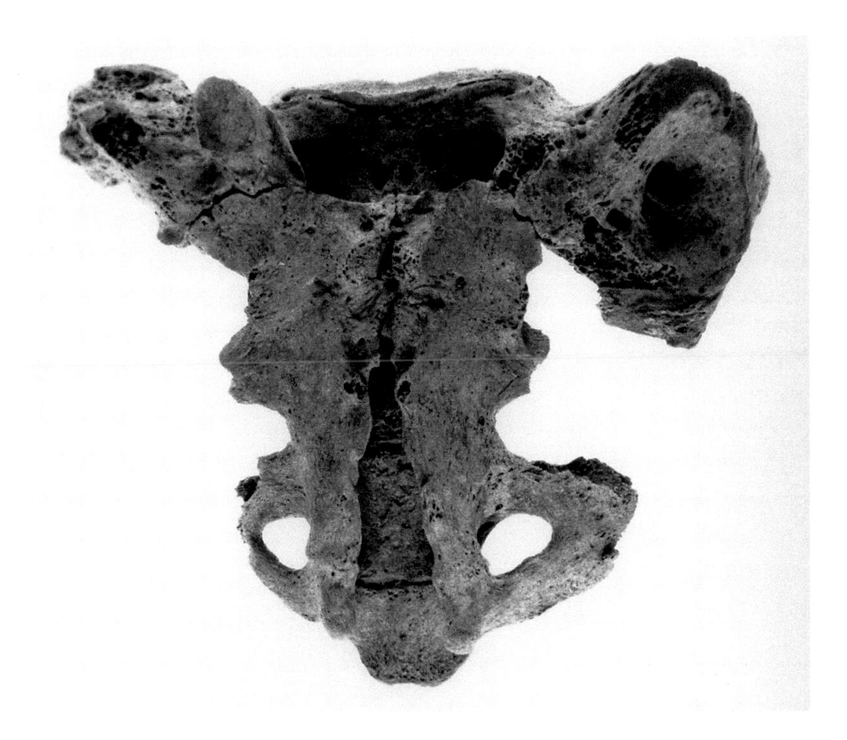

Figure 12.21. Spina bifida oculta, juvenile individual (burial 50)

advance any further interpretations because of the small number of cases present.

Spina bifida is characterized by the incomplete closure, fusion, or development of the neural arches of any vertebrae including the sacrum. A case of sacral spina bifida was observed on a juvenile individual from burial 50 (Fig. 12.21). In this case, the neural arch of the first sacral segment is divided at the midline of the spine and the arches of S2–S5 are not formed, leaving the sacral canal exposed.

Sacralization is the congenital condition that results in partial or complete fusion of the fifth lumbar vertebra to the first sacral vertebra. The vertebra can exhibit characteristics of a lumbar vertebra on one side and a sacral segment on the other. One case of sacralization was observed in Korytiani. An adult male (burial XVII) has six lumbar vertebrae, the sixth of which is fused with the sacrum.

TRACE ELEMENT ANALYSIS

Chemical research in bioanthropology involves the analysis of isotopes and trace elements. The applications of these analyses give us information about nutrition, weaning ages, paleopathology, and to some extent, patterns of migration.

The trace element method is based on the differential distribution of various elements within the sources of food used by human populations. In addition, anthropological interest extends to trace elements such as lead, mercury, and cadmium, which produce toxic effects in small doses. However, there were, and still are, many obstacles concerning this analysis, most of them relating to the problem of chemical diagenesis.

Geologists define diagenesis as the alteration of sediments after deposition. This term, adapted by some bioanthropologists, describes the postmortem alterations in the chemical constituents of bone following deposition in soil.[56] Diagenesis of bones is affected by intrinsic (density, microstructure, size, biochemistry of bone) and extrinsic (soil pH and texture, temperature, groundwater, microorganisms, pedogenic processes) factors.[57]

The soil of the Korytiani cemetery is calcareous clay with a high silica concentration and an alkaline pH. However, the pH and elemental concentration of each grave varied considerably. This variation affects the state of preservation of individual skeletons and consequently the effect of diagenesis. The location of the cemetery upon a mound encourages these differences; only by macroscopic inspection is it obvious that the skeletons buried on the east section (top of the mound) are better preserved than the skeletons of the west section (slope and foot of the mound).

In order to evaluate the extent of diagenesis in the skeletons, we examined the following:

1. The calcium/phosphorus ratio (Ca/P).
2. The differential elemental concentrations among individuals with various states of preservation, as determined by surface weathering and degree of fragmentation.
3. The differential elemental concentrations of skeletons in the west and east sections of the cemetery.
4. The soil elemental concentrations within and outside of the graveyard.

Skeletal calcium phosphate, commonly known as hydroxyapatite ($Ca_{10}(PO_4)_6(OH)_2$), contains about 18.5% P and 39.9% Ca. The expected Ca/P ratio of intact bone is 2.14. Diagenesis can be seen as a correction of this deviation.[58] The stoichiometric variations in the hydroxyapatite formula are slightly affected by diet[59] and potential deviations are due largely to the phenomena of heteroionic exchange, whereby various ions are incorporated into the bone by displacing its normal chemical constituents.

The Ca/P ratio (2.54, as shown in Table 12.4) in our sample reveals good integrity of the mineral matrix, but it is elevated in comparison to the stoichiometric ratio and to the Ca/P ratio from modern humans.[60] A Ca/P ratio could be raised by the presence of calcite, if CO_3 ions substitute for PO_4 ions. For our sample, the replacement of phosphate is also possible, because the phosphorus concentration is lower than the physiological levels. This suggests that diagenic processes are likely to have taken place because an elevated Ca/P helps diagenesis through the capacity of the CO_3 to increase solubility because of its weakening effect on the bonds in the apatite structure.[61]

The comparisons between the individuals in various states of preservation from different locations inside the cemetery, and analysis of soil concentration within and outside the cemetery showed statistically significant differences for certain elements. There is postmortem accumulation of Si and Al in all samples, and of Fe and Cu in some samples. For Mn, Cr, V, Ni, Li, Rb, and Co, we observed normal concentrations and slightly elevated

56. Lyman 1994.
57. von Endt and Ortner 1984, 247–253.
58. Sillen 1989, pp. 211–229.
59. Lambert and Weydert-Homeyer 1993, pp. 279–294.
60. Zachick and Tzaphlidou 2001, pp. 1090–1095.
61. Sillen 1989, pp. 211–229.

TABLE 12.4. MINERAL RATIOS FOR THE BONE SAMPLES (N = 116)

	Ca/P (mg/g)	*Sr/Ca (ppm)*	*Ba/Ca (ppm)*	*Zn/Ca (ppm)*
Mean	2.54	1.23	1.04	0.42
Median	2.57	1.22	1.03	0.41
Std. Dev.	0.19	0.22	0.21	0.22
Min.	1.16	0.59	0.38	0.59
Max.	2.85	2.17	1.77	2.17

concentrations in the less preserved bones. Because the differences between the physiological concentrations in human bones and the concentrations in this sample are very small, it is not clear whether those differences are due to postmortem contamination or intraindividual variation. For the rest of the elements, no evidence was found for postmortem accumulation or leaching of the bone.

For the dietary reconstruction, we examined the levels of strontium, barium, and zinc. Although Fe, Cu, Mg, and Mn are also used by some researchers[62] as diet markers, these were not included in our dietary analysis because of probable postburial alteration of their concentrations. Li, Rb, V, Ni, Ti, Co, Cs, Cr, and Mo exhibit almost normal concentrations compared with modern bones, but for these elements no connection has yet been made between their presence in humans and diet.

Measurements of skeletal strontium and Sr/Ca ratios have long been used for providing information on the proportion of meat and vegetable foods in the diets of ancient populations. This information is based on the well-documented reduction of Sr/Ca ratios in terrestrial food chains.[63] Mammals discriminate physiologically against strontium, resulting in the decrease of strontium levels in successively higher levels of the food chain from plants to herbivores to carnivores.[64] Therefore, the closer the value of the ratio comes to approaching 1, the higher the amount of vegetable food consumed by the individuals during their lifetime.[65] However, it has been suggested lately that the Sr/Ca for mixed diets is not linearly related to the plant/meat ratio and is more probably related to patterns of consumption of different plants (for example, browsing versus grazing), rather than the amount of meat in the diet.[66] Strontium is also used for demonstrating consumption of marine resources, although studies that have suggested this relationship were conducted on populations highly dependent on marine foods and it has not yet been investigated how the absence of marine foods might influence the Sr/Ca ratios of populations that are less or not at all dependent on them,[67] such as is the case at Korytiani.

The Korytiani population exhibits relatively high concentrations of Sr. The mean value is 463 ppm (as shown in Table 12.5), the minimum value is 230 ppm, and the maximum is 650 ppm. The Sr/Ca ratio (Table 12.4) is higher than 1 and does not vary much among the individuals.[68] It is probable that high strontium values reflect a substantial consumption of plants, although exact species definition is not possible.

62. Ezzo 1994a, pp. 1–34; Ezzo 1994b, pp. 606–621; Gilbert 1977, pp. 85–100; Sandford 1992, pp. 79–103; Sandford 1993, pp. 3–57.

63. Sillen and Kavanagh 1982, pp. 67–90.

64. Elias, Hirao, and Patterson 1982, pp. 2561–2580.

65. Schoeninger 1982, pp. 37–52.

66. Burton and Price 2000, pp. 159–171.

67. Sillen and Kavanagh 1982, pp. 67–90.

68. The mean Sr soil concentration from the area inside the cemetery is 137.5 ppm, while from the nearby region is 190 ppm.

TABLE 12.5. ELEMENT CONCENTRATIONS FOR THE BONE SAMPLES (*N* = 116)

	Mean	*Median*	*Std. Dev.*	*Range*	*Min.*	*Max.*
Ba	394.75	390	79.43	530	150	680
Sr	463.28	460	78.96	420	230	650
Zn	158.76	157.5	34.95	280	80	360
Mg	897.67	900	198.48	1,200	300	1,500
Mn	6.93	5	6.49	39	1	40
Fe	490.46	400	387.47	1,480	20	1,500
Cu	54.28	40	45.66	295	5	300
Pb	5.52	6	2.97	17	0	17
Co	0.73	0.65	0.30	1.7	0.2	1.9
Cd	0.16	0.2	0.15	1.5	0	1.5
Ti	7.93	7	4.79	37	2	39
Cr	11.75	10	6.12	42	3	45
Cs	9.77	9	2.69	16	0	16
V	11.79	12	2.51	18	4	22
Ni	9.41	9	4.23	24	3	27
Li	2.62	2.6	0.70	4	1	5
Rb	6.36	5.2	2.54	17.6	3.4	21
Mo	4.19	4	1.27	11	2	13
Ag	0	0	0	0	0	0
Ca%	37.89	38	2.11	25.2	18.4	43.6
Si	1,463.9	1,100	1,142.18	9,000	200	9,200
Al	571.27	420	469.57	4,000	100	4,100
K	136.35	110	85.57	450	40	490
Na	2,702.3	2,700	317.62	1,700	1,800	3,500
P%	14.99	15	0.89	8	13.4	21.4

Barium, like strontium, is chemically similar to calcium and enters bone in proportion to dietary levels. The intestinal epithelium discriminates against the absorption of barium even more strongly than it does against strontium.[69] Barium appears to be a useful indicator for separating marine from terrestrial diet because it is not abundant in seawater and consequently in marine resources.[70]

In this study, we found that Ba shows high values; aside from one individual that had Ba levels of 150 ppm, all individuals showed a mean of 395 ppm. The Ba/Ca ratio is as high as the Sr/Ca ratio. Barium values suggest no consumption of marine food, verifying our hypothesis of an absence of marine resources in the everyday diet—not inconsistent with the inland location of the settlement at Korytiani.

The use of zinc concentrations as indicators of past diets[71] and/or health conditions has become accepted in bone chemistry analysis despite the fact that a direct correlation between diet and zinc concentrations in human bones has not yet been fully proven. However, if Zn and Zn/Ca levels are combined with Sr/Ca and Ba/Ca values, this could give useful information about diet and health.[72] Zinc is largely derived from edible animal tissues. Milk and milk products release Zn from its mineral matrix in times of demand or undersupply. In addition, the absorption rate of Zn is inhibited by the presence of phytate cellulose, hemicellulose, and fibers.

69. Elias, Hirao, and Patterson 1982, pp. 2561–2580.

70. Burton and Price 1990, pp. 547–557; Burton and Price 1991, pp. 787–795; Gilbert, Sealy, and Sillen 1994, pp. 173–184.

71. Gilbert 1975.

72. Klepinger 1984, pp. 75–96.

These substances are commonly present in certain cereal proteins and legumes, so that diets rich in any of these antagonists ought to promote deficiency of this essential nutrient.

In our samples, Zn values are moderate, with a mean of 159 ppm (Table 12.5), a minimum of 80 ppm, and a maximum of 360 ppm. The Zn/Ca ratio (0.42 as shown in Table 12.4) is much lower than that of strontium and barium, and also the variability among the individuals is smaller in comparison to that of Ba and Sr. It could be assumed that the absorption rate for Zn in the intestines was inhibited by the presence of phytate cellulose and hemicellulose in the diet, as is suggested by the high strontium levels.

The data concerning elemental concentrations were grouped according to age and sex. All elemental concentrations and ratios besides Sr, Ba, Mg, Ni, and Ca/P showed significant deviations from the normal distribution. Therefore, for the comparisons between age and sex groups we used nonparametric tests (Kruskal-Wallis) to identify significant differences. In the case of the Sr, Ba, Mg, and Ni comparisons, we employed a one-way ANOVA statistical test.

No differences were recorded between male and females for the Ba/Ca, Sr/Ca, and Zn/Ca ratios. The comparisons indicate that the elemental differences between males and females are not significant, with the exception of the Mn, Cr, V, Ni, Si, and Al levels. As mentioned above, these elements are probably subject to postdepositional alterations and those differences probably are due to intra-individual diagenesis and not to dietary or occupational differences. The one-way ANOVA tests for Sr, Ba, Mg, and Ni did not exhibit statistically significant differences between males and females.

The data were also grouped according to age (adult and immature groups). Statistically significant differences were found only for Na and Pb. Lead levels have a higher mean concentration in the immature individuals. This is surprising, considering the fact that lead concentration is additive over the lifecourse. Thus, it would be expected that adults would exhibit higher concentrations of lead than immature individuals.[73] On the other hand, contemporary research on modern populations highlights the vulnerability of children to lead poisoning.[74]

Paleopathological Analysis Clarified Through Trace Element Analysis

The only pathological case related to trace element concentrations is an example of toxic levels in a probable adult female (burial 9) who exhibits 180 ppm of lead. Postburial contamination has been excluded because:

1. the individual was buried in a grave together with three other individuals, who exhibit normal lead values (7, 0, and 10 ppm, respectively);
2. all examined individuals (N = 115) exhibit expected lead levels with a mean of 5.52 ppm (Table 12.5) and none shows signs of postburial accumulation of lead;
3. two soil samples from inside the grave show 14 ppm of lead;

73. The small sample size for immature individuals may influence these results.

74. http://www.keepkidshealthy.com/welcome/lead/leadfacts.html (accessed August 17, 2007)

4. the lead value for soil inside the graveyard is 21 ppm and the value for the soil around the graveyard is 17 ppm;
5. no grave goods were found inside the grave, such as bronze and iron jewelry or clay and glass pots, that might have decomposed and contaminated the skeleton; and
6. after the first analysis yielded high lead levels, a second sample (x = 185 ppm) from another part of the femur was extracted to ensure that the first value was not a result of contamination in the lab or a measurement error.

As we have no reason to suspect that only this individual was contaminated postmortem (and it is unlikely that only one individual would have been contaminated), we suggest that the high 180 ppm value is the individual's lead burden during life. Most of the medical literature relates symptoms of lead toxicity with the level of lead in the blood or urine. We converted the bone lead levels into blood lead levels, using a formula[75] that demonstrates a relationship between skeletal lead and blood lead levels if the latter are related to years of exposure.[76]

The derived blood lead value of 80.1 μg/100 ml suggests that the individual probably suffered from moderate symptoms of lead intoxication such as abdominal pain, headache, joint pain, fatigue, anemia, neuropathy, irritability, and deficits in short-term memory and the ability to concentrate.[77] The individual also exhibits many pathological lesions:

1. A large cyst in the left maxillary sinus. The bone around the cyst is highly porotic and the cyst is not related to a periapical tooth abscess.
2. The right mandibular body, posterior to the second molar, is missing; the area exhibits a smooth rounded surface with signs of healing that may indicate another cyst in this area of the jaw (Fig. 12.22).
3. The fifth, sixth, and seventh cervical vertebrae and the first thoracic, and probably the second thoracic, vertebrae are ankylosed. The ankylosis occurs mainly between the vertebral arches and not between the vertebral bodies. There are no signs of osteoarthritic changes.
4. All preserved vertebrae are highly porous. On the second thoracic vertebra, in addition to the low density of the trabecular bone, there is a large cyst on the upper surface of the body.
5. The sternal epiphysis of the right clavicle has active periostitis.
6. The proximal epiphyses of both humeri exhibit lesions in the form of cysts with irregular borders that are distinct from the common form of osteoarthritic porosity.
7. The right pubic bone shows a periosteal reaction on the dorsal aspect and irregular loss of bone on the ventral surface. The right acetabulum exhibits a moderate degree of osteoarthritis.
8. In general, we noted a reduction in total bone volume caused by thinning of both the cortical and trabecular bone of the long bones. It is probable that besides the various local lesions the individual was suffering from severe osteoporosis.

75. Tibia Pb (μg/g) = 0.03 × blood Pb (μg/100ml) × years of exposure - 0.09 (after Corruccini et al. 1987, pp. 233–239).

76. Corruccini et al. 1987, pp. 233–239.

77. http://www.keepkidshealthy.com/welcome/lead/leadsymptoms.html (accessed August 17, 2007)

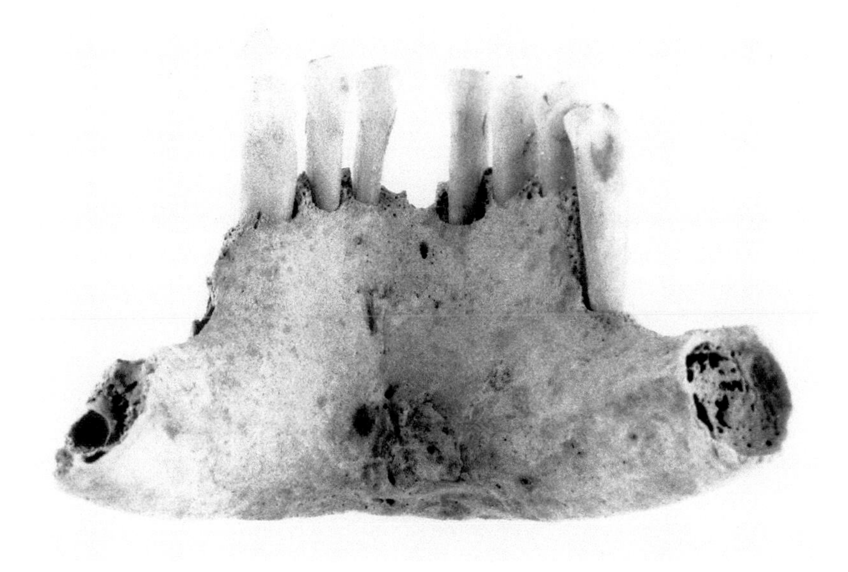

Figure 12.22. Mandible exhibiting a smooth rounded surface with signs of healing at the position of the right second molar, adult female (burial 9)

It is not certain whether the specific pathologic lesions have any connection with the high lead level. In the medical literature, lead poisoning effects are restricted to soft tissues.[78] The only connection between bone reaction and lead toxicity has recently been made by some new studies connecting osteoporosis and high lead levels.[79]

How did the individual get exposed to high lead levels? We can probably exclude occupational exposure because it is rather unusual for a woman to be working with metals. A possible explanation is that all of the lesions are connected with a kind of euplastic disease, and that the individual was treated with medicines containing lead. The complete diagnosis and medical evaluation of this individual is very difficult and further research is needed.

DISCUSSION

The analysis of the dental pathological lesions from Korytiani indicates an economy that was to a large extent agricultural: slight-to-moderate attrition, a high caries rate, and severe antemortem tooth loss, as well as slight-to-medium calculus deposits, suggest a diet of protein, starch, and carbohydrates. The high rate of caries, especially at the CEJ, provides support for the hypothesis of high carbohydrate consumption together with fruits and vegetables that tend to stagnate and decompose between the teeth. The slight to moderate, rather than severe, accumulation of calculus suggests higher grain and vegetable consumption relative to protein consumption.[80]

The above conclusions are in agreement with the results of the trace element analysis. For this medieval population, the high Sr/Ca, the high Ba/Ca, and the low Zn/Ca ratios should correspond to a diet rich in foods of plant origin and low in marine resources. Considering the available spectrum of food sources, an abundant supply of cereal grains and vegetables

78. The only macroscopically present sign for plumbism is the Burton's line, a blue-purplish line about the one-twentieth part of an inch in width, seen only on the gums, and used as a diagnostic feature only for clinical studies because no soft tissues are normally preserved in archaeological skeletal material, Pearce 2007; Nogue and Culla 2006.

79. Hu, Aro, and Rotnitzky 1995, pp. 105–110; Latorre et al. 2003, pp. 631–636.

80. Littleton and Frohlich 1993, pp. 427–447.

appears most probable. In addition, because phytates in some grains inhibit zinc absorption, the bones of wheat consumers should incorporate less zinc, a probable interpretation for the low Zn/Ca ratio in our sample. However, the consumption of meat and dairy products cannot be excluded as Zn is not completely absent and no severe nutritional or other kind of stress was indicated by the low prevalence of linear enamel hypoplasias and anemias. In the end, we should note that even plants themselves vary in their mineral content and, correspondingly, in their ability to affect dietary Sr/Ca;[81] so whether we define in our population the low or high consumption of meat or the high or low consumption of plants rich or poor in strontium levels remains unclear and only rough speculations are possible. Taking into account the ecological characteristics of the Korytiani region, which presents favorable conditions for agricultural activities, we assume that the medieval diet was multifaceted and contained varying amounts of food derived mainly from farming and horticulture and secondarily from livestock. Because of the geographical location of the inland settlement, it is unlikely that seafood contributed to the daily diet; therefore, we attribute the high strontium ratios more to plant consumption than to marine sources. This interpretation is further supported by the high barium levels.

The results of the paleopathological analysis show that the population had a relatively high frequency of degenerative spinal disease, medium levels of osteoarthritis and infectious diseases of the long bones, low rates of fractures and neoplasms, and no signs of metabolic diseases or anemia other than a few individuals with linear enamel hypoplasias. The distribution of osteophytosis and osteoarthritis along the spine and osteoarthritis of the long bones does not show a unique pattern that can be easily attributed to specific activities. It can instead be attributed to common biological factors, such as aging, and habitual physical activities. The presence of spinal arthritis on three juveniles and many young adults indicates that individuals experienced substantial mechanical stress early in life and consequently reflects intense labor from a young age.[82] The sexually dimorphic differences in the pattern of osteophytosis and osteoarthritis and changes among the different joints may indicate that males and females were not engaged in the same types of physical labor. In addition, the higher percentage of osteoarthritis in males probably suggests the more rigorous physical labor in which men were engaged.

In addition to typical osteoarthritic changes in the spine and long bones, this population exhibits some other skeletal characteristics recognized as skeletal markers of occupational stress.[83] These markers, in combination with the osteoarthritic changes, may suggest that many individuals at Korytiani were exposed to extracorporeal and external forces that were not only attributable to disorders like metabolic and hormonal imbalances, but also to physical stresses from routine everyday activities. Activities such as carrying heavy burdens involve supination and hyperextension of the arm, slinging and pitching, hyperflexion of the hip and knee with both hyperdorsiflexion of the foot at the tibiotalar and subtalar joints, and dorsiflexion of the foot at the tibiotalar joint. This kind of physical activity is very common in rural agricultural societies; the intraindividual differences documented here probably indicate that some individuals were engaged in more rigorous physical labor than other individuals from the same community.

81. Burton and Price 2000, pp. 159–171.

82. Roberts and Manchester 1995, p. 109.

83. Schmorl's disk herniation, ossification of the costal cartilage of the manubrium, enthesophyte formations on the fibula and the tibia, accessory sacroiliac facets, prolongation of the external side of the trochlear articular surface of the talus and irregular ossification of the *os trigonum* on the talus. The analysis of the stress markers is not included in the present study.

In general, the anthropological analysis of this skeletal sample suggests an agricultural population following a low-protein diet based primarily on carbohydrates and vegetables. The population was probably exposed to a certain level of physical stress during routine agricultural activities and the individuals show low fracture and infectious disease rates. It can be suggested that anthropological data presented above confirm the historical data, according to which the broader region of Thesprotia underwent a cultural decline during the 10th century A.D.[84]

84. C. Pitsakis (pers. comm.).

REFERENCES

Alt, K. W., F. Rosing, and M. Teschler-Nicola, eds. 1998. *Dental Anthropology, Fundamentals, Limits and Prospects,* Vienna.

Amos, M. D., P. A. Bennett, J. P. Matousek, C. R. Parker, E. Rothery, C. J. Rowe, and J. B. Sanders. 1975. *Absorption Spectroscopy: A Modern Introduction,* Victoria.

Aufderheide, A. C., and C. Rodríguez-Martín. 1998. *The Cambridge Encyclopedia of Human Paleopathology,* Cambridge.

Bridges, P. S. 1989. "Spondylolysis and its Relationship to Degenerative Joint Disease in the Prehistoric Southeastern United States," *American Journal of Physical Anthropology* 79, pp. 321–329.

Brooks, S. T., and J. M. Suchey. 1990. "Skeletal Age Determination Based on the *Os Pubis*: A Comparison of the Acsadi-Nemeskeri and Suchey-Brooks Methods," *Human Evolution* 5, pp. 227–238.

Brothwell, D. R. 1981. *Digging Up Bones,* 3rd ed., Ithaca, N.Y.

Burton, J. H., and T. D. Price. 1990. "The Ratio of Barium to Strontium as a Paleodietary Indicator of Consumption of Marine Resources," *JAS* 17, pp. 547–557.

———. 1991. "Paleodietary Applications of Barium in Bone," in *Proceedings of the 27th International Symposium on Archaeometry, Heidelberg, April 2–6, 1990,* ed. E. Pernicka and G. A. Wagner, Basel, pp. 787–795.

———. 2000. "The Use and Abuse of Trace Elements for Paleodietary Research," in *Biogeochemical Approaches to Paleodietary Analysis,* ed. S. H. Ambrose and M. A. Katzenberg, New York, pp. 159–171.

Corruccini, R. S., A. C. Aufderheide, J. S. Handler, and L. E. Wittmers. 1987. "Patterning of Skeletal Lead Content in Barbados Slaves," *Archaeometry* 29, pp. 233–239.

Davies, P. L., D. C. A. Picton, and A. G. Alexander. 1969. "An Objective Method of Assessing the Periodontal Condition in Human Skulls," *Journal of Periodontal Research* 4, pp. 74–77.

Efstratoglou, S. 1998. "Fthiotis (Greece)," in *Labour Situation and Strategies of Farm Women in Diversified Rural Areas of Europe, Research-Programme DEMETRA, Final Report,* Brussels, pp. 27–54.

Elias, R. W., Y. Hirao, and C. C. Patterson. 1982. "The Circumvention of the Natural Biopurification of Calcium Nutrient Pathways by Atmosphere Inputs of Industrial Lead," *Geochimica et Cosmochimica Acta* 46, pp. 2561–2580.

Elting, J., and W. Starna. 1984. "A Possible Case of Pre-Columbian Treponematosis from New York State," *American Journal of Physical Anthropology* 65, pp. 267–275.

Ezzo, J. A. 1994a. "Putting the Chemistry Back into Archaeological Bone Chemistry Analysis: Modeling Potential Paleodietary Indicators," *JAnthArch* 13, pp. 1–34.

———. 1994b. "Zinc as a Paleodietary Indicator: An Issue of Theoretical Validity in Bone-Chemistry Analysis," *AmerAnt* 59, pp. 606–621.

Gilbert, R. I. 1975. "Trace Element Analyses of Three Skeletal Amerindian Populations at Dickson Mounds" (diss. Univ. of Massachusetts).

———. 1977. "Applications of Trace Element Research to Problems in Archaeology," in *Biocultural Adaptation in Prehistoric America,* ed. R. L. Blakely, Athens, Ga., pp. 85–100.

Gilbert, C., J. Sealy, and A. Sillen. 1994. "An Investigation of Barium, Calcium, and Strontium as Paleodietary Indicators in the Southwestern Cape, South Africa," *JAS* 21, pp. 173–184.

Goodman, A. H., and G. J. Armelagos. 1989. "Infant and Childhood Morbidity and Mortality Risks in Archaeological Populations," *WorldArch* 21, pp. 225–243

Goodman, A. H., and J. C. Rose. 1991. "Dental Enamel Hypoplasias as Indicators of Nutritional Status," in *Advances in Dental Anthropology,* ed. M. Kelley and C. S. Larsen, New York, pp. 279–294.

Gregory, T. E. 2005. *A History of Byzantium*, Oxford.

Hillson, S. W. 1979. "Diet and Dental Disease," *WorldArch* 11, pp. 147–162.

———. 1996. *Dental Anthropology*, Cambridge.

Hu, H., A. Aro, and A. Rotnitzky. 1995. "Bone Lead Measured by X-ray Fluorescence: Epidemiological Methods," *Environmental Health Perspectives* 103, pp. 105–110.

Klepinger, L. L. 1984. "Nutritional Assessment from Bone," *Annual Review of Anthropology* 13, pp. 75–96.

Lambert J. B., and J. M. Weydert-Homeyer. 1993. "The Fundamental Relationship between Ancient Diet and the Inorganic Constituents of Bone as Derived from Feeding Experiments," *Archaeometry* 35, pp. 279–294.

Latorre, F. G., M. Hernandez-Avila, J. T. Orozco, C. A. Albores Medina, A. Aro, E. Palazuelos, and H. Hu. 2003. "Relationship of Blood and Bone Lead to Menopause and Bone Mineral Density among Middle-Age Women in Mexico City," *Environmental Health Perspectives* 111, pp. 631–636.

Leigh, R. W. 1925. "Dental Pathology of Indian Tribes of Varied Environmental and Food Conditions," *American Journal of Physical Anthropology* 8, pp. 179–199.

Littleton, J., and B. Frohlich. 1993. "Fish-Eaters and Farmers: Dental Pathology in the Arabian Gulf," *American Journal of Physical Anthropology* 92, pp. 427–447.

Lyman, R. L. 1994. *Vertebrate Taphonomy*, Cambridge.

Magdalino, P. 2002. "The Medieval Empire 780–1204," in *The Oxford History of Byzantium,* ed. C. Mango, Oxford, pp. 169–213.

Mann, R. W., and S. P. Murphy. 1990. *Regional Atlas of Bone Disease, a Guide to Pathologic and Normal Variation in Human Skeleton*, Springfield, Ill.

Meindl, R. S., and C. O. Lovejoy 1985. "Ectocranial Suture Closure: A Revised Method for the Determination of Skeletal Age at Death Based on the Lateral-Anterior Sutures," *American Journal of Physical Anthropology* 68, pp. 57–66.

———. 1989. "Age Changes in the Pelvis: Implications for Paleodemography," in *Age Markers in the Human Skeleton*, ed. M. Y. İşcan, Springfield, Ill., pp. 137–168.

Merbs, C. F. 1989. "Trauma," in *Reconstruction of Life from the Skeleton*, ed. M. Y. İşcan and K. A. R. Kennedy, New York, pp. 161–189.

———. 1995. "Incomplete Spondylolysis and Healing: A Study of Ancient Canadian Eskimo Skeletons," *Spine* 20, pp. 2328–2334.

———. 1996. "Spondylolysis and Spondylolisthesis: A Cost of Being an Erect Biped or a Clever Adaptation?" *Yearbook of Physical Anthropology* 39, pp. 201–228.

Miles, A. E. 1962. "Assessment of the Ages of a Population of Anglo-Saxons from Their Dentitions," *Proceedings of the Royal Society of Medicine* 55, pp. 881–886.

Molnar, S. 1971. "Human Tooth Wear, Tooth Function and Cultural Variability," *American Journal of Physical Anthropology* 34, pp. 175–190.

Molnar, S., and J. K. McKee. 1988. "Measurements of Tooth Wear among Australian Aborigines: II. Intrapopulational Variation in Patterns of Dental Attrition," *American Journal of Physical Anthropology* 76, pp. 125–136.

Molnar, S., J. K. McKee, and I. Molnar. 1983a. "Measurements of Tooth Wear among Australian Aborigines: I. Serial Loss of the Enamel Crown," *American Journal of Physical Anthropology* 61, pp. 51–65.

———. 1983b. "Tooth Wear Rates among Contemporary Australian Aborigines," *American Journal of Physical Anthropology* 62, pp. 562–565.

Murphy, T. 1959. "The Changing Pattern of Dentine Exposure in Human Molar Tooth Attrition," *American Journal of Physical Anthropology* 17, pp. 167–178.

Nogue, S., and A. Culla. 2006. "Burton's Line," *New England Journal of Medicine* 354, p. 21.

Ortner, D. J., and G. J. Putschar. 1981. *Identification of Pathological Conditions in Human Skeletal Remains*, Washington, D.C.

Papageorgopoulou, C. 2002. "Anthropological Research on Skeletal Material from a Byzantine Cemetery in Epirus Paleodontological, Paleopathological and Chemical Analysis as a Tool for Quantifying Diet and Health" (diss., Univ. of Florence).

Pearce, J. M. S. 2007. "Burton's Line in Lead Poisoning," *European Neurology* 57, pp. 118–119.

Richards, L. C. 1984. "Principal Axis Analysis of Dental Attrition Data from Two Australian Aboriginal Populations," *American Journal of Physical Anthropology* 65, pp. 5–13.

Roberts, C., and K. Manchester. 1995. *The Archaeology of Disease*, 2nd ed., Ithaca, N.Y.

Sandford, M. K. 1992. "A Reconsideration of Trace Element Analysis in Prehistoric Bone," in *Skeletal Biology of Past Peoples: Research Methods*, ed. S. R. Saunders and M. A. Katzenberg, New York, pp. 79–103.

———. 1993. "Understanding the Biogenic-Diagenetic Continuum: Interpreting Elemental Concentrations of Archaeological Bone," in *Investigations of Ancient Human Tissue: Chemical Analyses in Anthropology*, ed. M. Sandford and K. Langhorne, New York, pp. 3–57.

Sandford, M. K., and D. S. Weaver. 2000. "Trace Element Research in Anthropology: New Perspectives and Challenges," in *Biological Anthropology of the Human Skeleton*, ed. A. Katzenberg and S. Saunders, New York, pp. 329–350.

Savvidis, A. G. K. 1986. *Μελέτες Βυζαντινής Ιστορίας 11ου–13ου αιώνα*, Athens.

Scheur, L., and S. Black. 2000. *Developmental Juvenile Osteology*, San Diego.

Schoeninger, M. J. 1982. "Diet and the Evolution of Modern Human Form in the Middle East," *American Journal of Physical Anthropology* 58, pp. 37–52.

Sillen, A. 1989. "Diagenesis of the Inorganic Phase of Cortical Bone," in *The Chemistry of Prehistoric Bone*, ed. T. D. Price, Cambridge, pp. 211–229.

Sillen, A., and M. Kavanagh. 1982. "Strontium and Paleodietary Research: A Review," *Yearbook of Physical Anthropology*, 25, pp. 67–90.

Stewart, T. D. 1953. "The Age Incidence of Neural-Arch Defects in Alaskan Natives, Considered from the Standpoint of Etiology," *Journal of Bone and Joint Surgery* 35, pp. 937–950.

Tóth, K. 1970. *The Epidemiology of Dental Caries in Hungary*, Budapest.

Ubelaker, D. H. 1978. *Human Skeletal Remains: Excavation, Analysis, Interpretation*, 2nd ed., Washington, D.C.

von Endt, D. W., and D. J. Ortner. 1984. "Experimental Effects of Bone Size and Temperature on Bone Diagenesis," *JAS* 11, pp. 247–253.

y Edynak, G. 1989. "Yugoslav Mesolithic Dental Reduction," *American Journal of Physical Anthropology* 78, pp. 17–36.

Zachick, V., and M. Tzaphlidou. 2001. "Sex- and Age-Related Ca/P Ratio in Rib Bone of Healthy Humans," in *Proceedings of the 3rd International Symposium on Trace Elements in Human: New Perspectives, Athens, October 4–6, 2001,* ed. S. Ermidou-Pollet and S. Pollet, Athens, pp. 778–785.

CHAPTER 13

Bioarchaeological Analysis of the Human Osteological Material from Proskynas, Lokris

by Anastasia Papathanasiou, Eleni Zachou, and Michael P. Richards

Very little interdisciplinary biocultural research has been undertaken for Neolithic and Bronze Age Greece, although both periods have been extensively studied archaeologically. However, over the past 20 years there has been growing interest and a developing corpus of bioanthropological research into these periods. During the Neolithic, Greece was the setting for the earliest transition to agriculture and domestication in Europe, and in the following period—the Bronze Age—the region where the first urban (specialized and stratified) communities developed.

This chapter presents an interesting case study of a skeletal series from the site of Proskynas in central Greece that spans both time periods. The site dates from the Final Neolithic to the Mycenaean period, approximately 4500–1300 B.C., and has revealed a series of very well preserved human osteological remains. Analysis of these seeks to provide information about health, disease, and lifestyle using a number of commonly accepted health indicators observed on the osteological material, and to identify how successfully the population of Proskynas adapted to the stresses to which it was subjected. In addition, a reconstruction of past diet and economy will be attempted through paleodietary analysis.

THE SITE OF PROSKYNAS

Proskynas is located in central Greece, approximately 128 km north of Athens and 2 km from the sea. The site was excavated from 1996 to 2000 by the 14th Ephorate of Prehistoric and Classical Antiquities under the direction of Eleni Zachou. During the construction of the E-65 highway, building remains from all prehistoric periods were uncovered on a soft limestone hill near Rachi Proskyna Fthiotidos. On the flat hilltop there is a stratum in which four habitation phases were uncovered, ranging from Final Neolithic to the Mycenaean period. The major habitation period, with the best preserved remains, dates to the Early Bronze Age.

During the excavation of the buildings, 14 burials were also uncovered. Six of these form a group, while the rest are individual cases uncovered in the foundations of buildings or in open spaces. The dating of these burials is based solely on stratigraphic evidence because they were not associated with any chronologically diagnostic grave goods.

Two pit burials and one burial in a pot are dated to the Final Neolithic period because they were retrieved from a layer much deeper than the foundations of the Early Bronze Age buildings. Five other pit burials from foundation levels can also be attributed to the Final Neolithic.

The remaining six burials belong to the Middle Helladic or the Early Mycenaean period, and come from a burial circle 11 m in diameter and defined by large stones. One of the six burials is a grave outlined and covered by substantial slabs. The other five are simple pit burials, also covered by stones. The dead were placed in a flexed position and there were no offerings. Over the stones that cover the burial circle were found potsherds that are characteristic of the Middle Helladic period.

METHODOLOGY

Paleodemography and Paleopathology

The osteological series from Proskynas consists mainly of almost complete, articulated skeletons. The bone preservation ranges from fair to very good, which is rare for an open-air site of this period. All human remains were first mechanically cleaned with water, soft brushes, and delicate instruments in order to minimize post-excavation damage. Preservation and reconstruction were performed whenever needed.

The analysis of the material followed the standard procedures developed by Buikstra and Ubelaker for working on complete skeletons.[1] The study, recording, and photography of the material was done at the 14th Ephorate of Prehistoric and Classical Antiquities in Lamia where the material is currently stored.

In order to determine the health status of the Proskynas skeletal series, the analysis took into account a number of health and stress indicators in the osteological assemblage. The basic demographic parameters of the skeletal series (minimum number of individuals [MNI], age, sex, and stature) were determined first, as they provide the essential context for any further analysis. Stature and mean adult age are directly related to overall well-being and are used as evidence of chronic environmental and/or nutritional stress. Indicators of pathology were then scored for each individual. The following observations were made: porotic hyperostosis and cribra orbitalia for identification of nutritional deficiencies and deprivation; periosteal reaction for assessment of inflammatory response to infectious pathogens; osteoarthritis and musculoskeletal stress markers for the determination of the mechanical stress of the joints due to workload and physical activity; cranial injuries (depressed fractures) and postcranial fractures for indications of trauma; dental caries and antemortem tooth loss as indices for oral and general health in relation to food composition; and enamel defects (linear enamel hypoplasia) for physiological stress and growth disturbances in childhood. Finally, nonmetric traits were used to infer population variation, genetic affinities, and general relatedness of the individuals in the sample.

1. Buikstra and Ubelaker 1994.

The procedure for determining the MNI seemed initially to be straightforward as each burial represented one individual and was collected and stored in a separate box. However, care was taken to determine if all the skeletal elements in a container represented the same individual, on the basis of age at death, size, side, preservation, and the actual matching of two or more bones. The search yielded an additional individual, as skeletal elements of an infant were found in the same burial with the remains of another body.

Sex determination was performed following the methods outlined by Buikstra and Ubelaker, as well as other standard methods including patterns of robusticity and cranial and pelvic morphology.[2] Only adults with mature characteristics were sexed.

For this rather fragmentary sample, a number of different methods were used to estimate age at death. Meindl and Lovejoy's[3] method for ectocranial suture closures was used for crania belonging to adults while Lovejoy's[4] method for stages of dental wear was used for adult dentitions. The methods of Todd[5] and Brooks and Suchey[6] for analyzing morphological changes of the pubic symphyseal face were used for the adult *os coxa*. Ubelaker's[7] method for tooth formation and eruption was applied to subadult remains. McKern and Stewart's[8] method for epiphyseal fusion was used for aging subadult and young adult postcranial bones. Subadult age was also estimated by long bone diaphyseal length and iliac breadth from the available complete bones following Ubelaker.[9] Age at death estimates were used to construct a profile with five-year age increments.

Stature was estimated using all the available complete long bones according to formulae for white males and females,[10] and the stature of each individual was calculated as the average of the stature estimates from all the long bones of that individual. If the sex could not be determined, stature was derived as the average of the corresponding estimates for males and females.

Pathological conditions were identified following Ortner and Putschar,[11] Resnick,[12] and Buikstra and Ubelaker.[13] Observations were made under normal light conditions without the aid of microscopy. The pathological conditions were recorded by presence-absence, and the percentages reflect the observed over the observable, that is, the number of elements that display the characteristic over the total number of the same element. In the case of porotic hyperostosis, the degree of severity was also recorded. Additionally, a small number of bones were selected for radiographic analysis in order to determine if there were cases of osteopenia or growth arrest lines that could be related to anemic conditions or other nutritional deficiencies.

Nonmetric Variation

The collection of morphological nonmetric variation was limited to observation of cranial traits. Two discrete cranial traits were selected because there are published sources confirming a good understanding of their development, genetic control, high level of heritability, low sensitivity to pathological or environmental factors, and the existence of comparable

2. Buikstra and Ubelaker 1994; Ubelaker 1989; White 1991; Milner 1992; Krogman and İşcan 1986; Phenice 1969.
3. Meindl and Lovejoy 1985.
4. Lovejoy 1985.
5. Todd 1920, 1921.
6. Brooks and Suchey 1990.
7. Ubelaker 1989.
8. McKern and Stewart 1957.
9. Ubelaker 1989.
10. Trotter 1970.
11. Ortner and Putschar 1985.
12. Resnick 1995.
13. Buikstra and Ubelaker 1994.

data. These traits, metopism and extra ossicles, were recorded by presence-absence, following Hauser and DeStefano.[14] The percentages reflect the number of observed over the potentially observable cases.

Paleodietary Reconstruction

We measured the stable carbon and nitrogen isotopes of human bone collagen to provide additional information about the diet at Proskynas. Stable isotope analysis provides direct, specific, and detailed dietary information using the chemical analysis of the human remains themselves.[15] Many foods, or groups of foods, are isotopically distinct, and when consumed and incorporated into body tissues, their isotopic signature is retained, although usually altered to some extent. Analysis of the stable carbon and nitrogen isotopes of collagen, which is the major protein in bone, provides information about the sources of protein in diets.[16] Carbon isotope ratios (the ratio of ^{13}C to ^{12}C, presented as the $\delta^{13}C$ value) can distinguish between three major dietary protein categories:[17] diets based on marine resources, diets reliant on the consumption of plants (and animals that consume them) that use the C_3 photosynthetic pathway, and those dietary sources that use the C_4 pathway.[18] C_4 plants have less negative ^{13}C values, closer to that of the atmosphere, and they differ by about 14‰ from C_3 plants. Marine fish and mammals have more positive $\delta^{13}C$ values compared to terrestrial C_3 foods, and less positive values than terrestrial C_4 foods.[19]

The nitrogen stable isotope ^{15}N accumulates along the food chain, depending on the trophic level, with a stepwise increase in the ratio of ^{15}N to ^{14}N, the $\delta^{15}N$ value, of 2–4‰. Atmospheric nitrogen has a value of 0‰. Legumes, which get nitrogen from symbiotic bacteria in the soil, have values closer to atmospheric nitrogen, while non-leguminous plants have higher values following the pattern of the stepwise increase to herbivores and to carnivores. Marine environments have higher $\delta^{15}N$ values than terrestrial ones. Thus, nitrogen isotope ratios can be used to distinguish between different trophic levels and especially between marine and terrestrial protein consumption. Values for consumers of terrestrial foods generally are less positive than those for consumers of marine foods.[20]

Samples for isotopic analysis were taken from 13 individuals. The sample was taken from the rib and from the femur of each individual. Of the 13 samples, 10 yielded valid results. The analysis was performed at the Stable Isotope Laboratory, Department of Archaeological Sciences, University of Bradford, UK. Samples were prepared following procedures outlined in Richards and Hedges,[21] with the addition of an ultrafiltration step.[22]

RESULTS

Paleodemography and Paleopathology

Based on counts of skeletal elements, there is a minimum of 15 individuals (Table 13.1), consisting of seven adults and eight subadults. Of the individuals over 18 years of age for whom sex could be determined, five

14. Hauser and DeStefano 1989.
15. DeNiro and Epstein 1978, 1981; van der Merwe and Vogel 1978; Norr 1995; Ambrose and Norr 1993; Ambrose et al. 1997.
16. Ambrose and Norr 1993; DeNiro and Epstein 1978, 1981; Schoeninger, DeNiro, and Tauber 1983; Schoeninger 1989; Norr 1995.
17. Schoeninger and DeNiro 1984; Dupras, Schwarcz, and Fairgrieve 2001; Katzenberg 2000.
18. Ambrose 1987. C_3 pathway plants include most leafy plants, temperate grasses like wheat and rice, all tree and shrubs and their nuts and fruits; C_4 pathway plants include tropical grasses, maize, sugarcane, millet, some amaranths, and some chenopods.
19. Tauber 1981; Dupras, Schwarcz, and Fairgrieve 2001; Katzenberg 2000.
20. Schoeninger and DeNiro 1984; Ambrose 1987; Minagawa and Wada 1984; Dupras, Schwarcz, and Fairgrieve 2001; Katzenberg 2000.
21. Richards and Hedges 1999.
22. Brown, Nelson, and Southon 1988.

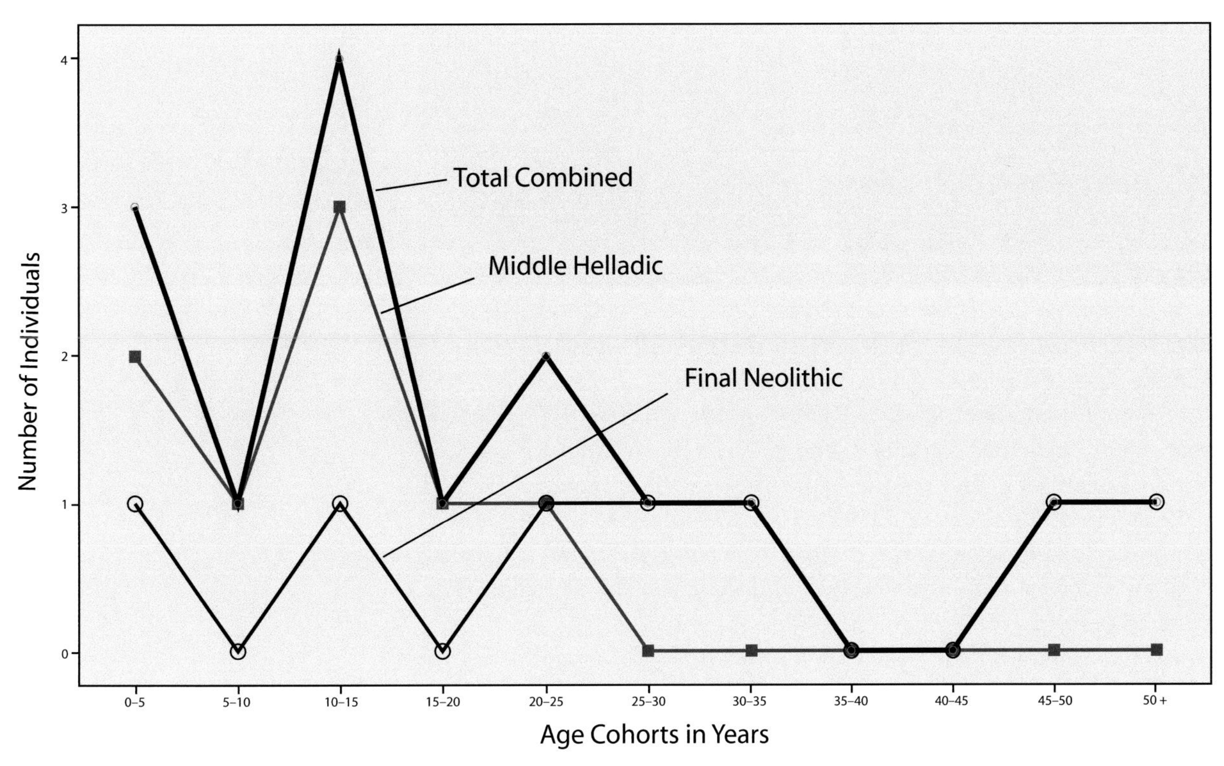

Figure 13.1. Proskynas: age at death

TABLE 13.1. MINIMUM NUMBER OF INDIVIDUALS (MNI) FROM PROSKYNAS

Burial Number	*Age*	*Sex*	*Period*
I	9 mos.		Final Neolithic
II	2.5		Middle Helladic
III	5		Middle Helladic
IV	14–15		Final Neolithic
V	10–15		Middle Helladic
VI	>50	M	Final Neolithic
VII	<30	I	Final Neolithic
VIII	10+		Middle Helladic
VIIIa	<0		Middle Helladic
IX	20–25	M	Final Neolithic
X	18	M?	Middle Helladic
XI	20–25	F	Middle Helladic
XII	14		Middle Helladic
XIII	30–35	M?	Final Neolithic
XIV	45	M?	Final Neolithic

are males or probable males, and one is a female or probable female. The remaining individual is of indeterminate sex. The mean adult age at death (for the individuals that could not be accurately aged, the midpoint of the age range was used) is 33.8 years. The ages range from newborn to 50 years. The life expectancy at birth is 19.17 years. The age profile for the total excavated skeletal series exhibits a peak at ages 10–15 (Fig. 13.1). The D_{30+}/D_{5+} index is 0.25, indicative of a large number of young individuals

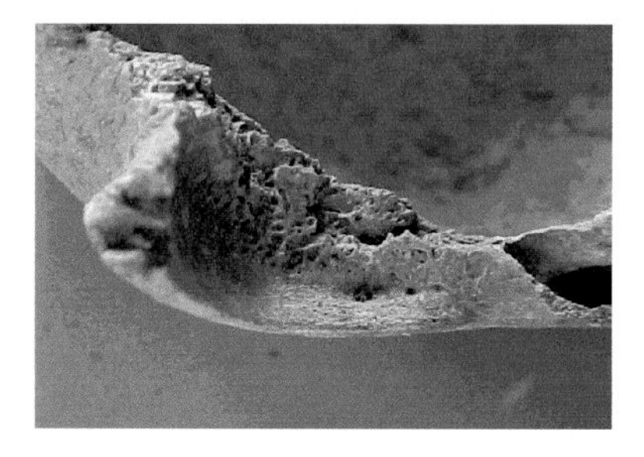

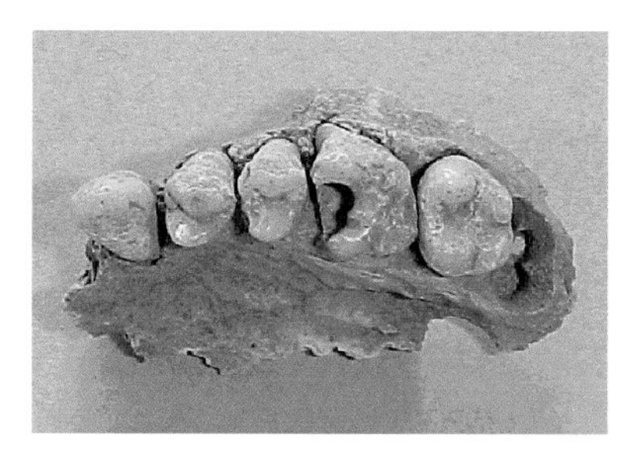

Figure 13.2 ***(left).*** **Cribra orbitalia on orbital roof.** Photo A. Papathanasiou

Figure 13.3 ***(right).*** **Dental caries.** Photo A. Papathanasiou

entering the skeletal series. The average stature for males is 167.5 cm (n = 3) and for females it is 152.5 cm (n = 1).

The most frequently observed pathological conditions in this skeletal series are cribra orbitalia (a pathological lesion on the roof of the eye socket, shown in Fig. 13.2) and porotic hyperostosis, with a frequency of 54.5 percent, or six out of eleven individuals, that exhibit one or both lesions in either a mild or severe form. Lesions were observed in one adult male and one adult female, a five-year-old, and three subadults ranging from 10 to 18 years of age.

Dental pathology at Proskynas is represented by dental caries, enamel hypoplasias, calculus, and antemortem tooth loss. Linear enamel hypoplasia, the lines of enamel deficiency observed on teeth due to disruption of the enamel formation process, has a prevalence of 6.4%, or 17 out of 264 teeth (ten canines, six incisors, and one premolar). Dental caries, the focal demineralization of teeth caused by organic acids produced by bacterial fermentation of dietary carbohydrates,[23] has a relatively low prevalence of 3.0%, or 8 out of 264 teeth, although it is present in five out of fifteen individuals (Fig. 13.3). Of the infected teeth, seven are molars and one is a premolar. Antemortem tooth loss, which results from dental infection, trauma, gingivitis, or periodontitis,[24] is present in only 1 individual out of 11 observable, or 6 of 264 teeth (2.3%). The majority of teeth lost premortem are molars. In contrast, calculus is relatively common and is present in three out of eleven individuals, or 27.3%. All cases are of the supragingival type. There are also two cases of periodontal disease (two out of eleven individuals, or 18.2%). Furthermore, the series exhibits a low prevalence of tooth maleruption, often due to insufficient space and impacted teeth,[25] at a prevalence of 4.2%, or 11 out of 264. Of these, eight incisors and one canine are crowded, and there are also two cases of third molar agenesis.

The most frequently observed postcranial pathological conditions are osteoarthritis and enthesopathies. Osteoarthritis is present in three out of eleven individuals, or 27.3% of the sample. It occurs mainly as vertebral osteoarthritis (Fig. 13.4), and it includes moderate-to-severe spinal osteophytosis and Schmorl's nodes, as well as lipping of hand phalanges. Enthesopathies, or ossified tendons (Fig. 13.5), are relatively common and severe and are observed in seven out of twelve individuals, or 58.3% of the sample. They are most frequently associated with muscles of the arm, the leg, and the hand digits. Musculoskeletal stress markers almost always involve the interosseous ligament of the hand phalanges. The linea aspera of the femur and the sternal and acromial ends of the clavicle follow in frequency, with the radioulnar joint, the acromion of the scapula, the proximal humerus, and the talus being less frequent sites.

23. Larsen 1997.
24. Hillson 1986.
25. Larsen 1997.

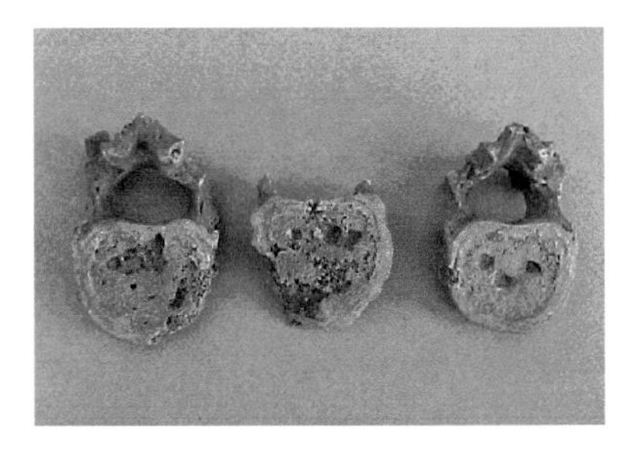

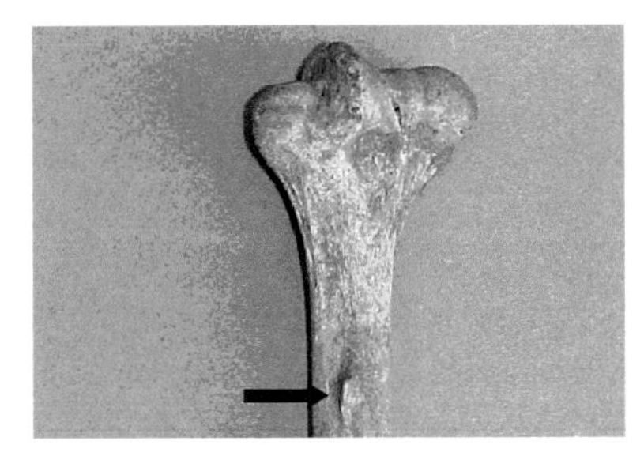

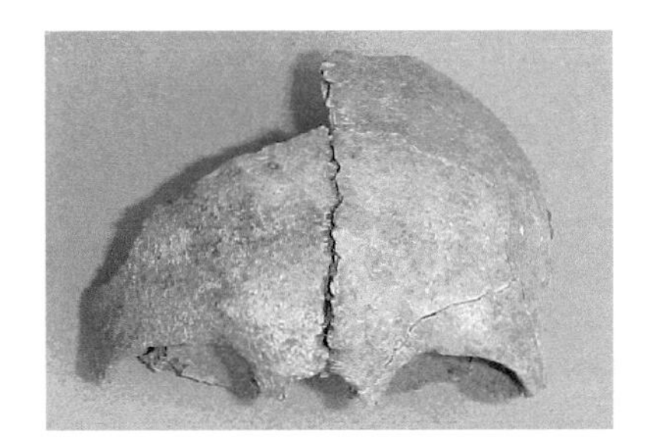

Figure 13.4 ***(left)*****. Vertebrae with Schmorl's nodes and spinal osteoarthritis.** Photo A. Papathanasiou

Figure 13.5 ***(center)*****. Humerus with ossified tendon insertion (marked by arrow).** Photo A. Papathanasiou

Figure 13.6 ***(right)*****. Unfused metopic suture.** Photo A. Papathanasiou

Finally, there is only one instance of periosteal reaction, on a tibia. In terms of trauma, there is a case of a healed depressed fracture on a parietal bone, but no postcranial trauma was observed.

Nonmetric Variation

The Proskynas sample exhibits unfused metopic sutures (Fig. 13.6) in two out of eight individuals and extrasutural bones again in two individuals.

Paleodietary Reconstruction

The results of the stable isotope analysis are presented in Figure 13.7 and Table 13.2. The mean $\delta^{13}C$ value is -19.7 ± 0.7 and the mean $\delta^{15}N$ value is 7.5 ± 1.6. The majority of the samples have values that indicate that the humans were consuming primarily terrestrial protein from C_3 ecosystems. The $\delta^{15}N$ values are best interpreted through comparison with contemporary fauna, but none was available from Proskynas. Therefore, our tentative conclusion, based on comparison with fauna from other sites

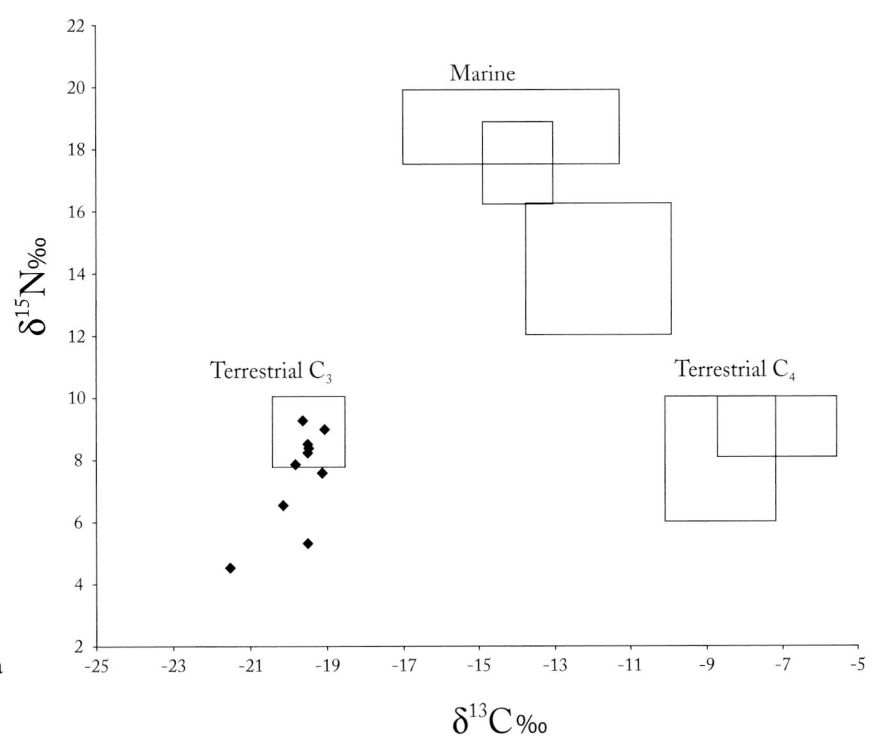

Figure 13.7. Stable carbon and nitrogen isotope analysis. Results from bone collagen, in comparison with isotopically distinct diets (data from Schoeninger, DeNiro, and Tauber 1983).

TABLE 13.2. STABLE ISOTOPE ANALYSIS RESULTS

Sample	*δ¹³C*	*δ¹⁵N*	*%C*	*%N*	*C:N*	*% Collagen Yield*
AP1	-19.62	9.25	48.82	17.35	3.28	2.9
AP2	-19.05	8.97	42.94	14.96	3.35	7.9
AP3	-20.13	6.52	41.17	14.06	3.42	3.9
AP4	-19.81	7.84	41.86	14.41	3.39	3.3
AP5	–	–	–	–	–	0.0
AP6	–	–	–	–	–	0.0
AP7	–	–	–	–	–	1.1
AP8	-19.49	5.30	42.65	15.79	3.16	4.9
AP9	-19.11	7.57	35.21	13.05	3.14	1.7
AP10	-19.49	8.22	39.72	14.61	3.17	5.0
AP11	-21.51	4.52	36.35	12.95	3.28	1.4
AP12	-19.46	8.37	37.95	13.88	3.18	1.9
AP13	-19.49	8.49	41.97	15.41	3.18	1.2

in prehistoric Greece, is that the humans consumed a mix of plant and animal protein. What is clear is that there was little, if any, consumption of the marine resources that were located near the site. The likely sources of dietary protein were then C_3 plants, such as wheat, barley, legumes, and fruits, and possibly dairy products and some meat. Until we have comparable faunal isotopic data, it is not possible to be more specific about the relative amounts of plant versus animal protein in the diets, but these data suggest that it is unlikely that large amounts of animal protein were consumed.

There is no clear evidence of a difference in isotopic values between the Neolithic and Middle Helladic periods, although the sample sizes are small. It is interesting to note that the two humans with quite low $\delta^{15}N$ values (AP8 and AP11, see Table 13.2) both date to the Middle Helladic period. These low values may indicate a diet with almost all of the protein coming from plant sources, or else the significant consumption of legumes, which, as discussed above, have lower $\delta^{15}N$ values than other plants.

DISCUSSION

The Proskynas sample of 15 individuals is certainly a very small one from which to draw general conclusions reflecting the entire living population from which it came. Certainly biological, cultural, and other selective biases are present in this skeletal series as in almost any archaeological sample. Nevertheless, it resembles other contemporary populations, as it is characterized by high fertility rates and comparable distributions of sex, stature, and age at death.

In terms of paleopathology, while always keeping in mind the limitations of the small sample, a number of observations can produce some insight into its health and dietary status. The most prevalent pathological conditions in this sample are cribra orbitalia and porotic hyperostosis.

Both pathological conditions are characterized by marrow expansion and thickening of the diploë; this is a result of increased production of red blood cells in response to either genetic or acquired anemias.[26] The increase in red blood cells and the consequent expansion of the diploë is more evident in children when the bones are more malleable, while the lesions in adults are usually healed. A few other conditions have a similar expression on the cranial bones, such as hereditary spherocytosis, cyanotic congenital heart disease, and polycythemia vera, but they are very uncommon and could not possibly account for the observed high prevalence.[27] Thus, the high prevalence in this population is indicative of either genetic or acquired anemia.

There are several reasons why genetic forms of anemia,[28] such as thalassemia and sickle cell anemia, are probably not implicated, including the absence of the characteristic pathological lesions on postcranial bones, the slight-to-moderate expression of the lesions, and the evidence of extensive healing and remodeling.[29] These considerations support the hypothesis that the anemic conditions observed in this skeletal series represent an acquired anemia caused by either deficient or inadequate diet, probably poor in iron as suggested by the paleodietary analysis, and/or environmentally induced circumstances such as high pathogen loads and parasitic infection.[30] Radiographic analysis, although based on an extremely small sample, correlated the anemic conditions with osteopenia and growth arrest episodes in postcranial bones. Iron deficiency anemia is not the only disease that is related to nutritional deficiencies. Two other major pathological conditions are scurvy (vitamin C deficiency) and rickets (vitamin D deficiency). Two or three of these diseases may occur in the same group of people or even the same individual in cases of malnutrition,[31] and differentiation between these pathological conditions is not always possible. It is probable that the Proskynas population was experiencing a chronic nutritional deficiency, but it is not possible to isolate a specific deficiency or even additional factors that might have resulted in these lesions.

Regarding dental health, the most common pathological conditions in archaeological populations, dental caries and enamel hypoplasias, exhibit a relatively low frequency in the Proskynas sample. The individuals with enamel defects likely experienced either a nutritional deficiency or infectious disease, or a combination of the two that caused growth disruption.[32] The relatively low prevalence of dental enamel hypoplasia in the Proskynas sample indicates that the stresses were not as common or severe as documented for many other prehistoric populations.[33] Furthermore, the observed low prevalence of dental caries contrasts with contemporary populations of the Near East that exhibit higher prevalences of dental caries.[34] Antemortem tooth loss is very low as well, implying a low dental trauma and infection rate.[35]

Osteoarthritis can be due to numerous causes. Age, genetic predisposition, obesity, activity, lifestyle, and environmental factors could contribute to its development.[36] In the skeletal series under study, we observe two types of osteoarthritis: (1) spinal osteophytosis and Schmorl's nodes, a direct consequence of spinal mechanical stress, and (2) a pattern of lipping of hand phalanges, probably due to excessive use of the hands and the fingers.

26. Ortner, Kimmerle, and Diez 1999; Ortner et al. 2001; Schultz 2001; Larsen 1997.
27. Ortner and Putschar 1985; Steinbock 1978.
28. Angel 1966.
29. Papathanasiou 1999, 2001; Papathanasiou, Larsen, and Norr 2000; Lagia 1993.
30. Stuart-Macadam 1985; Stuart-Macadam and Kent 1992.
31. Ortner, Kimmerle, and Diez 1999.
32. Goodman and Rose 1990, 1991.
33. Papathanasiou 2005; Smith, Bar-Yosef, and Sillen 1984; Teegen and Schultz 1997, 1998.
34. Larsen 1997; Smith, Bar-Yosef, and Sillen 1984; Cook 1999.
35. Papathanasiou 2005; Larsen 1997; Smith, Bar-Yosef, and Sillen 1984.
36. Roberts and Manchester 1995.

The observed enthesopathies are new bone formation at tendinous and ligament insertions as a result of increased muscle size or trauma. However, they cannot be directly correlated with specific occupations in the past.[37] In the Proskynas skeletal series, enthesopathies seem to follow a pattern and their distribution may indicate (1) excessive use of the hand digits, and (2) significant mechanical stress of the legs. Additionally, the morphology and geometry of the long bones, the curved shapes of the clavicle and the femur and especially the marked linea aspera of the femora, are also indicative of long bones that have been mechanically altered due to high activity and mobility levels.

Nonmetric Variation

In terms of nonmetric variation, metopism appears sporadically and infrequently in most populations, with some of the highest frequencies reaching 8%–14%.[38] Proskynas exhibits an unusually high frequency of metopism, considering that two cases are present. The genetic relatedness of these two individuals could be confirmed with future DNA analyses.

Comparisons between Subsamples

The Proskynas skeletal series originates from two different cultural periods, the Final Neolithic and the Middle Helladic, and so biologically it should be considered as consisting of at least two subpopulations. Comparison of these two subpopulations yields interesting results when the prevalence of pathological conditions is examined. The differences between the two subpopulations are the following: Most of the anemic conditions (four out of eight individuals [50%] as opposed to two out of seven [28.6%]) are observed in the Middle Helladic subpopulation, while most of the musculoskeletal stress markers (six out of seven [85.7%] compared to two out of eight [25%]) are observed in the Neolithic subpopulation. Also, linear enamel hypoplasia is more prevalent in the Neolithic (three out of seven [42.9%]) than in the Middle Helladic (one out of eight [12.5%]).

It must be noted that we observed different maturation rates on four out of the eight Proskynas subadults, where the dental age estimate points to an age of a year or more older than the estimate from the long bones. Because teeth are more resistant to environmental factors, this observation may indicate long-term stress for these individuals, possibly involving malnutrition and growth retardation. These same individuals also exhibit remodeling scars on the femoral neck.[39] These differential maturation rates between teeth and long bones are observed exclusively in the Middle Helladic sample. Also, an inversion of the adult to subadult ratio is observed between the two subpopulations (five adults to two subadults in the Neolithic versus one adult to seven subadults in the Middle Helladic), as clearly seen in the profiles of the two subpopulations (Fig. 13.1). This implies that the observed skeletal series are not representative of the living populations they come from, possibly due to burial practices such as differential treatment according to age within the framework of their beliefs about death. However, these differences are not statistically significant because the sample size is very small.

37. Roberts and Manchester 1995; Merbs 1983; Kennedy 1989.
38. Hauser and DeStefano 1989.
39. Angel 1964.

CONCLUSIONS

The Proskynas population of 15 individuals exhibits multiple skeletal and dental stress indicators, including a relatively high frequency of anemic conditions, arthritis, and musculoskeletal stress markers. This suggests a consistent pattern of population stress. These observations relate to the health, nutrition, and lifestyle of the population and indicate that this group may have been stressed by a long-term nutritional deficiency, most probably associated with an iron-poor diet. This finding appears to be supported by the isotopic analysis, which did not indicate a diet rich in animal meat (a good source of dietary iron). Iron deficiency could have resulted in both the observed anemic conditions and the growth arrest signs on the teeth, as well as in the differential maturation rate. In addition to dietary stresses, the Proskynas individuals probably experienced elevated mobility and physical activity stresses that resulted in the observed osteoarthritis and the severe enthesopathies. A subtle distinction can be made between the two subpopulations of the site, with the Final Neolithic subsample being more physically stressed and the Middle Helladic group being more physiologically stressed.

Overall, although the Proskynas skeletal series is a very small sample, it conforms to the trends of the Neolithic and Middle Helladic periods in Greece in terms of demographic parameters of age, stature, the frequencies of anemia and osteoarthritis, and trends in dental health.[40] It is characterized by the same terrestrial diet consisting mainly of cereals and pulses, and an agricultural lifestyle involving high levels of physical stress.

40. Papathanasiou 1999, 2005; Cohen and Armelagos 1984; Triantaphyllou 2001.

REFERENCES

Ambrose, S. H. 1987. "Chemical and Isotopic Techniques of Diet Reconstruction in Eastern North America," in *Emergent Horticultural Economies of the Eastern Woodlands,* ed. W. F. Keegan (Center for Archaeological Investigations: Southern Illinois University at Carbondale, Occasional Paper 7), Carbondale, Ill., pp. 87–107.

Ambrose, S. H., B. M. Butler, D. B. Hanson, R. L. Hunter-Anderson, and H. W. Krueger. 1997. "Stable Isotopic Analysis of Human Diet in the Marianas Archipelago, Western Pacific," *American Journal of Physical Anthropology* 104, pp. 343–361.

Ambrose, S. H., and L. Norr. 1993. "Experimental Evidence for the Relationship of the Carbon Isotope Ratios of Whole Diet and Dietary Protein to Those of Bone Collagen and Carbonate," in *Prehistoric Human Bone: Archaeology at the Molecular Level,* ed. J. B. Lambert and G. Grupe, New York, pp. 1–37.

Angel, J. L. 1964. "The Reaction Area of the Femoral Neck," *Clinical Orthopedics* 32, pp. 130–142.

———. 1966. "Porotic Hyperostosis, Anemias, Malarias, and Marshes in the Prehistoric Eastern Mediterranean," *Science* 153, pp. 760–763.

Brooks, S. T., and J. M. Suchey. 1990. "Skeletal Age Determination Based on the *Os Pubis*: A Comparison of the Ascadi-Nemeskeri and Suchey-Brooks Methods," *Human Evolution* 5, pp. 227–238.

Brown, T. A., D. E. Nelson, and J. R. Southon. 1988. "Improved Collagen Extraction by Modified Longin Method," *Radiocarbon* 30, pp. 171–177.

Buikstra, J. E., and D. H. Ubelaker, eds. 1994. *Standards for Data Collection from Human Skeletal Remains,* Fayetteville, Ark.

Cohen, M. N., and G. J. Armelagos, eds. 1984. *Paleopathology at the Origins of Agriculture,* Orlando.

Cook, D. H. 1999. "Skeletal Evidence for Nutrition in Mesolithic and Neolithic Greece: A View from Franchthi Cave," in *Paleodiet in the Aegean,* ed. S. J. Vaughan and W. D. E. Coulson, Oxford, pp. 99–104.

DeNiro, M. J., and S. Epstein. 1978. "Influence of Diet on the Distribution of Carbon Isotopes in Animals," *Geochimica and Cosmochimica Acta* 42, pp. 495–506.

———. 1981. "Influence of Diet on the Distribution of Nitrogen Isotopes in Animals," *Geochimica and Cosmochimica Acta* 45, pp. 341–351.

Dupras, T. L., H. P. Schwarcz, and S. I. Fairgrieve. 2001. "Infant Feeding and Weaning Practices in Roman Egypt," *American Journal of Physical Anthropology* 115, pp. 204–212.

Goodman, A. H., and J. C. Rose. 1990. "Assessment of Systematic Physiological Perturbations from Dental Enamel Hypoplasias and Associated Histological Structures," *Yearbook of Physical Anthropology* 33, pp. 59–110.

———. 1991. "Dental Enamel Hypoplasias as Indicators of Nutritional Status," in *Advances in Dental Anthropology,* ed. M. A. Kelly and C. S. Larsen, New York, pp. 279–294.

Hauser, G., and G. F. DeStefano. 1989. *Epigenetic Variants of the Human Skull,* Stuttgart.

Hillson, S. 1986. *Teeth,* Cambridge.

Katzenberg, M. A. 2000. "Stable Isotope Analysis: A Tool for Studying Past Diet, Demography, and Life History," in *Biological Anthropology of the Human Skeleton,* ed. M. A. Katzenberg and S. R. Saunders, New York, pp. 305–327.

Kennedy, K. A. R. 1989. "Skeletal Markers of Occupational Stress," in *Reconstruction of Life from the Skeleton,* ed. M. Y. İşcan and K. A. R. Kennedy, New York, pp. 129–160.

Krogman, W. M., and M. Y. İşcan. 1986. *The Human Skeleton in Forensic Medicine,* 2nd ed., Springfield, Ill.

Lagia, A. 1993. "Differential Diagnosis of the Three Main Types of Anaemia (Thalassaemia, Sickle Cell Anaemia, Iron Deficiency Anaemia) Based on Macroscopic and Radiographic Skeletal Characteristics" (M.Sc. diss., Univ. of Bradford).

Larsen, C. S. 1997. *Bioarchaeology: Interpreting Behavior from the Human Skeleton,* Cambridge.

Lovejoy, C. O. 1985. "Dental Wear in the Libben Population: Its Functional Pattern and Role in the Determination of Adult Skeletal Age at Death," *American Journal of Physical Anthropology* 68, pp. 47–56.
McKern, T. W., and T. D. Stewart. 1957. *Skeletal Age Changes in Young American Males* (Environmental Protection Research Division Technical Report EP-45), Natick, Mass.
Meindl, R. S., and C. O. Lovejoy. 1985. "Ectocranial Suture Closure: A Revised Method for the Determination of Skeletal Age at Death Based on the Lateral Anterior Sutures," *American Journal of Physical Anthropology* 68, pp. 57–66.
Merbs, C. 1983. *Patterns of Activity Induced Pathology in a Canadian Inuit Population* (Archaeological Survey of Canada Paper 119), Ottawa.
Milner, G. R. 1992. *Determination of Skeletal Age and Sex,* University Park, Pa.
Minagawa, M., and E. Wada. 1984. "Stepwise Enrichment of ^{15}N along Food Chains: Further Evidence and the Relation between δ^{15}N and Animal Age," *Geochimica et Cosmochimica Acta* 48, pp. 1135–1140.
Norr, L. 1995. "Interpreting Dietary Maize from Stable Isotopes in the American Tropics: The State of the Art," in *Archaeology in the Lowland American Tropics: Current Analytical Methods and Applications,* ed. P. W. Stahl, Cambridge, pp. 1–27.
Ortner, D. J., W. Butler, J. Cafarella, and L. Milligan. 2001. "Evidence of Probable Scurvy in Subadults from Archaeological Sites in North America," *American Journal of Physical Anthropology* 114, pp. 343–351.
Ortner, D. J., E. H. Kimmerle, and M. Diez. 1999. "Probable Evidence of Scurvy in Subadults from Archaeological Sites in Peru," *American Journal of Physical Anthropology* 108, pp. 321–331.
Ortner, D. J., and W. G. J. Putschar. 1985. *Identification of Pathological Conditions in Human Skeletal Remains,* 2nd ed., Washington, D.C.
Papathanasiou, A. 1999. "A Bioarchaeological Analysis of Health, Subsistence, and Funerary Behavior in the Eastern Mediterranean Basin: A Case Study from Alepotrypa Cave, Greece" (diss. Univ. of Iowa).
———. 2001. *A Bioarchaeological Analysis of Neolithic Alepotrypa Cave, Greece* (*BAR-IS* 961), Oxford.
———. 2005. "Health Status of the Neolithic Population of Alepotrypa Cave, Greece," *American Journal of Physical Anthropology* 126, pp. 377–390.
Papathanasiou, A., C. S. Larsen, and L. Norr. 2000. "Bioarchaeological Inferences from a Neolithic Ossuary from Alepotrypa Cave, Diros, Greece," *International Journal of Osteoarchaeology* 10, pp. 210–228.
Phenice, T. W. 1969. "A Newly Developed Visual Method of Sexing the *Os Pubis*," *American Journal of Physical Anthropology* 30, pp. 297–302.
Resnick, D. 1995. *Diagnosis of Bone and Joint Disorders,* Philadelphia.
Richards, M. P., and R. E. M. Hedges. 1999. "Stable Isotope Evidence for Similarities in the Types of Marine Foods Used by Late Mesolithic Humans at Sites along the Atlantic Coast of Europe," *JAS* 26, pp. 717–722.
Roberts, C. A., and K. Manchester. 1995. *The Archaeology of Disease,* 2nd ed., Ithaca, NY.
Schoeninger, M. J. 1989. "Reconstructing Prehistoric Human Diet," in *Chemistry of Prehistoric Human Bone,* ed. T. D. Price, New York, pp. 38–67.
Schoeninger, M. J., and M. J. DeNiro. 1984. "Nitrogen and Carbon Isotopic Composition of Bone Collagen from Marine and Terrestrial Animals," *Geochimica et Cosmochimica Acta* 48, pp. 625–639.
Schoeninger, M. J., M. J. DeNiro, and H. Tauber. 1983. "Stable Nitrogen Isotope Ratios of Bone Collagen Reflect Marine and Terrestrial Components of Prehistoric Human Diet," *Science* 220, pp. 1381–1383.
Schultz, M. 2001. "Paleohistopathology of Bone: A New Approach to the Study of Ancient Diseases," *American Journal of Physical Anthropology* 44, pp. 106–147.
Smith, P., O. Bar-Yosef, and A. Sillen. 1984. "Archaeological and Skeletal Evidence for Dietary Change during the Late Pleistocene/Early Holocene in the Levant," in *Paleopathology at the Origins of Agriculture,* ed. M. N. Cohen and G. J. Armelagos, Orlando, pp. 101–136.
Steinbock, R. T. 1978. *Paleopathological Diagnosis and Interpretation*, Springfield, Ill.
Stuart-Macadam, P. 1985. "Porotic Hyperostosis: Representative of a Childhood Condition," *American Journal of Physical Anthropology* 66, pp. 391–398.
Stuart-Macadam, P., and S. Kent, eds. 1992. *Diet, Demography, and Disease: Changing Perspectives on Anemia,* New York.
Tauber, H. 1981. "^{13}C Evidence for Dietary Habits of Prehistoric Man in Denmark," *Nature* 292, pp. 332–333.
Teegen, W. R., and M. Schultz. 1997. "Transverse Linear Enamel Hypoplasia in the PPNB Population of Nevalı Çori (Turkey)," *American Journal of Physical Anthropology* Supplement 24, pp. 226–227.
———. 1998. "Teeth as Tools in the Late PPNB Population from Nevalı Çori (Turkey)," *American Journal of Physical Anthropology* Supplement 26, p. 217.
Todd, T. W. 1920. "Age Changes in the Pubic Bone: I. The White Male Pubis," *American Journal of Physical Anthropology* 3, pp. 285–334.
———. 1921. "Age Changes in the Pubic Bone: II, III, IV," *American Journal of Physical Anthropology* 4, pp. 1–70.
Triantaphyllou, S. 2001. *A Bioarchaeological Approach to Prehistoric Cemetery Populations from Central and Western Greek Macedonia* (*BAR-IS* 976), Oxford.
Trotter, M. 1970. "Estimation of Stature from Intact Limb Bones," in *Personal Identification in Mass Disasters,* ed. T. D. Stuart, Washington, D.C., pp. 71–83.
Ubelaker, D. H. 1989. *Human Skeletal Remains: Excavation, Analysis, Interpretation,* 2nd ed., Washington, D.C.
van der Merwe, N. J., and J. C. Vogel. 1978. "^{13}C Content of Human Collagen as a Measure of Prehistoric Diet in Woodland North America," *Nature* 276, pp. 815–816.
White, T. D. 1991. *Human Osteology,* San Diego.

CHAPTER 14

Isotope Paleodietary Analysis of Humans and Fauna from the Late Bronze Age Site of Voudeni

by Eirini I. Petroutsa, Michael P. Richards, Lazaros Kolonas, and Sotiris K. Manolis

The reconstruction of diet in prehistoric Greek populations has traditionally relied on indirect methods of analysis. These include study of the food remains recovered from archaeological sites, changes in technology, artifacts used for food preparation, and textual and iconographic evidence.

Stable isotope analysis of human and faunal bone collagen can also provide dietary information.[1] This measure of the diet is supplementary to the more traditional methods of analysis mentioned above, giving direct information on sources of dietary protein.[2] Stable isotope analysis has been widely used to investigate dietary patterns in prehistoric human populations in the rest of the world, but there are still only a few isotopic studies of human and faunal remains from prehistoric Greece.

In this chapter we report on the carbon and nitrogen isotopic analysis of human and faunal bone collagen from the Late Bronze Age cemetery site of Voudeni.[3] We were particularly interested in searching for possible correlations between social status and the isotopic evidence of diet that cannot be inferred from the grave context.

THE SITE OF VOUDENI

Voudeni cemetery lies about 8 km from the city of Patras and is the best-studied Mycenaean cemetery in Achaia prefecture. It consists of 55 graves that date from the Late Helladic (LH) IIB period to the LH IIIA period. A number of graves exhibit house architecture, and most likely belonged to higher-status individuals. The majority of the graves are circular constructions and no special grave goods were included. There are also a few pit graves dating to the end of the LH II period.[4]

The skeletal material studied comes from the 55 tholos tombs excavated in the 1990s. The state of preservation is good for most of the material, with 75.2% (288/383) of the individuals identifiable to sex and age (175 males, 80 females, and 33 children).

1. For recent reviews, see Sealy 2001; Katzenberg 2000.
2. Ambrose and Norr 1993.
3. This project was funded through a Marie Curie training fellowship awarded to E. Petroutsa.
4. Further archaeological information about the site will be published by L. Kolonas.

METHODS

Age Estimation and Sex Determination

The estimation of juvenile age at death was based on dental development,[5] epiphyseal union,[6] and long bone length.[7] It was possible to assess age at death based on dentition, which is considered to be the most accurate method.[8] The terms used for the different juvenile age groups are infancy (from birth to one year), early childhood (from two to five years), late childhood (from six to twelve years), and adolescence (from thirteen to nineteen years).

The estimation of the adult age at death was based on the observation of age-related changes in the os coxae morphology (the pubic symphysis[9] and the auricular surface[10]) and the cranium (degree of cranial suture closure[11] and dental wear[12]). Sex was also determined from morphological assessment of the os coxae and the cranium following the methodology detailed in Buikstra and Ubelaker.[13]

Isotope Sample Preparation

Samples of femur were taken from 36 adult humans and 6 animals, and were prepared for isotope analysis in the Stable Isotope Laboratory, Department of Archaeological Sciences, at the University of Bradford, UK. Preparation methods follow the procedure described in Richards and Hedges[14] with some modifications, including an ultrafiltration step.[15] Approximately 300 to 400 mg of bone pieces or powder were demineralized in 0.5 M HCl (aq.) at 4°C for three to five days. Samples were rinsed with distilled water and then gelatinized in acidic solution (pH 3) at 75–83°C for 48 hours. The liquid fraction containing the gelatinized protein was isolated by filtration (5 μm filter then 30,000 KDa ultrafilter), evaporated to dryness, rehydrated, frozen, and lyophilized to produce the final collagen product. For each analysis, 900 to 1,000 mg portions of the extracted collagen were used. Samples were combusted in an automated carbon and nitrogen analyzer (Carlo Erba elemental analyzer) coupled with a continuous-flow isotope ratio-monitoring mass spectrometer (Europa Geo 20/20 mass spectrometer). Where possible, samples were run in duplicate. The $\delta^{13}C$ values were measured relative to the vPDB standard, and $\delta^{15}N$ values were measured relative to the AIR standard reference. The analytical error (1σ) for all samples is 0.2‰ for both $\delta^{13}C$ and $\delta^{15}N$. Based on the criteria outlined by DeNiro,[16] most of the collagen was very well preserved with acceptable C:N ratios, collagen yields, and %C and %N values in the combusted collagen. Only samples with values that fell within the acceptable ranges for these criteria are reported here.

RESULTS

The isotopic analysis of faunal material from the Voudeni site provides the baseline isotopic data against which the human values can be compared

5. Moorrees, Fanning, and Hunt 1963a, 1963b; Ubelaker 1989.
6. Ubelaker 1989; Scheuer and Black 2000.
7. Ubelaker 1989.
8. Hillson 1996.
9. Brooks and Suchey 1990; Buikstra and Ubelaker 1994.
10. Ubelaker 1989.
11. Meindl and Lovejoy 1985.
12. Brothwell 1981.
13. Buikstra and Ubelaker 1994.
14. Richards and Hedges 1999a.
15. Brown, Nelson, and Southon 1988.
16. DeNiro 1985.

TABLE 14.1. FAUNAL $\delta^{13}C$ AND $\delta^{15}N$ VALUES

Species	$\delta^{13}C$	$\delta^{15}N$	*C/N*
Bird	-16. 5	8.5	3.4
Sheep	-20.9	3.2	3.4
Sheep	-21.1	3.7	3.5
Sheep	-21.1	3.7	3.3
Turtle	-23.2	8.1	3.6
Turtle	-23.6	5.7	3.6

TABLE 14.2. HUMAN $\delta^{13}C$ AND $\delta^{15}N$ VALUES

Sample	$\delta^{13}C$	$\delta^{15}N$	*C/N*	*% Collagen*	*%C*	*%N*	*Sex*
V01	-20.5	6.7	3.4	4.4	43.5	14.9	F
V05	-20	9.3	3.6	0.3	34.8	11.4	F
V06	-19.7	7.7	3.4	0.8	34.5	12.0	F
V09	-20.1	7.9	3.5	3.6	36.6	12.2	M
V11	-20.1	7.9	3.6	1.0	43.2	14.2	F
V12	-20.2	8.2	3.4	2.3	24.4	8.5	?
V14	-20.1	8	3.4	2.2	42.2	14.6	F
V15	-20.2	7.9	3.4	8	44.9	15.4	?
V16	-20	8.4	3.4	6.1	45.1	15.7	M
V17	-19.9	9	3.3	1.5	41.6	14.6	M
V19	-20.1	8.6	3.5	0.33	39.6	13.3	F
V21	-20.2	7.3	3.5	0.2	10.0	3.4	?
V23	-20.2	9	3.4	1.0	10.4	3.6	M
V24	-20.5	9	3.4	1.0	20.2	6.9	M
V25	-19.8	8.6	3.3	2.7	30.8	10.9	M
V26	-20	8.8	3.4	0.8	15.9	5.4	M
V27	-19.6	8.1	3.4	0.3	10.7	3.7	?
V28	-19.9	8.7	3.3	2.1	27.7	9.7	F
V29	-19.7	8.3	3.4	0.3	14.4	4.9	M
V30	-20.4	7.4	3.5	0.3	15.9	5.3	?
V31	-20.3	8.4	3.4	1.6	18.8	6.5	M
V32	-19.9	7.7	3.5	0.7	16.4	5.5	M
V33	-19.8	9.6	3.4	1.4	18.2	6.3	M
V34	-20	8.7	3.4	0.5	11.7	4.0	M
AVERAGE	-20.1	8.3		1.8	27.2	9.3	
STD	0.2	0.7		2.0	12.8	4.4	

(Table 14.1). Unfortunately, as is common in most Bronze Age cemetery sites, there is relatively little faunal material available for analysis. We were able to measure the isotopic values of sheep, birds, and turtles.

The herbivore (sheep) bone collagen $\delta^{13}C$ values range from -21.1‰ to -20.9‰ and the $\delta^{15}N$ values from 3.2‰ to 3.7‰, as expected for animals in temperate terrestrial C_3 ecosystems.[17] Bird $\delta^{13}C$ and $\delta^{15}N$ values are -16.7‰ and 8.5‰, respectively.

The two turtles have isotopic values that are quite different from the other herbivores (i.e., the sheep), because their $\delta^{13}C$ values are more negative ($\delta^{13}C$ = -23.2‰, -23.6‰) and the $\delta^{15}N$ values are higher ($\delta^{15}N$ = 8.1‰ and 5.7‰).

In Figure 14.1, each individual animal is represented by a separate icon. This is done so that we can easily differentiate them not only by species but by their environmental context, potentially providing useful information about varying human diets.

The isotopic results for the humans are given in Table 14.2 and are plotted alongside the faunal data in Figure 14.1. The human $\delta^{13}C$ values have a narrow range (-20.5‰ to -19.6‰). The human $\delta^{15}N$ values have a wider range (6.7‰ to 9.6‰).

17. Schoeninger and DeNiro 1984.

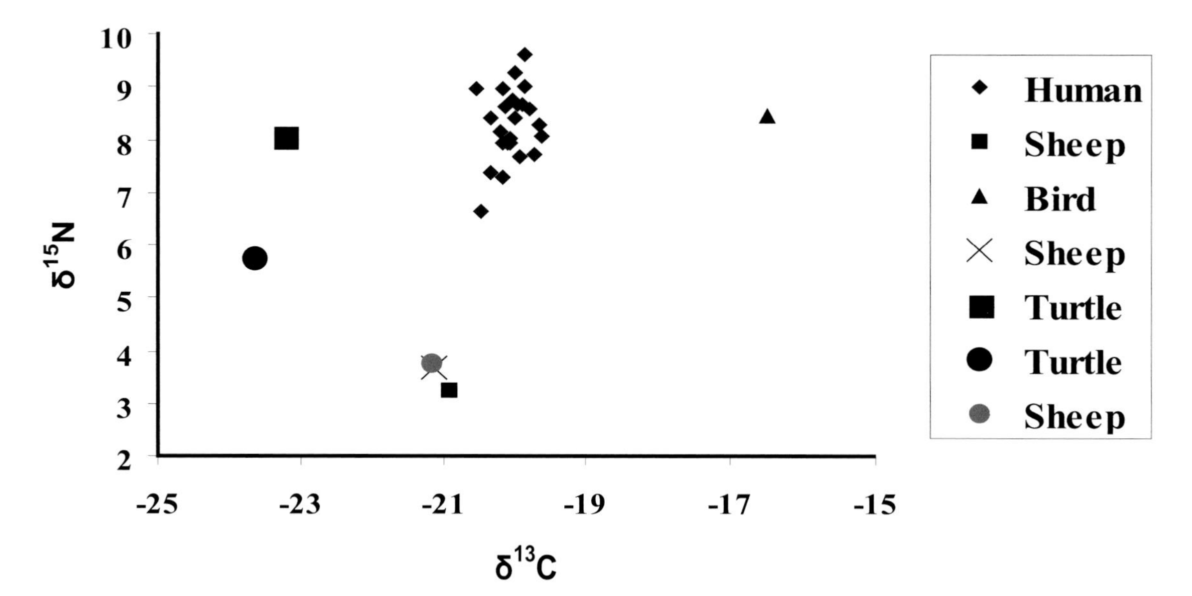

Figure 14.1. Stable carbon and nitrogen isotope values ($\delta^{13}C$, $\delta^{15}N$) from human and faunal bone collagen from Voudeni

DISCUSSION

The sheep bone collagen $\delta^{13}C$ and $\delta^{15}N$ values were in the range expected for animals in temperate terrestrial C_3 ecosystems.[18] Bird $\delta^{13}C$ and $\delta^{15}N$ values are quite different and indicate the consumption of marine resources.[19] The two turtles have isotopic $\delta^{13}C$ values consistent with a C_3 plant diet, and also indicative of the consumption of freshwater plants, which often have more negative $\delta^{13}C$ values than terrestrial plants from the same region due to the uptake of water dissolved CO_2. This is supported by higher $\delta^{15}N$ values compared to the sheep. This is due to the consumption of plants with higher $\delta^{15}N$ values, and the most likely explanation is that the turtles consumed more freshwater plants, which can have higher $\delta^{15}N$ values than terrestrial plants.

The human $\delta^{13}C$ values indicate a diet where the majority of protein was obtained from terrestrial C_3 plants and/or animals.

All human $\delta^{15}N$ values are higher than the sheep herbivore values. Using the sheep as a comparison, we can interpret these data as indicating that all these individuals consumed animal protein (meat and/or milk), with some individuals consuming more plants (those with lower $\delta^{15}N$) and some obtaining most of their protein from animal sources (those with high $\delta^{15}N$ values). There is no evidence in the carbon and nitrogen isotopic values for any significant consumption of marine foods.

The average isotopic values for seven individuals identified as female are $\delta^{13}C = -20.1 \pm 0.5‰$ and $\delta^{15}N = 8.1 \pm 1.1‰$. The average isotopic values of twelve individuals identified as males are $\delta^{13}C = -20.0 \pm 0.2‰$ and $\delta^{15}N = 8.6 \pm 0.5‰$. There is no statistically significant difference in the isotopic results between males and females, as shown in a plot of the data (Fig. 14.2). The two lowest $\delta^{13}C$ and $\delta^{15}N$ values are for two females, and the isotopic results indicate that they had relatively little animal protein in their diet. The highest $\delta^{15}N$ value, indicating a diet where the protein is largely coming from animals, is for a male. Therefore, despite these outliers,

18. Schoeninger and DeNiro 1984.
19. Chisholm, Nelson, and Schwarcz 1982; Schoeninger, DeNiro, and Tauber 1983.

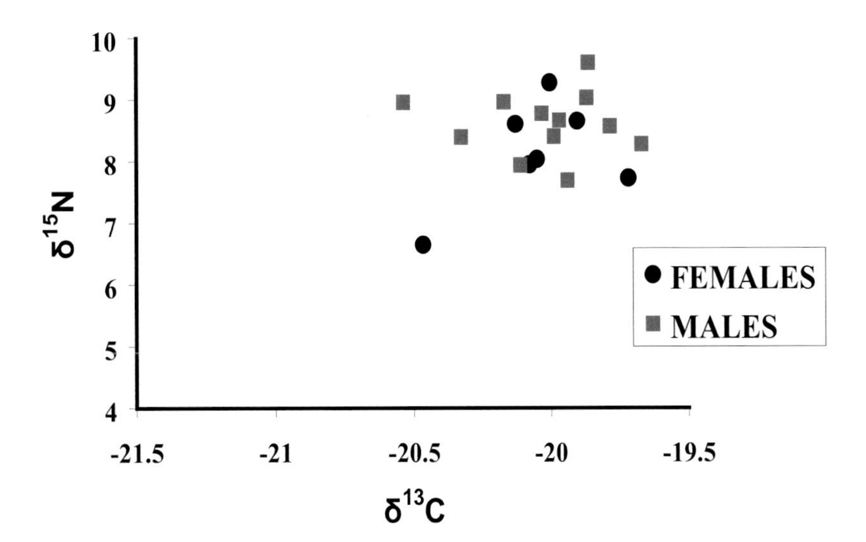

Figure 14.2. Comparison of stable carbon and nitrogen isotope values of males and females from Voudeni

it generally appears, in terms of isotopic data, that there was no difference between males and females in sources of dietary protein.

The isotopic results from Voudeni are similar to those found by other researchers for Late Bronze Age humans. Richards and Hedges[20] found very similar isotopic values for humans from the Late Bronze Age site of Armenoi, Crete, where there was also no difference between males and females, or between differently sized (and therefore, by inference, different status) graves. Similar data from two Late Bronze Age sites in the Peloponnese[21] also show that the humans had an isotopically homogenous diet, although in those studies it was not always possible to associate individuals with specific grave contexts or obtain secure sex information. In all three of these Late Bronze Age sites, the isotopic evidence did not show any significant input of marine foods.

CONCLUSIONS

We undertook stable carbon and nitrogen isotopic analysis of human and faunal bone collagen from the Late Bronze Age cemetery of Voudeni. We found that the population did not consume any significant amounts of marine protein despite the close proximity of the site to the sea. Instead, we found that the bulk of the dietary protein was derived from C_3 terrestrial sources, with most people having a mix of animal (meat and/or milk) and plant protein. We could not find any evidence of status (based on grave context) or gender differences in isotopic values. This study supports a pattern observed in other studies of prehistoric Greek populations: that the Bronze Age Aegean diet contained very little marine food.

20. Richards and Hedges 1999b, 2008.

21. Richards and Vika 2008.

REFERENCES

Ambrose S. H., and L. Norr. 1993. "Experimental Evidence for the Relationship of the Carbon Isotope Ratios of Whole Diet and Dietary Protein to Those of Bone Collagen and Carbonate," in *Prehistoric Human Bone: Archaeology at the Molecular Level,* ed. J. B Lambert and G. Grupe, New York, pp. 1–37.

Brooks, S. T., and J. M. Suchey. 1990. "Skeletal Age Determination Based on the *Os Pubis:* A Comparison of the Ascadi-Nemeskeri and Suchey-Brooks Methods," *Human Evolution* 5, pp. 227–238.

Brothwell, D. R. 1981. *Digging Up Bones: The Excavation, Treatment, and Study of Human Skeletal Remains,* 3rd ed., Ithaca, N.Y.

Brown, T. A., D. E. Nelson, and J. R. Southon. 1988. "Improved Collagen Extraction by Modified Longin Method," *Radiocarbon* 30, pp. 171–177.

Buikstra J. E., and D. H. Ubelaker, eds. 1994. *Standards for Data Collection from Human Skeletal Remains,* Fayetteville, Ark.

Chisholm, B. S., D. E. Nelson, and H. P. Schwarcz. 1982. "Stable-Carbon Isotope Ratios as a Measure of Marine versus Terrestrial Protein in Ancient Diets," *Science* 216, pp. 1131–1132.

DeNiro, M. J. 1985. "Post-Mortem Preservation and Alteration of in vivo Bone Collagen Isotope Ratios in Relation to Paleodietary Reconstruction," *Nature* 317, pp. 806–809.

Hillson, S. 1996. *Dental Anthropology,* Cambridge.

Katzenberg, M. A. 2000. "Stable Isotope Analysis: A Tool for Studying Past Diet, Demography, and Life History," in *Biological Anthropology of the Human Skeleton,* ed. M. A. Katzenberg and S. R. Saunders, New York, pp. 305–327.

Meindl, R. S., and C. O. Lovejoy. 1985. "Ectocranial Suture Closure: A Revised Method for the Determination of Skeletal Age at Death Based on the Lateral-Anterior Sutures," *American Journal of Physical Anthropology* 68, pp. 57–66.

Moorrees, C. F. A., E. A. Fanning, and E. E. Hunt. 1963a. "Formation and Resorption of Three Deciduous Teeth in Children," *American Journal of Physical Anthropology* 21, pp. 205–213.

———. 1963b. "Age Variation of Formation Stages for Ten Permanent Teeth," *Journal of Dental Research* 42(6), pp. 1490–1502.

Richards, M. P., and R. E. M. Hedges. 1999a. "Stable Isotope Evidence for Similarities in the Types of Marine Foods Used by Late Mesolithic Humans at Sites along the Atlantic Coast of Europe," *JAS* 26, pp. 717–722.

———. 1999b. "How Chemical Analysis of Human Bones Can Tell Us the Diets of People Who Lived in the Past," in *Flavours of Their Times: Food and Drink in Minoan and Mycenaean Times,* ed. Y. Tzedakis and H. Martlew, Athens, pp. 210–214, 216–217, 222–223, 226–227, 230–231, 246–247.

———. 2008. "Evidence of Past Human Diet at the Sites of the Neolithic Cave of Gerani; the Late Minoan III Cemetery of Armenoi; Grave Circles A and B at the Palace Site of Mycenae; and Late Helladic Chamber Tombs," in *Archaeology Meets Science: Biomolecular Investigations in Bronze Age Greece,* ed. Y. Tzedakis, H. Martlew, and M. K. Jones, Oxford, pp. 220–230.

Richards, M. P., and E. Vika. 2008. "Stable Isotope Results from New Sites in the Peloponnese: Cemeteries from Sykia, Kalamaki, and Spaliareika," in *Archaeology Meets Science: Biomolecular Investigations in Bronze Age Greece,* ed. Y. Tzedakis, H. Martlew, and M. K. Jones, Oxford, pp. 231–235.

Scheuer, L., and S. Black. 2000. *Developmental Juvenile Osteology,* San Diego.

Schoeninger, M., and M. J. DeNiro. 1984. "Nitrogen and Carbon Isotopic Composition of Bone Collagen from Marine and Terrestrial Animals," *Geochimica et Cosmochimica Acta* 48, pp. 625–639.

Schoeninger, M., M. J. DeNiro, and H. Tauber. 1983. "Stable Nitrogen Isotope Ratios of Bone Collagen Reflect Marine and Terrestrial Components of Prehistoric Human Diet," *Science* 220, pp. 1381–1383.

Sealy, J. 2001. "Body Tissue Chemistry and Palaeodiet," in *Handbook of Archaeological Sciences,* ed. D. R. Brothwell and A. M. Pollard, Chichester, pp. 269–279.

Ubelaker, D. H. 1989. *Human Skeletal Remains: Excavation, Analysis, Interpretation,* 2nd ed., Washington, D.C.

CHAPTER 15

Population Mobility at Frankish Corinth: Evidence from Stable Oxygen Isotope Ratios of Tooth Enamel

by Sandra J. Garvie-Lok

A number of years ago, the American School of Classical Studies conducted an intensive excavation of one area of Frankish Corinth. The resulting information has provided valuable insights into life in the city during the Frankish occupation. In particular, the recovery of a large number of burials has offered us the chance to investigate health, diet, and other aspects of daily life in those times.[1] The work reported here represents the preliminary results of a stable oxygen isotope study of human remains from the burials. Its goal is to improve our understanding of residential mobility in the populations of Frankish Corinth by identifying immigrants within these populations. This research forms part of an ongoing study examining diet, mobility, and other aspects of daily life at Corinth through stable carbon, nitrogen, oxygen, and strontium isotope analysis.[2]

HISTORICAL CONTEXT

When the Franks[3] invaded the Morea in A.D. 1204, they found at Corinth "a formidable castle, the finest of all Romania . . . the town lies below on the plain, well enclosed, indeed, with towers and walls."[4] The town finally fell to Geoffroy de Villehardouin after a five-year siege. At this time, Corinth became integrated into a feudal system in which the original Greek population was ruled by the linguistically and culturally distinctive Franks. Perhaps due to the hardships of the siege, Corinth appears to have lost a significant proportion of its population, and remained relatively small and poor until the middle of the 13th century.[5] At this time, Corinth appears

1. Williams, Barnes, and Snyder 1997; Williams et al. 1998; Garvie-Lok 2002; Barnes 2003.

2. Other results of this study are reported in Garvie-Lok 2002. The initial stages of this work were done while I was a J. Lawrence Angel Fellow of the Wiener Laboratory. Funding was also provided by the Social Sciences and Humanities Research Council of Canada, the University of Calgary, and the University of Alberta. Thanks are due to Charles K. Williams II for permission to study the Corinth material; to Ethne Barnes and Art Rohn for their cooperation, advice, and feedback at various stages of the work; and to Sherry Fox of the Wiener Laboratory for her support and encouragement during the preparation of this paper.

3. The term *Frank* refers, more or less, to Western Europeans. Although the Franks of Greece shared a common feudal Western European culture, they were disparate in language, origins and customs.

4. *Chronicle of the Morea*, pp. 110–111. The whole of the Latin Aegean was referred to as "Romania."

5. Williams 2003.

to have enjoyed a sudden increase in prosperity and importance; this may have been due to political and economic changes in the Frankish Mediterranean that caused a new influx of Frankish residents to the town.[6] This brief era of prosperity was ended abruptly in A.D. 1312 by a Catalan raid, perhaps in conjunction with an earthquake.[7] Corinth does not seem to have recovered its former economic status after this time. From the mid-14th century onward, it experienced several changes of rule, culminating in the Ottoman conquest of the region in the mid-15th century.

THE BURIALS

The 1989–1996 excavations at Corinth recovered a large number of human remains. The study of these remains and their context has shown that these may be separated into three distinctive burial groups representing three different eras at Corinth. One of the principal ways in which these groups differ is the apparent proportion of locally born individuals to immigrants.

The Room 4 Burials

The first group of burials dates to Corinth's period of prosperity. It is located in a former Byzantine monastery that was converted for use as a cemetery in the mid-13th century. After conversion, the monastery church served as a burial chapel and some of the surrounding rooms as burial areas.[8] This cemetery saw the burial of at least 200 individuals, most of whom were interred in one large room of the structure designated by the excavators as Room 4.[9] Analysis of the stratigraphy and associated finds indicates that these burials date to the second half of the 13th century.[10] The cemetery is likely associated with the activities taking place in a large building nearby.[11] Built in the 1260s or 1270s, this building includes a number of design elements typical of northern Europe, suggesting that it was erected and used by members of the Frankish community rather than Greek residents of the town.[12] Its layout and associated finds suggest that it was a hospice set up by the Frankish community of Corinth "to serve the sick, the poor, and pilgrims passing through Corinth to and from the Near East."[13] An analysis of the paleopathology of these remains by Ethne Barnes has revealed several individuals with signs of severe chronic or acute health problems, which would be consistent with the mission of a hospice to serve the sick.[14] Her analysis of genetic markers of the bones and teeth shows clusters of specific genetic markers in some areas of the cemetery, suggesting the presence of family plots and thus the use of the cemetery by a local long-term population. However, there is also evidence for genetic variability consistent with the immigration of outside individuals into the community.[15] Given the above interpretation, the Room 4 burial group might be expected to include a high proportion of individuals born outside the Corinth area. These individuals could potentially have come from a wide variety of locales, including not only various parts of western Europe but also the Frankish-controlled regions of the eastern Mediterranean.

6. Williams et al. 1998, p. 262.
7. Williams, Barnes, and Snyder 1997, pp. 41–42.
8. Williams and Zervos 1996, pp. 19–21; Williams, Barnes, and Snyder 1997, p. 261.
9. Williams et al. 1998, p. 239
10. Williams et al. 1998, p. 245.
11. Williams and Zervos 1996, p. 39; Williams et al. 1998, p. 261.
12. Williams, Barnes, and Snyder 1997, p. 31
13. Williams and Zervos 1995, p. 3.
14. Williams, Barnes, and Snyder 1997; Williams et al. 1998; Barnes 2003.
15. Williams et al. 1998, pp. 242–243.

The Ruined Church Burials

The second burial group postdates the events of A.D. 1312. In this period, the hospice was no longer in operation.[16] Burials in the nearby monastery buildings continued but in a different pattern. The Room 4 area was not used; instead, bodies were buried in and around the church.[17] This burial group dates to the 14th century.[18] Barnes's analysis of genetic variations of the skeleton suggests that the Ruined Church burials represent a population with lower genetic variability than the population represented by the Room 4 burials.[19] Given this information, the fact that the hospice was no longer in operation and the prospect that Corinth was less attractive to immigrants after its prosperity declined, the Ruined Church group might be expected to include fewer non-locally born individuals than the earlier Room 4 burial group.

The South Court Burials

The final group of burials was recovered from an open area to the south of the former monastery. Although this places them close to some of the Ruined Church burials, the two groups are separated clearly by burial style. While individuals in the two burial groups described above were buried with their heads propped up with tiles or stones to face the east and the hands crossed over the body, individuals in this later group were buried with their heads unsupported, sometimes turned to face the south, and their hands at their sides.[20] Stratigraphic analysis suggests that the South Court burials postdate those of the Ruined Church group.[21] Analysis of skeletal traits shows differences between this group and the earlier two groups suggesting the movement of individuals from an outside population into the area.[22] The stable carbon and nitrogen isotope values of the remains support such an interpretation, as they include some values that are highly atypical of Greece in this period and are also much more variable than would be expected in a single community.[23] Based on these findings, the South Court burial group, like the Room 4 burial group, might be expected to include a number of individuals not born in the Corinth area.

THE RESEARCH PROBLEM: DETECTING IMMIGRANTS AT CORINTH

As reviewed above, the evidence available at this time suggests that some of the communities who buried their dead in the cemeteries of Frankish Corinth included a substantial number of immigrants. There is reason to suspect that the proportion of these immigrants varied over time. However, this evidence is largely drawn from cultural practices and from analysis of genetic variations of the skeleton. These provide information on an individual's cultural affiliations and genetic heritage, but are silent on the question of his or her geographic origin. The direct identification of immigrants in the three burial groups would be extremely useful. It could, for example, provide independent evidence for the hypothesis that the

16. Williams, Barnes, and Snyder 1997, p. 42.
17. Williams and Zervos 1991, pp. 38–39; Williams, Barnes, and Snyder 1997, p. 245.
18. Barnes 2003, p. 436.
19. E. Barnes (pers. comm.).
20. Williams and Zervos 1991, p. 39; Barnes 2003.
21. Williams and Zervos 1991, p. 40.
22. E. Barnes (pers. comm.).
23. Garvie-Lok 2002. The Room 4 and Ruined Church burial groups were analyzed in the same study and produced no unusual values. As dietary $\delta^{13}C$ and $\delta^{15}N$ values are determined by diet rather than place of origin, the South Court values should be seen as suggesting rather than confirming the presence of immigrants; similarly, the relative uniformity of the Room 4 and Ruined Church values does not confirm the absence of immigrants.

Room 4 burial group includes a mixture of locally born individuals and immigrants from a variety of geographic locales. The stable oxygen isotope analysis reported here represents a first step toward the identification of immigrants in the cemeteries of Frankish Corinth.

Theoretical Background

Since the first application of the method in 1991,[24] stable oxygen isotope analysis has been used to study residential mobility in archaeological populations from a number of areas.[25] These applications have proven the value of the technique, providing insights on issues such as patterns of immigration in urban ethnic enclaves[26] and the source of recruitment for military units.[27] Although powerful, stable oxygen isotope analysis has some important limitations that are best understood through a brief review of its theoretical basis.

In the ecological and archaeological literature stable oxygen isotope ratios are expressed as $\delta^{18}O$ values. The $\delta^{18}O$ value of a substance expresses its $^{18}O/^{16}O$ ratio in terms of per mil (‰) departure from one of two international standards (PDB or SMOW), with higher values indicating a higher ratio of ^{18}O to ^{16}O in the sample.

The stable oxygen isotope analysis of archaeological human remains is performed on the mineral fraction (apatite) of tooth enamel and bone.[28] This apatite contains oxygen in the form of phosphate and carbonate.[29] In mammals, bone and tooth apatite $\delta^{18}O$ values are determined by the $\delta^{18}O$ value of body water, which is determined in turn by the $\delta^{18}O$ values of the organism's drinking water and food.[30] Because of differences in factors such as water metabolism, the exact relationship between body water $\delta^{18}O$, water $\delta^{18}O$, and food $\delta^{18}O$ varies from species to species.[31] As a result, the apatite $\delta^{18}O$ values of two species in a given region can differ considerably.[32] However, when $\delta^{18}O$ variation within a given species is considered, a clear regional pattern is seen. This pattern reflects regional differences in surface water and plant $\delta^{18}O$, which are generally higher in warmer climates.[33] As a result, apatite $\delta^{18}O$ values for mammals of a given species vary significantly between regions, with higher values typical of warmer climates.[34]

24. Schwarcz, Gibbs, and Knyf 1991.

25. E.g., Stuart-Williams et al. 1996; White et al. 1998, 2000, 2002, 2004a, 2004b; Dupras and Schwarcz 2001; White, Longstaffe, and Law 2001; Budd et al. 2003, 2004; Spence et al. 2004.

26. White et al. 2004a.

27. White et al. 2002; Spence et al. 2004.

28. Although oxygen is present in bone collagen, $\delta^{18}O$ analysis of archaeological bone collagen is not feasible due to the likelihood of contamination (Hedges, Stevens, and Richards 2004).

29. Due to in vivo differences and differences in the techniques used to prepare them for analysis, phosphate and carbonate $\delta^{18}O$ analyses of a bone apatite sample will produce different values. However, the separation between these values is consistent (Bryant et al. 1996b; Iacumin et al. 1996b).

30. Longinelli 1984; Luz, Kolodny, and Horowitz 1984; Luz and Kolodny 1989.

31. E.g., Luz, Kolodny, and Horowitz 1984; Luz and Kolodny 1989; Kohn, Schoeninger, and Valley 1996; for a model of the factors in this relationship, see Kohn 1996.

32. E.g., Bocherens et al. 1996; Kohn, Schoeninger, and Valley 1996; Sponheimer and Lee-Thorp 2001.

33. This variation results from regional variation in precipitation $\delta^{18}O$ values; for a recent review of precipitation $\delta^{18}O$, see Bowen and Wilkinson 2002.

34. Longinelli 1984; Luz, Kolodny, and Horowitz 1984; Levinson, Luz, and Kolodny 1987; Luz and Kolodny 1989; Luz, Cormie, and Schwarcz 1990.

If an individual mammal migrates from its natal region to an area with a significantly different climate, its tooth and bone $\delta^{18}O$ values will differ from the tooth and bone $\delta^{18}O$ values of the local population. In the case of bone, this difference will be slowly erased by turnover. However, as tooth enamel is not altered by the body after it is formed, its $\delta^{18}O$ value remains unchanged, preserving a permanent record of the individual's $\delta^{18}O$ value at the time of enamel formation.[35] Thus, the enamel $\delta^{18}O$ difference between the immigrating individual and the local population will persist regardless of the time since migration.

As described above, the theoretical basis for the use of $\delta^{18}O$ analysis to study residential mobility is straightforward. However, a number of complicating factors must be kept in mind. The first is possible alteration of apatite $\delta^{18}O$ values of archaeological bones and teeth after burial. This potential problem has long been recognized[36] and is handled in a number of ways. Some researchers restrict their work to tooth enamel, on the logic that the high crystallinity and low organic content of this substance render it fairly impervious to diagenetic alteration.[37] Others, while acknowledging the resistance of tooth enamel to alteration, point out that such alteration is unlikely but not impossible.[38] Due to this concern as well as the desire to work with bone as well as enamel, they have tried other approaches. As the phosphate in bone and tooth apatite is generally agreed to be less prone to diagenesis than carbonate, one solution is to perform phosphate $\delta^{18}O$ analysis.[39] This must be weighed against the fact that phosphate $\delta^{18}O$ analysis is significantly more difficult and expensive than carbonate $\delta^{18}O$ analysis. Another solution is to monitor samples for diagenetic alteration using indicators such as sample crystallinity and carbonate content.[40] All of these approaches have their merits and there is as yet no generally agreed standard.

The second limiting factor of the technique is that it works by identifying individuals whose apatite $\delta^{18}O$ values appear nonlocal in comparison to those of others buried in the area. This leads to two problems. First, a difference between immigrants and the locally born is seen only if local $\delta^{18}O$ values in the immigrants' place of origin differ significantly from those of their new residence. Thus, movement between areas with similar climates will not be detected. Second, because many locales in an area have similar $\delta^{18}O$ signatures, it is not possible to use apatite $\delta^{18}O$ to confirm a specific point of origin for a given individual; one can only conclude that that individual's $\delta^{18}O$ value is consistent with that point of origin. However, apatite $\delta^{18}O$ can be used to rule out a potential point of origin, providing that the range of local human $\delta^{18}O$ values for that place is known. In practice, $\delta^{18}O$ analysis becomes more powerful as more work is done on local $\delta^{18}O$ values for various sites in a region. When a number of such local values are known, researchers can begin to compare the nonlocal $\delta^{18}O$ values seen at their site to the $\delta^{18}O$ values typical of other sites in the region. Depending on the research problem in question, the ability to rule out a given point of origin for an individual, or to say at least that the individual *could* have come from a given place, can be extremely valuable.[41] When research in a region begins, however, potential points of origin may only be discussed in

35. E.g., Fricke and O'Neil 1996; Stuart-Williams and Schwarcz 1997.
36. Stuart-Williams et al. 1996.
37. Budd et al. 2003, 2004.
38. E.g., White et al. 2000.
39. White, Longstaffe, and Law 2001.
40. E.g., Stuart-Williams et al. 1996; White et al. 1998, 2000, 2002, 2004a, 2004b.
41. E.g., White, Longstaffe, and Law 2001, 2004; White et al. 2004a.

the broadest sense, based on the general relationship between mammalian $\delta^{18}O$ values and local temperature, humidity, and altitude.

A final factor to be considered is potential $\delta^{18}O$ variability within a population due to nursing infants and individual differences in diet. Nursing has a known effect on infant $\delta^{18}O$ values: until an infant is weaned, the $\delta^{18}O$ of its tissues will be higher than that of the general population.[42] As a result, the tooth enamel that was formed while an individual was a nursing infant should show a higher $\delta^{18}O$ value than the enamel formed after that individual was weaned. This difference can be used to great effect in the study of weaning age in past populations,[43] but it is a complicating factor when considering mobility, as younger-forming teeth can produce higher enamel $\delta^{18}O$ values that could potentially be mistaken as a sign of nonlocal origin.[44] Variation in individual $\delta^{18}O$ values within human populations may also be caused by differences in diet. Members of a population may have access to a variety of water sources such as wells whose $\delta^{18}O$ value may differ from that of local surface water. Consumption of traded foods and beverages with nonlocal $\delta^{18}O$ values may also vary. While it is difficult to confirm the effects of such variation in archaeological populations, several researchers have noted that they should be borne in mind.[45]

Materials and Methods

For this analysis, teeth were taken from a number of adult individuals from each of the three burial groups. In order to minimize damage to the skeletal collection, sampling was limited to single-rooted teeth that could be removed without damage to the alveolar bone; premolars were selected when possible, but the canine was sampled for one individual. In all, 30 samples are being studied. At this point, however, results are only available for 17 individuals. Information on these individuals is provided in Table 15.1. Sex and age at death estimates were performed by Ethne Barnes in the course of her broader study of the remains using methods standard in the osteological literature.[46]

Tooth enamel was prepared for analysis using a modified version of the protocol used in our laboratory to prepare bone samples for carbonate analysis.[47] After the outer surface of the enamel was removed with a Dremel hand tool, the enamel was separated from the underlying dentine, rinsed with purified water, and allowed to dry. The dried enamel was powdered with an agate mortar and pestle, and 50 mg was placed in a disposable 2.5 ml centrifuge tube. To remove organic contaminants, the sample was treated with 2% sodium hypochlorite for 24 hours, then rinsed to neutrality with purified water. Following this, the sample was leached to remove its most soluble fraction by adding 2 ml (0.04 ml/mg of sample) of 0.1 M acetic acid. The acid was allowed to react with the powder for two hours; at the one hour point, the vial was exposed to vacuum for five minutes to encourage complete exposure of the crystal surfaces to the acid solution. The sample was then rinsed to neutrality with purified water and lyophilized. To extract CO_2 for carbonate $\delta^{18}O$ analysis, samples were combined with 100% anhydrous phosphoric acid under vacuum at 25°C until the reaction was complete. The resulting CO_2 gas was cryogenically purified and analyzed

42. Luz and Kolodny 1989; Bryant et al. 1996a; Bryant and Froelich 1996; see Wright and Schwarcz 1998 for review.

43. E.g., Wright and Schwarcz 1998, 1999.

44. White et al. 2000.

45. E.g., Iacumin et al. 1996a; Wright and Schwarcz 1998; White et al. 2000.

46. Williams, Barnes, and Snyder 1997; Williams et al. 1998; Barnes 2003.

47. Garvie-Lok et al. 2004.

TABLE 15.1. CORINTH SAMPLES BY SEX, AGE, AND TOOTH POSITION

Grave	*Sex*	*Age*	*Tooth Sampled*
Room 4			
1991-5	?	16–17	RP_2
1995-7b	female	30–35	LP_2
1995-9c	female	ca. 18	RP_2
1996-12	male	40–45	LP_2
1996-18	male	25–35	RP^2
1996-20	female	>40	RP_1
1996-32	male	30–35	RP_2
1996-35	female	22–24	RP_1
1996-36	female	22–24	LP_2
Ruined Church			
1989-6	female	25–30	RP_1
1989-30	male	45–50	LUC
1990-7	female	25–35	RP_2
South Court			
1989-31	male	23–24	RP_2
1990-2	male	25–35	LP_1
1990-48	male	40–45	RP_2
1991-11	?	14–16	RP_1
1995-4	male	35–39	LP_1

on a Finnigan MAT 252 mass spectrometer at the University of Alberta Department of Earth and Atmospheric Sciences.

The enamel samples analyzed in this study were not directly assessed for diagenetic changes. However, prior stable isotope work on bone from these burials[48] used several preservation indicators including bone collagen content, bone collagen C/N ratio, and FTIR measures of crystallinity and carbonate content. The Corinth burials were judged to be well preserved according to these criteria. Because enamel is considerably more resistant than bone to the effects of diagenesis, these results argue strongly that the Corinth enamel samples are also well preserved.

Results

Sample $\delta^{18}O$ values are reported in Table 15.2. All $\delta^{18}O$ values are reported relative to the SMOW standard, which is more commonly used in the literature on archaeological mobility. This analytical method also produces enamel carbonate $\delta^{13}C$ values; as these produced some results of interest for the current discussion, they are also presented here. Figure 15.1 presents the results in visual form. Taken as a whole, the 17 individuals have $\delta^{18}O$ values ranging from 25.4‰ to 29.2‰, with a mean $\delta^{18}O$ value of 26.5 ± 1.0‰, and $\delta^{13}C$ values ranging from -12.8‰ to -1.9‰, with a mean $\delta^{13}C$ value of -11.1 ± 2.9‰. However, these figures include two clear outliers among

48. Garvie-Lok 2002.

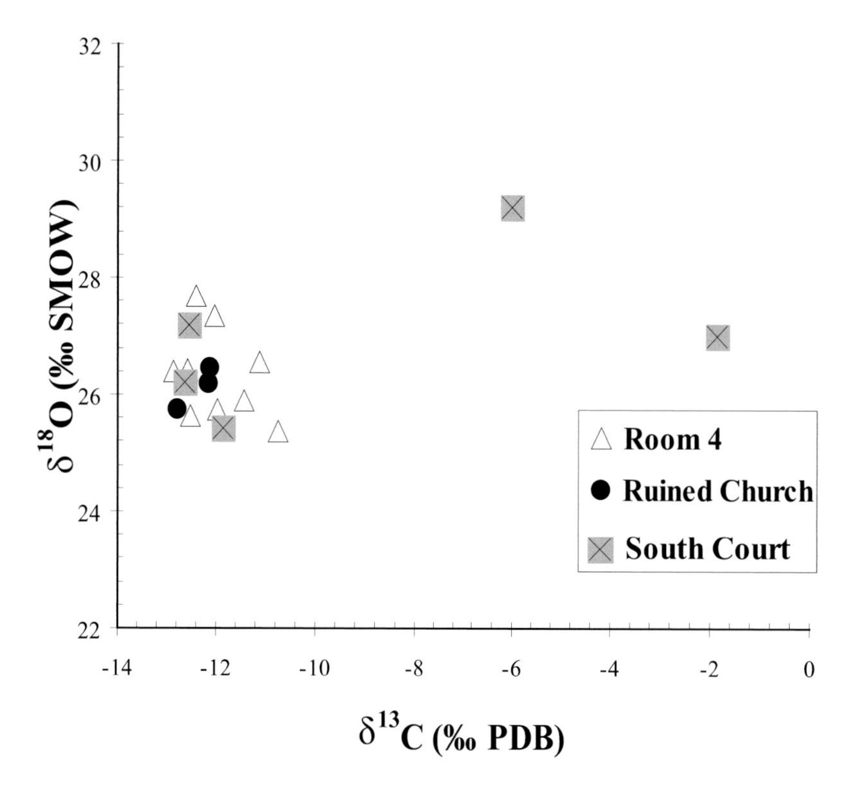

Figure 15.1. Corinth tooth enamel $\delta^{18}O$ and $\delta^{13}C$ values

the South Court group. When these are excluded, sample $\delta^{18}O$ ranges from 25.4‰ to 27.7‰, with a mean $\delta^{18}O$ value of 26.3 ± 0.7‰, and sample $\delta^{13}C$ values range from -12.8‰ to -10.7‰, with a mean $\delta^{13}C$ value of -12.1 ± 0.6‰.

Discussion

As a first step in the interpretation of these values, it is interesting to compare them to the human enamel carbonate $\delta^{18}O$ values one would expect given Corinth's climate. A rough comparison of this sort is possible, as data on modern rainfall $\delta^{18}O$ are available and there is a known relationship between drinking water $\delta^{18}O$ and tooth enamel $\delta^{18}O$ in humans.[49] While modern rainfall $\delta^{18}O$ data for Corinth have not been located, the average yearly value for Athens is -6.0‰ and that for Patras is -5.9‰; Corinth, lying between them and at a similar elevation, is likely similar.[50] The expected enamel carbonate $\delta^{18}O$ values for these rainfall values are 25.3‰ and 25.4‰; these fall at the low end of the Corinth $\delta^{18}O$ values. This is an extremely rough approximation, as rainfall $\delta^{18}O$ at Corinth in Frankish times may have differed somewhat from rainfall $\delta^{18}O$ today. However, it does show that the $\delta^{18}O$ values found for most of the enamel samples are reasonable given Corinth's modern climate.

Apart from a single outlying value, the $\delta^{18}O$ values show relatively little variation, with a total range of 2.2‰. This is in agreement with the 2‰ $\delta^{18}O$ range of variation that has been proposed for human communities not affected by migration or trade in food and drink.[51] Given this agreement, and the fit to Corinth's climate, it is tempting to argue that these 16 individuals all grew up in Corinth. However, this is not necessarily so,

49. Modern rainfall $\delta^{18}O$ values cited here are drawn from The International Atomic Energy Agency's database of precipitation stable isotope values (IAEA/WMO). Following Budd et al. (2003, 2004), the conversion used to calculate expected tooth enamel $\delta^{18}O$ from drinking water $\delta^{18}O$ is that established by Levinson, Luz, and Kolodny (1987).

50. IAEA/WMO. All rainfall $\delta^{18}O$ values cited here were taken from yearly average precipitation $\delta^{18}O$ values; where more than one year was recorded, the average of all years was calculated.

51. White et al. 2004a.

TABLE 15.2. CORINTH TOOTH ENAMEL $\delta^{18}O$ AND $\delta^{13}C$ VALUES

Grave	*$\delta^{18}O$ (‰ SMOW)*	*$\delta^{13}C$ (‰ PDB)*
Room 4		
1991-5	26.4	-12.6
1995-7b	25.6	-12.5
1995-9c	27.3	-12.0
1996-12	27.7	-12.4
1996-18	26.4	-12.8
1996-20	25.9	-11.4
1996-32	25.7	-12.0
1996-35	26.5	-11.1
1996-36	25.4	-10.7
Ruined Church		
1989-6	26.4	-12.1
1989-30	25.7	-12.8
1990-7	26.2	-12.1
South Court		
1989-31	29.2	-6.0
1990-2	27.0	-1.9
1990-48	25.4	-11.8
1991-11	27.2	-12.5
1995-4	26.2	-12.6

as many locales in the Mediterranean will have $\delta^{18}O$ values similar to that of Corinth. For example, the rainfall $\delta^{18}O$ value for Bari, Italy, is -5.8‰. Thus, the main $\delta^{18}O$ group may include individuals who did not, in fact, grow up in Corinth.

The Room 4 and Ruined Church Burials

As the data to this point include only three of the Ruined Church burials and these show no apparent differences from the Room 4 burials, these burial groups will be considered together. Despite reasons to suspect that the population of Frankish Corinth included a high proportion of nonlocal individuals, especially in the period when the hospice was in operation, these burials show modest $\delta^{18}O$ and $\delta^{13}C$ variation. With $\delta^{18}O$ values ranging from 25.4‰ to 27.7‰, they could easily be interpreted as a single community, all locally born. The group's $\delta^{13}C$ values, ranging from -12.8‰ to -10.7‰, would agree with a single community primarily dependent on C^3 grains such as wheat but also consuming some millet.[52] This finding is in agreement with the dietary data previously gathered on this population.[53]

As remarked above, these burials could include individuals who grew up elsewhere in the Mediterranean. Thus, the $\delta^{18}O$ values do not contradict the other evidence for mobility at Frankish Corinth, but do not add to it either. A wider sample of the group might locate individuals with nonlocal $\delta^{18}O$ values. However, strontium isotope analysis, which does not depend on climatic differences to distinguish between locales, is a more promising alternative.

The South Court Burials

Three of the five South Court burials show stable isotope values similar to those of the earlier burials. However, two individuals stand out. The first is set apart by both $\delta^{18}O$ and $\delta^{13}C$ values. At 29.2‰, this individual's enamel carbonate $\delta^{18}O$ value is quite high and suggests a childhood spent in a warm, arid region; it is comparable to bone carbonate $\delta^{18}O$ values documented for the Dakhleh Oasis in Egypt.[54] This individual also has a higher $\delta^{13}C$ value than most of the others analyzed; at -6.0‰, it suggests a diet more evenly balanced between C^3 and C^4 sources. As C^4 plants are tolerant of hot, dry conditions, greater C^4 dependence tends to be seen in warmer areas. Finally, in prior work on bone collagen from these burials, this individual was found to have a $\delta^{15}N$ value almost 4‰ higher than those of most adults at the site.[55] This too is often seen in individuals from a hot and arid environment.[56] Thus, in this case, all appears to fit quite well: enamel $\delta^{18}O$, enamel $\delta^{13}C$, and collagen $\delta^{15}N$ all suggest an origin in a warmer, drier climate than Corinth. It is worth noting that the collagen $\delta^{15}N$ was made on a rib fragment. The retention of the high $\delta^{15}N$ value in a bone that turns over quite rapidly suggests that the individual did not move in childhood, but rather some time in the last few years before death.

The second South Court burial that stands apart from the others has an enamel $\delta^{18}O$ value similar to that of most individuals at the site, but a $\delta^{13}C$ value departing even farther from the majority of values. At -1.9‰,

52. C^3 plants include most crop plants and have an average $\delta^{13}C$ value around -26‰, while C^4 plants include only a few important human food plants (most notably maize, millet, sorghum, and sugarcane) and have an average $\delta^{13}C$ value around -13‰ (Bender 1971). As maize was unknown in the Old World at this time, the C^4 grain in question would likely be millet.

53. Garvie-Lok 2002.

54. Dupras and Schwarcz 2001.

55. Garvie-Lok 2002.

56. See Dupras and Schwarcz (2001) for review.

the $\delta^{13}C$ value of this individual is consistent with a diet based almost completely on a C^4 staple. This individual also has a higher collagen $\delta^{15}N$ value than other adults in the site, although not as high as the burial described above. These values also suggest immigration to Corinth, but from a different sort of area. In this case, it appears that the foreign-born individual grew up in an area where dietary customs differed significantly from that of the Corinth population but whose climate was not all that different from Corinth's.

CONCLUSIONS

The results of this analysis, though based on a small sample, have added to our understanding of immigration patterns in the population of Frankish Corinth. Events at Corinth in this period give reason to suspect that the population included a significant number of people who were not originally born in the area. Studies of the material culture and human remains recovered during excavations at Frankish Corinth strengthen this suspicion. However, to the extent that can be determined from this small $\delta^{18}O$ and $\delta^{13}C$ study, most of the population may in fact have been born locally. With the exception of two outliers, enamel $\delta^{18}O$ and $\delta^{13}C$ values for the individuals studied vary within a relatively narrow range and are consistent with those expected for a group of people born in the Corinth area. However, it is important to recognize that many places in the Aegean are likely to have drinking water $\delta^{18}O$ values that are similar to that of Corinth. It is thus possible that these burials do include some individuals who were born elsewhere in the Aegean and whose presence was not detected through $\delta^{18}O$ analysis.

The potential of stable isotope analysis to detect nonlocal individuals in Greek cemeteries is demonstrated by the presence of two outliers in the South Court burial group. The high enamel $\delta^{18}O$ value of one individual, coupled with high enamel $\delta^{13}C$ and collagen $\delta^{15}N$, suggests that he originated from an area with a climate significantly different from Corinth's, and may not have moved from there to Corinth until relatively shortly before his death. The enamel $\delta^{18}O$ value is similar to values documented for Roman Egypt; along with the high $\delta^{13}C$ and $\delta^{15}N$ values, this suggests an origin in a hot, dry area (such as the Middle East or North Africa) rather than the cooler climates typical of much of Europe. In the case of the other individual identified as an immigrant, the combination of $\delta^{18}O$ and $\delta^{13}C$ values suggests a childhood spent in a climate not too different from that of Corinth, but in a community whose subsistence practices were quite different from those seen at Corinth. Thus, the results clearly indicate two different regions of origin for the two nonlocal individuals detected in the South Court group. This suggests that the community living at Corinth in this late period may have been quite diverse.

REFERENCES

Barnes, B., 2003. "The Dead Do Tell Tales," in *Corinth: The Centenary: 1896–1996* (*Corinth* XX), ed. C. K. Williams II and N. Bookidis, Princeton, pp. 435–443.

Bender, M. M. 1971. "Variations in the $^{13}C/^{12}C$ Ratios of Plants in Relation to the Pathway of Photosynthetic Carbon Dioxide Fixation," *Phytochemistry* 10, pp. 1239–1244.

Bocherens, H., P. L. Koch, A. Mariotti, D. Geraads, and J.-J. Jaeger. 1996. "Isotopic Biogeochemistry (^{13}C, ^{18}O) of Mammalian Enamel from African Pleistocene Hominid Sites," *Palaios* 11, pp. 306–318.

Bowen, G. J., and B. Wilkinson. 2002. "Spatial Distribution of $\delta^{18}O$ in Meteoric Precipitation," *Geology* 30, pp. 315–318.

Budd, P., C. Chenery, J. Montgomery, J. Evans, and D. Powlesland. 2003. "Anglo-Saxon Residential Mobility at West Heslerton, North Yorkshire, UK from Combined O- and Sr-Isotope Analysis," in *Plasma Source Mass Spectrometry: Applications and Emerging Technologies,* ed. G. Holland and S. D. Tanner, Cambridge, pp. 198–208.

Budd, P., A. Millard, C. Chenery, S. Lucy, and C. A. Roberts. 2004. "Investigating Population Movement by Stable Isotope Analysis: A Report from Britain," *Antiquity* 78, pp. 127–141.

Bryant, J. D., and P. N. Froelich. 1996. "Comment on 'Oxygen Isotope Composition of Human Tooth Enamel from Medieval Greenland: Linking Climate and Society,'" *Geology* 24, pp. 477–478.

Bryant, J. D., P. N. Froelich, W. J. Showers, and B. J. Jenna. 1996a. "A Tale of Two Quarries: Biologic and Taphonomic Signatures in the Oxygen Isotope Composition of Tooth Enamel Phosphate from Modern and Miocene Equids," *Palaios* 11, pp. 396–408.

Bryant, J. D., P. L. Koch, P. N. Froelich, W. J. Showers, and B. J. Jenna. 1996b. "Oxygen Isotope Partitioning between Phosphate and Carbonate in Mammalian Apatite," *Geochimica et Cosmochimica Acta* 60, pp. 5145–5148.

Chronicle of the Morea = H. E. Lurier, trans., *Crusaders as Conquerors: The Chronicle of Morea,* New York, 1964.

Dupras, T. L., and H. P. Schwarcz. 2001. "Strangers in a Strange Land: Stable Isotope Evidence for Human Migration in the Dakhleh Oasis," *JAS* 28, pp. 1199–1208.

Fricke, H. C., and J. R. O'Neil. 1996. "Inter- and Intra-Tooth Variation in the Oxygen Isotope Composition of Mammalian Tooth Enamel Phosphate: Implications for Palaeoclimatological and Palaeobiological Research," *Palaeogeography, Palaeoclimatology, Palaeoecology* 126, pp. 91–99.

Garvie-Lok, S. J. 2002. "Loaves and Fishes: A Stable Isotope Reconstruction of Diet in Medieval Greece" (diss. Univ. of Calgary).

Garvie-Lok, S. J., T. L. Varney, and M. Katzenberg. 2004. "Acetic Acid Treatment of Bone Carbonate: The Effects of Treatment Time and Solution Concentration," *JAS* 31, pp. 763–776.

Hedges, R. E. M., R. E. Stevens, and M. P. Richards. 2004. "Bone as a Stable Isotope Archive for Local Climatic Information," *Quaternary Science Reviews* 23, pp. 959–965.

Iacumin, P., H. Bocherens, A. Mariotti, and A. Longinelli. 1996a. "An Isotopic Palaeoenvironmental Study of Human Skeletal Remains from the Nile Valley," *Palaeogeography, Palaeoclimatology, Palaeoecology* 126, pp. 15–30.

———. 1996b. "Oxygen Isotope Analysis of Co-Existing Carbonate and Phosphate in Biogenic Apatite: A Way to Monitor Diagenetic Alteration of Bone Phosphate?" *Earth and Planetary Science Letters* 142, pp. 1–6.

IAEA/WMO. International Atomic Energy Agency/World Meteorological Organization, Global Network of Isotopes in Precipitation: The GNIP Database, http://www.naweb.iaea.org/napc/ih/GNIP/IHS_GNIP.html (accessed July 31, 2008).

Kohn, M. J. 1996. "Predicting Animal $\delta^{18}O$: Accounting for Diet and Physiological Adaptation," *Geochimica et Cosmochimica Acta* 60, pp. 4811–4829.

Kohn, M. J., M. J. Schoeninger, and J. W. Valley. 1996. "Herbivore Tooth Oxygen Isotope Compositions: Effects of Diet and Physiology," *Geochimica et Cosmochimica Acta* 60, pp. 3889–3896.

Levinson, A. A., B. Luz, and Y. Kolodny. 1987. "Variations in Oxygen Isotopic Composition of Human Teeth and Urinary Stones," *Applied Geochemistry* 2, pp. 367–371.

Longinelli, A. 1984. "Oxygen Isotopes in Mammalian Bone Phosphate: A New Tool for Paleohydrological and Paleoclimatological Research?" *Geochimica et Cosmochimica Acta* 48, pp. 385–390.

Luz, B., A. B. Cormie, and H. P. Schwarcz. 1990. "Oxygen Isotope Variations in Phosphate of Deer Bones," *Geochimica et Cosmochimica Acta* 54, pp. 1723–1728.

Luz, B., and Y. Kolodny. 1989. "Oxygen Isotope Variation in Bone Phosphate," *Applied Geochemistry* 4, pp. 317–323.

Luz, B., Y. Kolodny, and M. Horowitz. 1984. "Fractionation of Oxygen Isotopes between Mammalian Bone-Phosphate and Environmental Drinking Water," *Geochimica et Cosmochimica Acta* 48, pp. 1689–1693.

Schwarcz, H. P., L. Gibbs, and M. Knyf. 1991. "Oxygen Isotope Analysis as an Indicator of Place of Origin," in *Snake Hill: An Investigation of a Military Cemetery from the War of 1812,* ed. S. Pfeiffer and R. F. Williamson, Toronto, pp. 263–268.

Spence, M. W., C. D. White, F. J. Longstaffe, and K. R. Law. 2004. "Victims of the Victims: Human Trophies Worn by Sacrificed Soldiers from the Feathered Serpent Pyramid, Teotihuacan," *Ancient Mesoamerica* 15, pp. 1–15.

Sponheimer, M., and J. A. Lee-Thorp. 1999. "Oxygen Isotopes in Enamel Carbonate and Their Ecological Significance," *JAS* 26, pp. 723–728.

———. 2001. "The Oxygen Isotope Composition of Mammalian Enamel Carbonate from Morea Estate, South Africa," *Oecologia* 126, pp. 153–157.

Stuart-Williams, H. Le Q., and H. P. Schwarcz. 1997. "Oxygen Isotope Determination of Climatic Variation Using Phosphate from Beaver Bone, Tooth Enamel, and Dentine," *Geochimica et Cosmochimica Acta* 61, pp. 2539–2550.

Stuart-Williams, H. Le Q., H. P. Schwarcz, C. D. White, and M. W. Spence. 1996. "The Isotopic Composition and Diagenesis of Human Bone from Teotihuacan and Oaxaca, Mexico," *Palaeogeography, Palaeoclimatology, Palaeoecology* 126, pp. 1–14.

White, C. D., F. J. Longstaffe, and K. R. Law. 2001. "Revisiting the Teotihuacan Connection at Altun Ha: Oxygen-Isotope Analysis of Tomb F-8/1," *Ancient Mesoamerica* 12, pp. 65–72.

———. 2004. "Exploring the Effects of Environment, Physiology and Diet on Oxygen Isotope Ratios in Ancient Nubian Bones and Teeth," *JAS* 31, pp. 233–250.

White, C. D., M. W. Spence, F. J. Longstaffe, and K. R. Law. 2000. "Testing the Nature of Teotihuacán State Representation at Kaminaljuyú Using Phosphate Oxygen-Isotope Ratios," *Journal of Anthropological Research* 56, pp. 535–558.

———. 2004a. "Demography and Ethnic Continuity in the Tlailotlacan Enclave of Teotihuacan: The Evidence from Stable Oxygen Isotopes," *JAnthArch* 23, pp. 385–403.

White, C. D., M. W. Spence, F. J. Longstaffe, H. Le Q. Stuart-Williams, and K. R. Law. 2002. "Geographic Identities of the Sacrificial Victims from the Feathered Serpent Pyramid, Teotihuacan: Implications for the Nature of State Power," *Latin American Antiquity* 13, pp. 217–236.

White, C. D., M. W. Spence, H. Le Q. Stuart-Williams, and H. P. Schwarcz. 1998. "Oxygen Isotopes and the Identification of Geographical Origins: The Valley of Oaxaca versus the Valley of Mexico," *JAS* 25, pp. 643–655.

White, C. D., R. Storey, F. J. Longstaffe, and M. W. Spence. 2004b. "Immigration, Assimilation, and Status in the Ancient City of Teotihuacan: Stable Isotopic Evidence from Tlajinga 33," *Latin American Antiquity* 15, pp. 176–198.

Williams, C. K., II. 2003. "Frankish Corinth: An Overview," in *Corinth: The Centenary: 1896–1996* (Corinth XX), ed. C. K. Williams II and N. Bookidis, Princeton, pp. 423–434.

Williams, C. K., II, E. Barnes, and L. M. Snyder. 1997. "Frankish Corinth, 1996," *Hesperia* 66, pp. 1–47.

Williams, C. K., II, L. M. Snyder, E. Barnes, and O. H. Zervos. 1998. "Frankish Corinth, 1997," *Hesperia* 67, pp. 222–281.

Williams, C. K., II, and O. H. Zervos. 1991. "Corinth, 1990: Southeast Corner of Temenos E," *Hesperia* 60, pp. 1–58.

———. 1995. "Frankish Corinth, 1994," *Hesperia* 64, pp. 1–60.

———. 1996. "Frankish Corinth, 1995," *Hesperia* 65, pp. 1–39.

Wright, L., and H. P. Schwarcz. 1998. "Stable Carbon and Oxygen Isotopes in Human Tooth Enamel: Identifying Breastfeeding and Weaning in Prehistory," *American Journal of Physical Anthropology* 106, pp. 1–18.

———. 1999. "Correspondence between Stable Carbon, Oxygen and Nitrogen Isotopes in Human Teeth Enamel and Dentine: Infant Diets at Kaminaljuyú," *JAS* 26, pp. 1159–1170.

CHAPTER 16

Porotic Hyperostosis in Neolithic Greece: New Evidence and Further Implications

by Eleni Stravopodi, Sotiris K. Manolis, Stavros Kousoulakos, Vassiliki Aleporou, and Michael P. Schultz

The nature of Neolithization in Greece is an area of intense ongoing debate in prehistoric research. Most scholars emphasize the major impact of the agricultural revolution on demographic structure, socioeconomic institutions, and the health status of indigenous societies.[1] The association of disease with cultural development and elaboration is an established theme in the literature, and the direction of paleopathological research has been shaped by this overarching archaeological paradigm.

In this context, the study of the bone pathology known as porotic hyperostosis has been a particularly popular, yet controversial, subject for anthropological research in Greek prehistory. Porotic hyperostosis is defined as an osseous disorder manifested as pitting and/or porosis on the cranial vault (cribra cranii) and orbital areas (cribra orbitalia) with, occasionally, expanded lesions on the postcranial skeleton. The epidemic spread of porotic hyperostosis has been linked to the advent and adoption of Neolithic lifeways.

The prevailing argument for this linkage between pathology and sociocultural change suggests that the agricultural revolution is associated with an abrupt decline in health—not only in Greece,[2] but also in other world regions.[3] Factors involved include changes in settlement patterns with consequent population density increases resulting in less hygienic conditions, the dietary shift from protein to higher carbohydrate intake, and the occupation of extreme environmental settings.

Porotic hyperostosis is identified as the most common pathology found during the Neolithic in Greece.[4] A range of different diseases have been referred to as a cause of porotic hyperostosis, including genetic anemias (sickle cell and thalassemia) and the malarias that are often endemic in marshy and wetland environments.[5]

During the last decade, paleopathological research in prehistoric Greece has favored the argument that porotic hyperostosis might also serve as the marker of dietary iron deficiency in the Greek Neolithic.[6] It has been suggested that the major dietary shifts in the Neolithic from the abundance of iron-rich animal protein to a poor quality vegetarian diet that is high in carbohydrates and low in iron account for iron deficiency anemia. Iron deficiency anemia is favored over genetic anemias as a primary cause

1. Angel 1966; *Lerna* II.
2. Manolis and Stravopodi 2003; Stravopodi, Manolis, and Neroutsos 1997; Stravopodi and Manolis 2000.
3. Larsen 1998; Meiklejohn and Zvelebil 1991.
4. Angel 1966.
5. Angel 1966.
6. Manolis and Stravopodi 2003; Papathanasiou 2000; Stravopodi, Manolis, and Neroutsos 1997; Triantaphyllou 1999.

of porotic hyperostosis because of two observations: the lesions caused by genetic anemias do not develop a severe level of morphological expression (characterized by hyperplasia of marrow), and the association of cranial porotic hyperostosis with postcranial lesions, described as characteristic of genetic anemias,[7] is actually observed in very few archaeological cases.

Although most research has suggested that porotic hyperostosis is a marker of the Neolithic lifestyle, the intensification of research on newly discovered human skeletal collections in prehistoric sites and a reassessment of the process of Neolithization in Greece,[8] together with a new understanding of the complexity of the biochemical processes involved in iron intake,[9] contest this conventional interpretation. This challenge in the Greek context fits well with recent findings from adjacent regions such as the Levant.[10]

New evidence from the analysis of anthropological remains in Greece suggests that porotic hyperostosis shows significant morphological variability per site.[11] These results show that the actual pathology is more complicated than previously thought, and suggest that there is not a simple association between an epidemic increase in the occurrence of porotic hyperostosis and sociocultural change.

In many Neolithic societies in central Greece, where the inhabitants have been classified from archaeological data as full-time agriculturalists, and where porotic hyperostosis is therefore expected to occur in high frequencies, porotic hyperostosis is not found at all.[12] In contrast, societies of earlier and later periods with a balanced diet and protein intake often show significant prevalence of porotic hyperostosis. Additionally, while anemia seems to be one of the etiological factors, new research suggests that other diseases, such as scurvy, rickets, inflammatory disorders, and nonspecific infections, can be identified as alternative causal factors.

In other research, the idea of clear-cut boundaries between intensified agricultural practices and hunting-gathering in Greek prehistory is also being seriously challenged.[13] New evidence from skeletal and isotopic research,[14] in combination with paleobotanical and faunal data, supports the argument that mixed economies and shifting reliance on agriculture characterize the Neolithic. Broad variability in economic structures among Neolithic groups, with some systems resulting in surpluses rather than paucity of food, has been documented.[15]

The debate about the character, etiology, and sociocultural profile of porotic hyperostosis and the need to revisit concepts and methodologies is characteristic of the wider development of paleoepidemiological studies of prehistoric Greece. A number of new research projects focus on innovative laboratory methods, borrowed from both the biological and geological sciences. This recent work has significantly increased the importance of paleopathological analysis as an integral tool in archaeological investigations.

The present project aims to investigate the identity, biogeography, and possible etiology of porotic hyperostosis in prehistoric Greece, using a new methodological approach. Paleohistopathological examination is introduced as a reliable analytical technique that is based upon the differential diagnosis of diseased bone. The working hypothesis of this project is that the oft-proposed cause-and-effect relationship between porotic hyperostosis and Neolithic social changes is overemphasized. Porotic hyperostosis is

7. Ortner and Putschar 1981; Stuart-Macadam 1991, 1998; Hershkovitz et al. 1997; Schultz 2001.
8. Kotsakis 2000, p. 177.
9. Ryan 1997.
10. Hershkovitz 2004.
11. Stravopodi, in prep.
12. Stravopodi, in prep.
13. Kotsakis 2000; Kyparissi-Apostolika 2000.
14. Cook 2002.
15. Cook 2002; Grmek 1989; Kyparissi-Apostolika 2000.

rather defined as a morphological feature with a multifactorial etiology that exhibits a broad range of expression across different regions and periods during Greek prehistory.

MATERIALS AND METHODS

The sample under study includes 182 individuals from systematically excavated contexts in northern, central, eastern, and southern Greece (Table 16.1). The sites, some in caves and others in open areas, range from sea coast to mid-altitude inland, mountainous, and marshy settings. Most of the remains consist of scattered bone; very few of the specimens come from articulated skeletons. The minimum number of individuals (MNI) is determined from the articulated skeletons and from the anatomical element counts of the scattered bones.[16]

Our porotic hyperostosis project includes human remains from 11 sites, although analysis of the data from all sites is not completed. To date, none of the Voudeni sample has been studied so the individuals are not yet included in our sample. For Ayios Haralabos on Crete, only information on specimens selected for special analysis (e.g., histology) is provided.[17] Selected specimens from published data exposing lesions of porotic hyperostosis are also to be included in the study materials.[18]

The time periods represented by our samples include the Mesolithic (MNI = 4), Neolithic (MNI = 83), and Bronze Age (MNI = 95) periods. Although some sites might be dated more precisely to subphases (for example, Late Neolithic or Middle/Late Bronze Age), the small sample sizes necessitate the use of broader chronological categories. The samples are characterized by differential preservation, lack of homogeneity within and between sites,[19] and unbalanced statistical sizes. They thus should be regarded as case study samples rather than population units.

TABLE 16.1. STUDY SITES

Site	*Area of Greece*	*Setting*	*Period (Sample Size)*
Theopetra	Central	Mid-altitude, cave	Mesolithic (*N* = 4), Neolithic (*N* = 29)
Tharrounia	Eastern	Island, cave	Neolithic (*N* = 28)
Kouveleiki	Southern	Mid-altitude, cave	Neolithic (*N* = 9)
Kastria	Southern	Mountainous, cave	Neolithic (*N* = 15), Bronze Age (*N* = 17)
Lakonia	Southern	Coastal, cave	Neolithic (*N* = 2)
Lamia	Central	Valley, cave	Bronze Age (*N* = 7)
Perachora	Southern	Marshy lake, cave	Bronze Age (*N* = 52)
Kalogerovrisi	Eastern	Mid-altitude, open air	Bronze Age (*N* = 15)
Skyros	Eastern	Coastal, open air	Bronze Age (*N* = 2)
Orfeas	Southern	Mountainous cave	Bronze Age (*N* = 2)
Voudeni	Southern	Low altitude, open air	Bronze Age (*N* = 383)

Note: Sample sizes in the table are total identified for each site. Combined totals studied for this project are Mesolithic *N* = 4, Neolithic *N* = 83, Bronze Age *N* = 95. (Because the Voudeni sample has not yet been studied, the individuals are not included in the Bronze Age sample.)

16. As described by Buikstra and Ubelaker 1994.
17. P. McGeorge (pers. comm.).
18. Papathanasiou 2000; Triantaphyllou 1999.
19. The fieldwork at the Perachora and Orfeas caves is in progress.

Individuals are divided into six age groups, namely, fetal (<birth), infants (<3), children (3–12), adolescents (12–20), young adults (20–30), and adults (30+).[20] Not all age groups are represented at all sites, although subadults tend to be the largest cohorts.

For sex determination, morphological criteria from the examination of the gross specimens are used.[21] However, at this stage of the study, as we are dealing with mostly fragmentary cranial elements and an insufficiency of adult specimens, we can only make general statements about sexual differences in the prevalence of porotic lesions.

The research design involves three levels of analysis:

1. Macroscopic analysis: The presence and development of specific features of vault and orbital lesions are determined:[22] the morphology of the pits and/or pores, their regional development (outer and/or inner bone table), local distribution and frequency pattern per anatomical area, the degree of severity (slight, medium, severe), the morphology of the outer/inner/diplöic table, and the vascular activity of the exocranial and endocranial tables.[23] The correlation with lesions in other parts of the skeleton is also attempted although a systematic analysis of the skeleton is impossible because of the fragmentary nature of the samples.
2. Radiological examination: The radiological examination is conducted in order to identify any lesions on the areas under study.[24] An X-ray examination is undertaken of all complete skulls and selected cranial elements. This includes healthy as well as macroscopically detected diseased examples, as lesions are not always visible on the gross bone.
3. Endoscopic, scanning electron microscopy, and histological analyses: The tentative nature of the results of both the macroscopic and radiological methods, together with the limited number of samples and the complexity of porotic hyperostosis, point to the need for undertaking more precise analyses.

Endoscopic, scanning electron microscopy (SEM), and histological analyses represent three new methods being tested in order to establish the morphological profile of porotic hyperostosis. Endoscopic analysis is a non invasive technique that complements other methods. The internal or endocranial surfaces of the skull base and vault are viewed using low-power microscopy to locate the diagnostic clinical lesions.[25] Endoscopic investigation is applied to all specimens. The laboratory processing of the images produced and the diagnosis are still in progress.[26] Scanning electron microscopy, in combination with other microscopic methods, is helpful for examining the structure of the bone surface and the bone tissue microstructure. It can provide details of the distinction between pathology and the mere diagenesis of the bone. Using higher magnification in a back-scattered electron mode (BSE-mode in SEM), the internal structures and degree of mineralization of inorganic substances can be studied.[27] In paleopathological research, the controversy over discriminating true pathology and pseudopathology from diagenetic alteration[28] is ongoing. In this study, selected specimens suspected of exhibiting overlapping or undiagnostic lesions are being

20. Buikstra and Ubelaker 1994; Schultz 2003.
21. As described by Buikstra and Ubelaker 1994.
22. Using a descriptive system established by Buikstra and Ubelaker 1994; Carli-Thiele 1996; Ortner and Putschar 1981; Schultz 2001, 2003.
23. Hershkovitz et al. 1997; Lewis 2004; Schultz 2001, 2003.
24. The first X-rays were undertaken by Sherry Fox at the X-ray Unit of the Wiener Laboratory, American School of Classical Studies at Athens. Currently, they are conducted by S. Schultz at the facilities of the Anatomy Department, University of Göttingen.
25. Schultz 2001, 2003.
26. By Schultz at the University of Göttingen, Department of Anatomy.
27. Schultz 2001.
28. J. Buikstra (pers. comm.); Hanson and Buikstra 1987; Schultz 2001.

subjected to SEM examination for comparison with the alternative techniques. Finally, histological analysis involves the production of a series of undecalcified ground thin sections, 50μ and 70μ thick, from healthy and diseased dry vault and orbital areas. The techniques most appropriate for poorly preserved bones are applied.[29] The thin sections are examined in plane[30] and polarized light, and in polarized light using a red first order (quartz) as a compensator.[31]

Poor preservation of ancient bone due to highly acidic soil and geological turbulence is commonly encountered in Greek archaeological sites. Sixteen specimens exhibiting different levels of preservation were subjected to alternative histological protocols[32] in order to produce readable thin sections. This is a time-consuming methodology, as a good thin section takes four to six weeks to prepare. In advanced diagenesis, when the dry bone is brittle or high in moisture, it may take longer.

The diagnostic lesions for porotic hyperostosis on hard tissue are identified[33] in order to validate the gross examination results and to provide a differential diagnosis. At the same time, it is our goal to establish a data bank of pathological ancient bone tissue in Greece, such that samples are large enough to pursue statistical analyses of correlations among environmental, cultural, and biological variables.

The paleohistopathological investigation of porotic hyperostosis is necessary because in addition to the challenges listed above, there is the methodological issue of producing standardized data for the classification of porotic hyperostosis. Interpretations tend to still be highly subjective, although the methods employed in this study are designed to increase our ability to accurately identify pathological individuals and to document the variability of porotic hyperostosis expressions.

RESULTS AND DISCUSSION

The macroscopic study is still in progress for most of the samples. In the present stage of the project, it is not possible to perceive any patterning of the pathology due to region or environmental conditions. Priority is being given to the testing of the technical protocols applied in the microstructural analysis of the specimens. However, our preliminary results indicate some trends and provide new insights into the profile of porotic hyperostosis. These are shared below.

Macroscopic Analysis

The morphology and distribution of vault and orbital lesions by anatomical location that we have found are similar to those reported by other researchers working on Greek prehistoric collections.[34] Vault lesions are mainly observed in the region adjacent to the lambdoid suture (28%) (Fig. 16.1) and the frontal (38%) (Fig. 16.2) and parietal bones (72%) (Fig. 16.3). These lesions are mainly manifested as a pitting and/or porosis, occurring in different densities. The raying pattern of the lesions is not frequently observed.

Orbital lesions are always bilateral, with few cases of expanded lesions at the superior part of the orbital area toward the region near pterion (3%)

29. Modified and developed by Schultz 2001.

30. Oldfield 1998; Schultz 2001.

31. Schultz 2003.

32. Kousoulakos (pers. comm.); Maat, Van Den Bos, and Aarents 2001; S. Stout (pers. comm.); Schultz 2001.

33. For more information on microstructure and histopathology of bone, see Kousoulakos 1997; Schultz 2001; Stout 1992.

34. Manolis and Stravopodi 2003; Papathanasiou 2000; Triantaphyllou 1999.

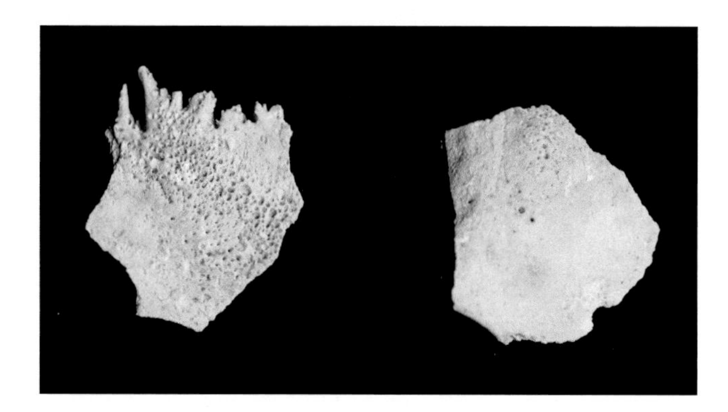

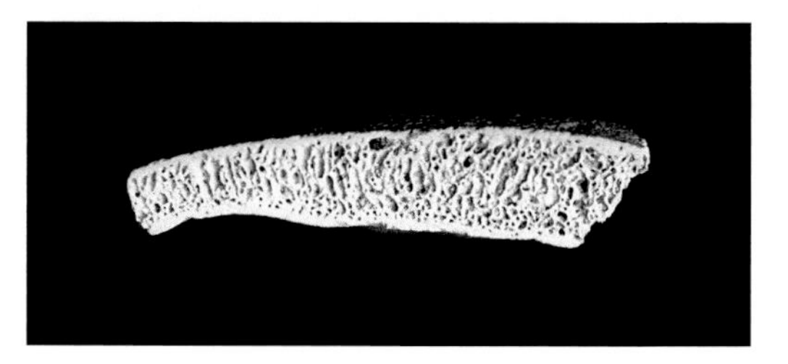

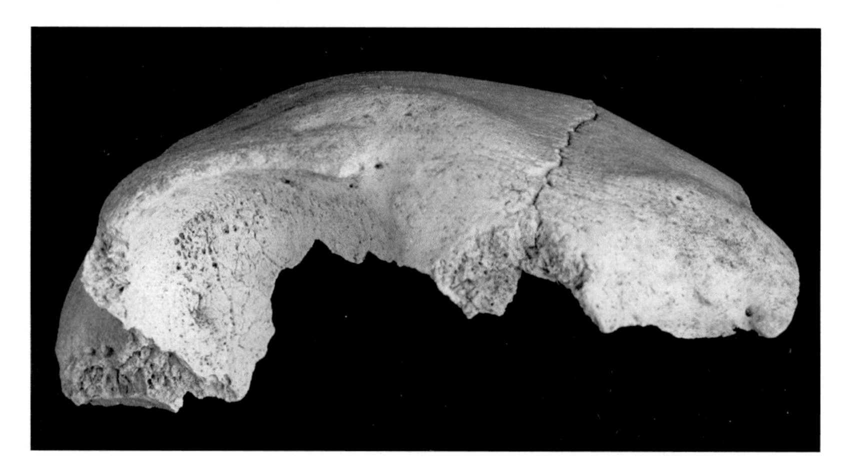

Figure 16.1 ***(top left).*** **Peracora cave, Bronze Age: endocranial table, pitting and hypervascularity.** Scale 1:1. Photo E. Stravopodi

Figure 16.2 ***(top right).*** **Kouveleiki cave, Neolithic: frontal parts, pitting, porosis, striations.** Scale 1:1. Photo E. Stravopodi

Figure 16.3 ***(middle left).*** **Perachora cave, Bronze Age: parietal, porosis, and pitting.** Scale 1:1. Photo E. Stravopodi

Figure 16.4 ***(bottom).*** **Perachora cave, Bronze Age: cribra orbitalia, lesions on the superior part.** Scale 1:1. Photo E. Stravopodi

Figure 16.5 ***(middle right).*** **Perachora cave, Bronze Age: parietal, porosis on outer table. Hyperplasia of marrow of the diploë, severe case.** Scale 1:1. Photo E. Stravopodi

(Fig. 16.4). The lesions are confined to the anterior orbit in most of the cases (64%). Yet these percentages will probably be modified when the study is completed.

In very few cases, the hypertrophy of the vault bone marrow (2%) and orbital lesions (1%) can be observed (Fig. 16.5), with occasional thinning of the outer bone table. The rarity of cases where this is severe is similar to the results of other studies in Greece.[35] This further suggests that the argument for genetic anemias being the primary cause of porotic hyperostosis in prehistoric Greece[36] has to be reevaluated.

Although it is not possible, for the present, to establish a demographic profile within a site or between sites, a preliminary assessment of the

35. Papathanasiou 2000.

36. Angel 1966. However, in his later studies, he reconsidered that.

TABLE 16.2. NEOLITHIC SAMPLE: DISTRIBUTION OF VAULT AND ORBITAL LESIONS BY AGE

Age Group	*Vault Lesions %*	*Orbital Lesions %*
Infants (<3)	0	0.6
Children (3–12)	16	19
Adolescents (12–20)	21	18
Young adults (20–30)	12	17

TABLE 16.3. BRONZE AGE SAMPLE: DISTRIBUTION OF VAULT AND ORBITAL LESIONS BY AGE

Age Group	*Vault Lesions %*	*Orbital Lesions %*
Infants (<3)	1	3
Children (3–12)	36	35
Adolescents (12–20)	47	59
Young adults (20–30)	35	29

prevalence of porotic hyperostosis by age categories is possible. During the Neolithic and Bronze Age, the lesions of porotic hyperostosis, both in the vault and orbital area, are mostly observed in adolescent individuals of 12–20 years old (Tables 16.2, 16.3). In the Neolithic, there is also a trend toward greater prevalence of the disease in the young adult age category. The younger ages have a lower prevalence, but this is possibly a biased result. Any suggestion concerning survival/mortality curves and their association with the disease is not yet possible. Although it is tempting to argue that mild episodes occurred during childhood, the inclusion of more data may alter this preliminary finding.

Despite the fact that the estimation of sex is very tentative in the studied samples, we made an initial attempt to correlate the lesions with biological sex. However, our study does not show any significant trend for greater prevalence in either sex. In some samples there is a heightened prevalence of the disease in male individuals (at Tharrounia cave, 65% of the males are affected), while in other sites females seem more susceptible to the disease (at Kastria cave, 52% of the females are affected).

The preliminary data indicate that there is a trend towards more frequent and severe lesions from the Neolithic to the Bronze Age (Tables 16.2 and 16.3) and fewer during the Neolithic. The diseased cases therefore do not fit the "expected" pattern of high frequencies of porotic hyperostosis for Neolithic societies in Greece.[37]

The above pattern also agrees with the high percentage of lesions reported for other Bronze Age sites such as Lerna[38] and Asine.[39] Despite the prevailing arguments about a declining trend in the occurrence of porotic hyperostosis in the Bronze Age, the pathology still occurs in a large number of sites. It is not unexpected that a high prevalence of the disease is observed in the Bronze Age because, according to epidemiological modeling, it takes time for a disease to spread on a population level and to be identified as an epidemiological outbreak. It is certainly the case that there

37. Angel 1966, 1977; Papathanasiou 2000; Stravopodi, Manolis, and Neroutsos 1997.
38. *Lerna* II, pp. 77, 110.
39. *Lerna* II, pp. 2, 33.

is evidence for the presence of porotic hyperostosis in many individuals during the Neolithic,[40] but whether the pathology and the causal factors can be viewed as endemic can still be questioned.

On the other hand, although it is not yet feasible to postulate a specific geographical patterning for porotic hyperostosis, there does seem to be a predictable relationship between specific environmental conditions and the pathology itself. It seems to be most prevalent in low altitude and coastal areas (such as Perachora) across time periods, while there is either absence or low frequency of lesions in higher altitude and arid regions (for example, at Kouveleiki, Tharrounia, and Theopetra). Notably, there is a significant correlation between extreme environmental settings, such as the marshy lake site at Perachora, and the more severe expressions and prevalence of the disease. The pattern at Perachora is similar to that of the Bronze Age people in Lerna,[41] another marshy site. The parasite load and infections caused by endemic malaria are the probable causes of porotic hyperostosis, reflecting anemic responses in these populations. Our histological investigation provides solid evidence for this scenario, because there are distinctive lesions caused by different agents that have been identified.[42]

In Ayios Haralabos cave in Crete, situated in a mountainous area, the prevalence of porotic hyperostosis is very high (mainly in subadults),[43] an "unexpected" result according to the general geographical pattern proposed above. However, the cave was used as a long-term ossuary and the metropolitan residential area is not found in the vicinity.[44] Thus, it may be that the individuals originate from a different, low altitude or coastal environment. Conversely, it should be noted that in a few cases including modern epidemiological studies, the same environment might be associated with high or low frequencies of porosis, because so many other factors may be involved (e.g., gene flow, selective biological and/or cultural mechanisms, or geographical barriers).

Any significant association between the prevalence of porotic hyperostosis and "poor diet"—low in protein intake and mainly based on carbohydrates—is difficult, if not impossible, to document. This may be due to broader dietary sufficiency than previously proposed, as documented by the recent paleodietary evidence for a mixed and balanced diet of both animal and plant protein intake during prehistory.[45]

In terms of the morphology of porotic hyperostosis, we find no strong relationship between the presence of vault and orbital lesions in individuals, even though the former are thought to be an earlier expression that then develops into cribra orbitalia. However, this initial expression is not easily detectable, even if more precise techniques than macroscopic inspection are used.

In a few specimens of the Neolithic/Bronze (Orfeas) and Bronze Age periods a parallel development of vault lesions in the outer and inner bone tables is observed. The histological examination of these specimens is very critical for the differential diagnosis of porotic hyperostosis, given the fact that the presence of lesions in the internal lamina is not of anemic origin but suggests inflammatory disorders.[46]

In the few cases of articulated skeletons (two in Kastria, four in Tharrounia, and two in Kouveleiki cave) no postcranial lesions are observed that are indicative of genetic anemias. The same holds true for the scattered

40. Angel 1966; Papathanasiou 2000; Stravopodi, Manolis, and Neroutsos 1997.

41. *Lerna* II.

42. Schultz 2001, 2003.

43. P. McGeorge (pers. comm.).

44. P. Betancourt (pers. comm.).

45. Cook 2002; E. Petroutsa (pers. comm.).

46. Schultz 2001, 2003.

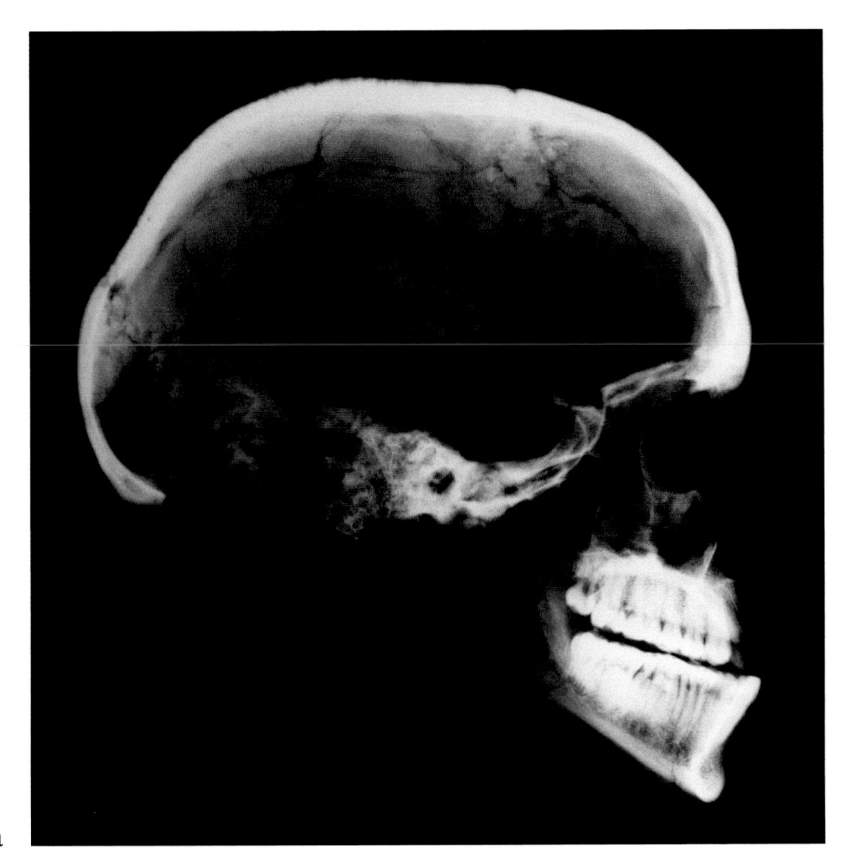

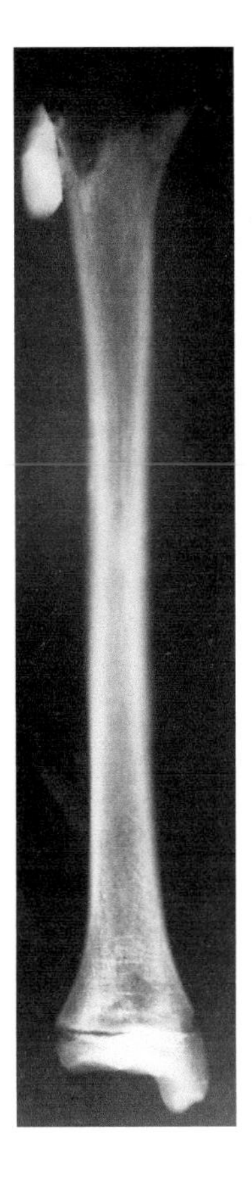

Figure 16.6. Theopetra cave, Mesolithic: (a) pseudopathological lesions appearing as the "hair-on-end" pattern, facial changes; (b) radiological picture of "Harris lines." Radiographs S. Fox

bone remains, except for one Perachora specimen that exhibits a severe inflammation of a tibia, with secondary lesions. This individual will be subjected to histological analysis.

Radiological Examination

The Mesolithic skull from Theopetra cave shows a pronounced "hair-on-end pattern" along with a thickening of the occipital region and facial and maxillary changes (Fig. 16.6:a). The postcranial skeleton shows evidence of "Harris lines" on the long bones (Fig. 16.6:b). The clinical picture points to anemia. However, it is a pseudopathological feature—a postmortem erosion on the sagittal suture. The endoscopic investigation does not show evidence of any pathological lesions, and the histological analysis is still in progress.

The 12 complete skulls from the Neolithic and Neolithic/Bronze Age sites (two from Kouveleiki cave, four from Tharrounia cave, two from Perachora, and four from Kastria cave) exhibit no radiographically detectable porotic lesions despite the fact that they showed evidence of the disease in the gross examination.

Scanning Electron Microscopy (SEM)

Pilot samples from the Perachora and Kouveleiki caves were analyzed using SEM. Osteoclastic activity and lamellar microstructure changes are observable.[47] In addition, mineralization processes in the inorganic portion of the bone can be seen.

47. Schultz, in prep.

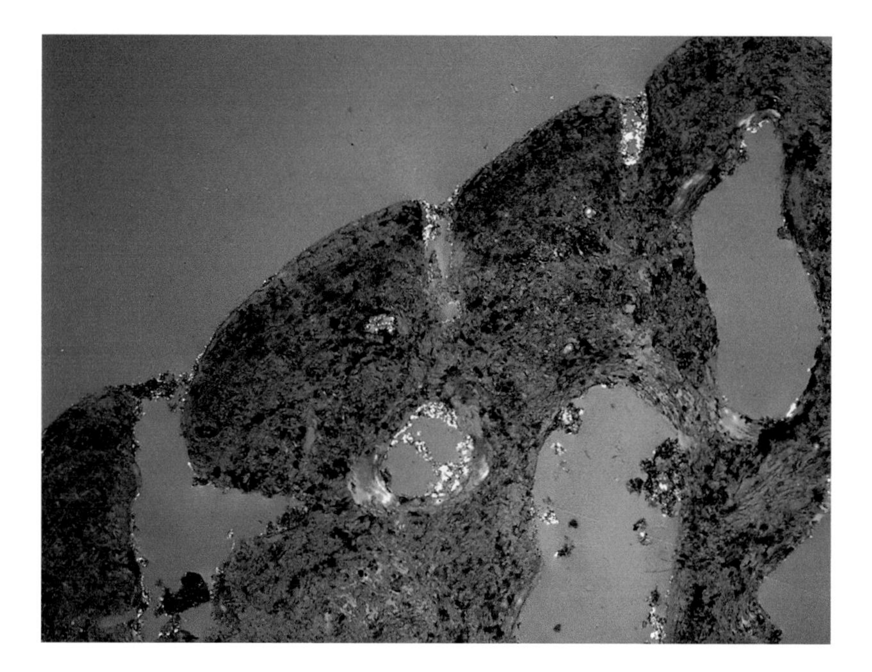

Figure 16.7. Peracohra cave, Bronze Age: parietal, outer lamina, diploë. Thin ground section, 50μ. Photo M. P. Schultz

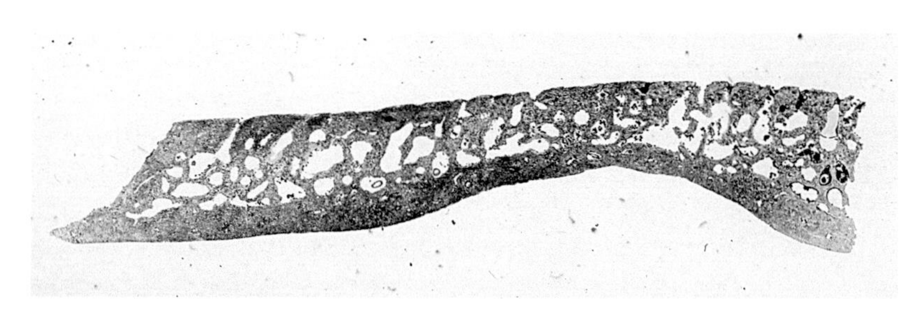

Figure 16.8. Perachora cave, Bronze Age: vault area, right parietal. Obliteration and thinning of the external lamina. Overgrowth of diploë, parallel trabeculae. Thin ground section, polarized light (red first order; Schultz 2003). Photo M. P. Schultz

Microscopic Analysis

Thirteen sections, 50μ and 70μ thick, from both healthy and diseased vault and orbit areas yielded interesting results when examined under plane-polarized light (an example from Perachora is shown in Fig. 16.7).[48] In the Perachora cases where the vault lesions were scored as severe in the gross examination, the hard tissue microscopy reveals the same severe lesions:[49] overgrowth, thinning, and direction change of the diploë in combination with the obliteration of the outer lamina (Fig. 16.8). These features are all characteristic of anemia.[50] In other Perachora specimens, there is destruction of the internal lamina with altered areas indicative of inflammatory diseases.[51]

From the examination of the thin sections of orbital areas, it seems that the destruction of the external lamina is possibly caused by inflammatory and diagenetic factors, and that these are often hard to distinguish.[52] In very few cases, the characteristic diplöe enlargement and parallel orientation of its trabeculae was recognized as evidence of a mild anemic episode, often overlapping with inflammatory episodes.

In a few histological sections the tissue morphology presents unidentifiable lesions that are possible nonspecific infections of the endocranial

48. By Schultz at the Department of Anatomy, University of Göttingen.
49. For further details on bone histology and histopathology, see Kousoulakos 1997; Martin 1991; Schultz 2001.
50. Schultz 2003.
51. Schultz, in prep.
52. Schultz 2001, 2003.

lamina, while often diagenesis destroyed certain areas with vestiges of disease. Interestingly, in Kouveleiki cave, where the macroscopic analysis yielded mild to severe stages of porotic hyperostosis, the hard tissue examination indicates advanced diagenesis with no sign of disease. This was verified by the micromorphological analysis as well.[53] In about 35% of the cases, the vault and orbital lesions are of clear-cut diagenetic, rather than pathological, origin. This is a significant finding and strongly suggests the necessity of histological examination in conjunction with gross examination in order to avoid false-positive diagnoses.

CONCLUSIONS

These preliminary results suggest a variable pattern of porotic hyperostosis prevalence through time and across space in Greece. It also seems important to bear in mind that a combination of different factors within a site may cause porotic hyperostosis. Another critical finding of this research project is that accurate macroscopic identification of porotic lesions is seriously compromised by diagenetic processes and the possible "healthy" appearance of bone that exhibits microscopic or histological disease alteration.

In keeping with the recent results from the isotopic analysis of bones,[54] the present data do not provide evidence that iron deficiency anemia, resulting solely from diet and/or in association with other agents, can be recognized as the predominant cause of porotic hyperostosis in Greek prehistoric populations. With regard to this question, there are unsolved problems to be considered. These include the following:

1. The profile of the intrasite variability in the dietary behavior of the individuals is missing because of the sampling problems.[55]
2. During the transition to the Neolithic, the change in the ratio of nutrients from animal and plant protein that could be responsible for dietary deficiency disease detectable in bone is still under investigation.[56]
3. The relation between iron intake and infectious episodes resulting in pathologies is unclear. The interaction of iron with other nutrients and its biological characteristics are issues of ongoing research. Many components of the diet may inhibit or enhance iron absorption. Marginal iron supply may be identified as a critical factor for the prevalence of dietary deficiency and consequent porosis. As a matter of fact, there is a concern that excessive iron supplementation could lead to decreased resistance to infection and promote gastrointestinal illness.[57]
4. Regional comparisons of genetic, paleodietary, and mobility pattern data demonstrate a more complex relationship between Mesolithic and Neolithic "populations" and considerable heterogeneity in the southeastern Mediterranean. The possibility of an exogenous origin for some Neolithic "populations" or "outlier" individuals has implications for the evaluation of the prevalence of porotic hyperostosis in Greek Neolithic societies.

53. T. Karkanas (pers. comm.).

54. See Papathanasiou 2004

55. The individuals within a site adopt various dietary habits because of different access to resources, personal preferences, health, sex, and age.

56. New data support that in many Neolithic societies, especially in harsh ecological settings, the groups continue to be mobile pastoralists, practicing agriculture only circumstantially.

57. Ryan 1997.

In addition, it is relevant that our preliminary histological investigations showed no indication of diet deficiency episodes as they are expressed in the hard tissue,[58] while in a number of cases, overlapping lesions of anemic and inflammatory processes were identified.[59] It has to be reemphasized that no type of anemia can be specifically identified in histological analysis; this can only be accomplished through DNA sequencing. Therefore, the serious consequences of iron deficiency anemia in prehistoric populations in Greece, although they cannot be ruled out, are not easily detected from the present data. The low frequencies of anemia and scurvy do not suggest a model of over-generalized nutritional deficiencies in the Greek Neolithic, while the microscopic analysis reveals a large number of diagenetic lesions.

Environmental stressors, however, such as parasitic infections (especially in areas with endemic malaria) are more likely to account for the varying morphotypes of porotic hyperostosis (both in the vault and orbital areas) as a marker of anemic and/or infectious diseases. There is a consistent prevalence of mild lesions in the subadult groups in these extreme environmental settings. Along these lines, it is worth reporting that in Pseira, a Bronze Age island site in Crete, signs of porotic hyperostosis as a consequence of an inflammatory disorder have been documented.[60] This study is pertinent to our work because it involves the extraction and sequencing of ancient DNA from diseased samples in order to determine the cause of porotic hyperostosis on the molecular level. The mechanism of balanced polymorphism, where the heterozygous individuals develop anemia as a protection against malaria in marshy areas, is not supported by the present data since that type of anemia can only be identified through molecular analysis.

The present data provide evidence that anemias, although present throughout Greek prehistory, are only one of the causal factors for porotic hyperostosis. In the light of new research on the etiology of iron-deficiency anemia[61] and scurvy[62] as genetic, and not purely nutritional, disorders, a molecular analysis of the diseased bones is essential.

In accordance with recent findings in clinical medicine on the "anemia of inflammation," the paradox of the symbiosis between bacteria and humans, and the evolution of infectious diseases, it is self-evident that the significance of any inferences presented here need to be further evaluated with more data and a synthesis of the results from the different analytical techniques.

POSTSCRIPT

Since this manuscript was submitted, a wealth of new information has been published.[63] The new data support and strengthen the working hypothesis presented in this paper.

58. Osteopenia of the diplöic table, specific changes in the endocranial lamina, involvement of all laminae, subdural hematoma, see Schultz 2001.
59. Schultz, in prep.
60. Arnott 2003, pp. 155, 163.
61. Centers for Disease Control and Prevention 2002.
62. Sotiriou et al. 2002.
63. Bailey, Whittle, and Cummings 2005; Fleming and Bacon 2005; Ganz 2006; Delanghe, Langlois, De Buyzere, and Torck 2007; Olivares, Pizarro, and Ruz 2007; Camberlein et al. 2008; Guillem et al. 2008; *ScienceDaily* 2008.

REFERENCES

Angel, J. L. 1966. "Porotic Hyperostosis, Anemias, Malarias, and Marshes in the Prehistoric Eastern Mediterranean," *Science* 153, pp. 760–763.

———. 1977. "Human Skeletons," in *Kephala: A Late Neolithic Settlement and Cemetery,* ed. J. C. Coleman, Princeton, pp. 133–156.

Arnott, R. 2003. "The Human Skeletal Remains," in *The Pseira Cemetery* 2: *Excavation of the Tombs* (*Pseira* VII), ed. P. P. Betancourt and C. Davaras, Philadelphia, pp. 153–164.

Bailey, D., A. Whittle, and V. Cummings. 2005. *(Un)Settling the Neolithic*, Oxford.

Buikstra, J. E., and D. H. Ubelaker, eds. 1994. *Standards for Data Collection from Human Skeletal Remains,* Fayetteville, Ark.

Camberlein, E., G. Zanninelli, L. Détivaud, A. R. Lizzi, F. Sorrentino, S. Vacquer, M.-B. Troadec, E. Angelucci, E. Abgueguen, O. Loréal, P. Cianciulli, M. E. Lai, and P. Brissot. 2008. "Anemia in β-Thalasemia Patients Targets Hepatic Hepcidin Transcript Levels Independently of Iron Metabolism Genes Controlling Hepcidin Expression," *Haematologica* 93, pp. 111–115.

Carli-Thiele, P. 1996. *Spuren von Mangelerkrankungen an steinzeitlichen Kinderskeleten* (Fortschritte in der Paläopathologie und Osteoarchäologie 1), Göttingen.

Centers for Disease Control and Prevention. 2002. "Iron Deficiency: United States, 1999–2000," *Morbidity and Mortality Weekly Report* 51, pp. 897–899.

Cook, D. C. 2002. "Skeletal Evidence for Nutrition in Mesolithic and Neolithic Greece: A View from Franchthi Cave," *Paleodiet in the Aegean,* ed. S. Vaughan and W. Coulson, Oxford, pp. 99–105.

Delanghe, J. R., M. R. Langlois, M. L. De Buyzere, and M. A. Torck. 2007. "Vitamin C Deficiency and Scurvy Are Not Only a Dietary Problem But Are Codetermined by the Haptoglobin Polymorphism," *Clinical Chemistry* 53, pp. 1397–1400.

Fleming, R. L., and B. R. Bacon. 2005. "Orchestration of Iron Homeostasis," *New England Journal of Medicine* 352(17), pp. 1741–1744.

Ganz, T. 2006. "Hepcidin and Its Role in Regulating Systemic Iron Metabolism," *Hematology* 1, pp. 29–35.

Grmek, M. D. 1989. *Diseases in the Ancient Greek World,* Baltimore.

Guillem, F., S. Lawson, C. Kannengiesser, M. Westerman, C. Beaumont, and B. Grandchamp. 2008. "Two Nonsense Mutations in the TMPRSS6 Gene in a Patient with Microcytic Anemia and Iron Deficiency," *Blood* 112, pp. 2089–2091.

Hanson, D., and J. Buikstra, 1987. "Histomorphological Alteration in Buried Human Bone from the Lower Illinois Valley: Implications for Paleodietary Research," *JAS* 14, pp. 549–563.

Hershkovitz, I. 2004. "Has the Transition to Agriculture Reshaped the Demographic Structure of Prehistoric Populations? New Evidence from the Levant," *American Journal of Physical Anthropology* 124, pp. 315–329.

Hershkovitz, I., B. Rothchild, B. Latimer, O. Dutour, G. Leonetti, C. Green Wald, C. Rothchild, and L. Jellema, 1997. "Recognition of Sickle-Cell Anemia in Skeletal Remains of Children," *American Journal of Physical Anthropology* 104, pp. 213–226.

Kotsakis, K. 2000. "Η αρχή της Νεολιθικής στην Ελλάδα," in *Theopetra Cave: Twelve Years of Excavation and Research 1987–1998. Proceedings of the International Conference, Trikala, November 6–7, 1998,* ed. N. Kyparissi-Apostolika, Athens, pp. 173–181.

Kousoulakos, S. 1997. *Εισαγωγή στην Αναπτυξιακή Βιολογία και Ιστολογία,* Athens, pp. 348–371.

Kyparissi-Apostolika, N. 2000. "Η ανασκαφή του σπηλαίου Θεόπετρας," in *Theopetra Cave: Twelve Years of Excavation and Research 1987–1998. Proceedings of the International Conference, Trikala, November 6–7, 1998,* ed. N. Kyparissi-Apostolika, Athens, pp. 17–37.

Larsen, C. S. 1998. Post Pleistocene Human Evolution Bioarchaeology of the Agricultural Transition. Poster presented at the 14th International Congress of Anthropological and Ethnological Sciences, Williamsburg, Virginia, July 26–August 1, 1998. http://www.cast.uark.edu/local/icaes/conferences/wburg/posters/cslarsen/larsen.html (accessed July 31, 2008).

Lerna II = J. L. Angel. *The People,* Princeton 1971.

Lewis, M. 2004. "Endocranial Lesions in Non-Adult Skeletons: Understanding Their Etiology," *International Journal of Osteoarchaeology* 14, pp. 82–97.

Maat, G., R. Van Den Bos, and M. J. Aarents. 2001. "Manual Preparation of Ground Sections for the Microscopy of Natural Bone Tissue: Update and Modification of Frost's 'Rapid Manual Method,'" *International Journal of Osteoarchaeology* 11, pp. 366–374.

Manolis, S., and E. Stravopodi, 2003. "An Assessment of the Human Skeletal Remains in the Mesolithic Deposits of Theopetra Cave: A Case Study," in *The Greek Mesolithic: Problems and Perspectives* (British School at Athens Studies 10), ed. N. Galanidou and C. Perlès, London, pp. 207–217.

Martin, D. 1991. "Bone Histology and Paleopathology: Methodological Considerations," in *Human Paleopathology: Current Synthesis and Future Options,* ed. D. Ortner and A. Aufderheide, Washington, D.C., pp. 55–60.

Meiklejohn, C., and M. Zvelebil. 1991. "Health Status of European Populations at the Agricultural Transition and the Implications of Adoption of Farming," in *Health in Past Societies: Biocultural Interpretations of Human Skeletal Remains in Archaeological Contexts* (*BAR-IS* 567), Oxford, pp. 115–129

Oldfield, B. 1998. "An Introduction to Light Microscopy," in *Cell Biology,* ed. J. Celis, New York, pp. 5–15.

Olivares, M., F. Pizarro, and M. Ruz. 2007. "New Insights about Iron Bioavailability Inhibition by Zinc," *Nutrition* 23, pp. 292–295.

Ortner, D., and W. Putschar. 1981. *Identification of Pathological Conditions in Human Skeletal Remains,* Washington, D.C.

Papathanasiou, A. 2000. "Bioarcheological Inferences from a Neolithic Ossuary from the Alepotrypa Cave, Diros, Greece," *International Journal of Osteoarchaeology* 10, p. 210.

———. 2004. "Stable Isotope Analysis in Neolithic Greece and Possible Implications on Human Health," *International Journal of Osteoarchaeology* 13, pp. 314–324.

Ryan, A. 1997. "Iron Deficiency Anemia in Infant Development: Implications for Growth, Cognitive Development, Resistance to Infection, and Iron Supplementation," *Yearbook of Physical Anthropology* 40, pp. 25–52.

Schultz, M. 2001. "Paleohistopathology of Bone: A New Approach the Study of Ancient Diseases," *Yearbook of Physical Anthropology* 44, pp. 106–147.

———. 2003. "The Differential Diagnosis of Ancient Disease at the Microscopic Level" (paper, Göttingen, 2003).

ScienceDaily. 2008. "A Genetic Cause for Iron Deficiency," April 15, 2008, http://www.sciencedaily.com/releases/2008/04/080413161038.htm (accessed September 29, 2008).

Sotiriou, S., S. Gispert, J. Cheng, Y. Wang, A. Chen, S. Hoogstraten-Miller, G. F. Miller, O. Kwon, M. Levine, S. H. Guttentag, and R. L. Nussbaum. 2002. "Ascorbic-Acid Transporter Slc23a1 Is Essential for Vitamin C Transport into the Brain and for Perinatal Survival," *Nature Medicine* 8, pp. 514–517.

Stout, S. 1992. "Methods of Determining Age at Death Using Bone Microstructure," in *Skeletal Biology of Peoples: Research Methods,* ed. S. Saunders and A. Katzenberg, New York, pp. 21–37

Stravopodi, E., and S. Manolis, 2000. "Το βιοαρχαιολογικό προφίλ των ανθρωπολογικών ευρημάτων του σπηλαίου Θεόπετρας: ένα πιλοτικό πρόγαμμα στον Ελλαδικό χώρο," in *Theopetra Cave: Twelve Years of Excavation and Research 1987–1998. Proceedings of the International Conference, Trikala, November 6–7, 1998,* ed. N. Kyparissi-Apostolika, Athens, pp. 95–109.

Stravopodi, E., S. Manolis, and T. Neroutsos. 1997. "Ανθρωπολογική μελέτη των σκελετικών καταλοίπων από το σπήλαιο των Λιμνών," in *Το σπήλαιο των Λιμνών στα Καστριά Καλαβρύτων* (Εταιρεία Πελοποννησιακών Σπουδών 7), ed. N. Kyparissi-Apostolika, Athens, pp. 415–456.

Stuart-Macadam, P. 1991. "Porotic Hyperostosis: Changing Interpretations," in *Human Paleopathology: Current Synthesis and Future Options,* ed. D. Ortner and A. Aufderheide, Washington, D.C., pp. 36–40.

———. 1998. "Iron-Deficiency Anemia: Exploring the Difference," in *Sex and Gender in Paleopathological Perspective,* ed. A. L. Grauer and P. Stuart-Macadam, Cambridge, pp. 45–64.

Triantaphyllou, S. 1999. "A Bioarchaeological Approach to Prehistoric Cemetery Populations from Western and Central Greek Macedonia" (diss. Univ. of Sheffield).

CHAPTER 17

The Application of mt-DNA Analysis to the Investigation of Kinship from Skeletal Remains

by Maria Georgiou, George D. Zouganelis, Chara Spiliopoulou, and Antonis Koutselinis

The Monastery of Koudouma was founded in the mid-19th century by the brothers Evmenios and Parthenios. The two founders died in the early 20th century and have been celebrated ever since. The monks saved the remains of Parthenios, but the whereabouts of the remains of his brother were unknown. Hence, when a skeleton was discovered during recent renovations, the monks questioned whether it might belong to Evmenios. In order to resolve this question, a femur from each skeleton was sent to our laboratory for mt-DNA analysis.

Human mt-DNA has become a useful tool in forensic and archaeological investigations. Its polymorphic nature, maternal inheritance, and good preservation are characteristics that have, combined with its sequence information, enabled investigators to identify missing persons,[1] war casualties,[2] individuals involved in mass disasters,[3] and criminal cases.[4] In addition, researchers have analyzed mt-DNA from past populations[5] and they have even undertaken evolutionary studies involving Neanderthal skeletal remains.[6]

Perhaps one of the most publicized cases involving mt-DNA analysis was the identification and assignment of kinship to the Romanov family.[7] Mt-DNA was useful for matching the genetic material of the Tsarina and her three daughters to Prince Philip, His Royal Highness the Duke of Edinburgh. Additionally, Tsar Nicholas was identified by comparison of mt-DNA from the bones in the grave with that from bone samples of Georgij Romanov.[8] They exhibited the same sequence as the Tsar. In a similar manner, the mt-DNA information from Anna Anderson Manahan, who had claimed to be the missing Royal Duchess Anastasia, showed that the claim was false.[9] The above cases clearly indicate the power of mt-DNA in the identification and establishment of kinship in historical bone samples.

1. Ginther, Issel-Tarver, and King 1992, pp. 135–138.
2. Holland et al. 1992, pp. 542–553.
3. Boles, Snow, and Stover 1995, pp. 349–355.
4. Butler and Levin 1998, pp. 158–162.
5. Kolman and Tuross 2000, pp. 5–23.
6. Krings et al. 1997, pp. 19–30.
7. Gill et al. 1994, pp. 130–135.
8. Ivanov et al. 1996, pp. 417–420.
9. Stoneking et al. 1995, pp. 9–10.

MATERIALS AND METHODS

The following precautions were taken to avoid contamination: Glassware used in this study was soaked overnight in 1 N HCl, rinsed with double-distilled (dd) water, autoclaved at 135°C, and finally baked at 100°C for 12 hours. Disposable plasticware was γ irradiated (when the manufacturers said this was possible) and autoclaved between uses. Solutions were prepared fresh and exposed to UV light two to three hours before use. Ancient DNA isolation was done in a dedicated room. All other steps of the analysis, PCR setup, the Polymerase chain reaction (PCR) itself, and molecular analysis were done in separate dedicated rooms. The handler wore both sterile latex gloves and a face mask.

To prepare the bone specimens, bone plugs (2 × 5 cm) were cut with a machine saw from each femur and placed in sterile 50 ml tubes. The samples were stored at -20°C. They were then washed with double-distilled water and Decon (5% solution) and dried in a clean UV laminal flow for 24 hours. The outer layer was sandblasted with a sterile sandblaster dish and pulverized with a flamed drill. The powder was collected in sterile ceramic dishes and stored in sterile tubes at -20°C. At all times appropriate precautions were taken to avoid contamination.

In order to isolate the DNA, the bone powder (0.5 g) was mixed with 750 µl extraction buffer (0.5 M ethylenediaminenetetracetic acid [EDTA] [pH 8], 0.5 mg/ml proteinase K [DNA free], and 0.5% sodium dodecyl sulphate [SDS]). The mix was incubated with simultaneous mixing at 52°C for 24 hours and at 37°C for another 24 hours. The samples were centrifuged at 14,000 RPM for five minutes and the supernatant was washed with silica gel columns (Qiagen). The DNA was bound at the silica gel and was eluted with 50 µl water (DNA free). The DNA was quantified with a spectrophotometer at A_{260}/A_{280}.

The appropriate primers for the PCR reaction and the technique for sex determination (Table 17.1) are provided below: The sample volume was 5 µl (3.5 ng DNA). The deoxyribonucleotide mix (consisting of dATP, dCTP, dGTP, and dTTP) were all in final concentration of 200 µM each. The PCR buffer, 10 µM Tri-Cl pH 8.3, contains 50 µM KCl, 2.5 µM $MgCl_2$, and 0.01% gelatin. PCR primers specific for amelogenin (Table 17.1) were utilized in concentrations of 20 pmol/reaction each. Taq-DNA-polymerase (2.5 units [U]/reaction) was added and double-distilled water was also added to attain a final volume of 50 µl.

For every reaction the appropriate thermocycler program was used (Table 17.2) which is dependent upon the primers. It should be pointed out that in every PCR an internal control without DNA (blank), a sample derived from an isolation without any bone powder (mock extraction), and positive controls of known sex were included. These measures are necessary for ensuring the authenticity of the results.

A multiplex PCR was developed in our laboratory in order to verify that the samples contain no high molecular weight (HMW) DNA due to modern contamination. The reaction was set up in the same manner as the sex determination. The primer mix used contained equimolar amounts of primers (20 pmol/each) and is presented in Table 17.3. The thermocycler program is listed in Table 17.4. A mock extraction, a PCR blank, and PCR positive controls were included as previously described.

PCR for amplification of mt-DNA fragments: We used two primer pairs (Table 17.5) in order to amplify two overlapping fragments of hypervariable region I (HVI) of mt-DNA from each sample. The PCR setup was analogous to the ones previously described. The thermocycler program is the same one shown in Table 17.4.

TABLE 17.1. PRIMERS FOR SEX DETERMINATION

X and Y Chromosomes	
Amelogenin A	5'-CCC TGG GCT CTG TAA AGA ATA GTG-3'
Amelogenin B	5'-ATC AGA GCT TAA ACT GGG AAG CTG-3'

Source: Faerman et al. 1995, pp. 327–332.

TABLE 17.2. PCR CYCLER PROGRAM FOR SEX DETERMINATION

X and Y chromosomes (Amelogenin A and Amelogenin B)	
94°C for 2 minutes	1 cycle
60°C for 10 seconds	10 cycles
65°C for 1 minute	
95°C for 10 seconds	30 cycles
60°C for 10 seconds	
72°C for 1 minute	30 cycles

Source: Evison, Fieller, and Smille 1999, pp. 1–28.

TABLE 17.3. PRIMER MIX FOR MULTIPLEX PCR

Primer	*Primer Sequence*
L 16055	5'-GAAGCAGATTTGGGTACCAC-3'
H 16139	5'-TACTACAGGTGGTCAAGTAT-3'
H 16218	5'-TGTGTGATAGTTGAGGGTTG-3'
H 16303	5'-TGGCTTTATGTACTATGTAC-3'
H 16410	5'-GCGGGATATTGATTTCACGG-3'
H 264	5'-TTATGATGTCTGTCTGGAAAG-3'

TABLE 17.4. THERMOCYCLER PROGRAM FOR MULTIPLEX AND SINGLE PCR FOR MT-DNA

95°C for 5 minutes	1 Cycle
95°C for 1 minute	
52°C for 1 minute	
72°C for 1 minute	40 Cycles
72°C for 10 minutes	1 Cycle

TABLE 17.5. PRIMERS FOR HVI REGION OF MT-DNA AMPLIFICATION

Region of mt-DNA	*Primer Name*	*Sequence*
HVI A	L 16055	5'-GAAGCAGATTTGGGTACCAC-3'
	H 16303	5'-TGGCTTTATGTACTATGTAC-3'
HVI B	L 16209	5'-CCATGCTTACAAGCAAGT-3'
	H 16410	5'-GCGGGATATTGATTTCACGG-3'

Polyacrylamide gel electrophoresis and visualization of PCR products were performed. PCR products (10 μl) were loaded in an 8% polyacrylamide gel. After the electrophoresis, the gel was stained with EtBr (0.5 mg/ml) for 20 minutes and visualized by UV light. In order to assess our results, we ran a special DNA marker (50 base pairs ladder or φX 174 marker) alongside the sample.

For DNA sequencing, the amplification products were purified from primer dimers with the use of Minelute PCR purification kit (Qiagen). The purified products were sequenced in a 310 ABI PRISM automatic sequencer with the use of Big Dye Terminator Cycle Sequencing Ready Reaction Mix volume 3.1 according to the manufacturer's instructions. The sequencing primers used for each reaction corresponded to each PCR amplification product and are listed in Table 17.5.

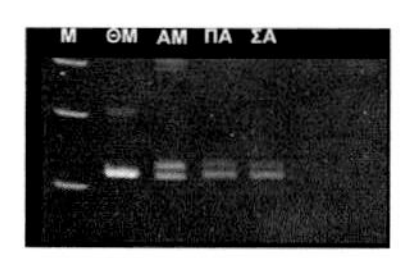

Figure 17.1. Polyacrylamide gel electrophoresis of PCR products from DNA isolated from Parthenios bone and from his putative brother for sex determination. Lane M: Molecular weight marker (50 bp ladder); Lane ΘM: Female positive control; Lane AM: Male positive control; Lane ΠΑ: Male-Parthenios; Lane ΣΑ: Male-putative brother. Photo G. Zouganelis

RESULTS

As a first step in this analysis, the authors determined the sex of the samples using molecular techniques. In Figure 17.1, the results of the amelogenin sexing test are shown. Both samples exhibited bands of 106 and 112 base pairs as expected for the male genotype. As a next step in this analysis, the authors developed and applied a multiplex PCR in order to verify that the samples contained no high molecular weight (HMW) DNA. It is well established that contamination of forensic and recent archaeological specimens results in amplification of products that are usually higher than 300 base pairs.[10] The results of multiplex PCR showed that products of 123 base pairs and 287 base pairs could be amplified from bone DNA samples, whereas products of 123 base pairs, 287 base pairs, 412 base pairs, and 876 base pairs were amplifiable from HMW DNA (Fig. 17.2). This contrast verifies the fact that no HMW DNA is present and that bone DNA is degraded as expected in respect to the time elapsed, temperature, moisture, and several other preservation conditions.

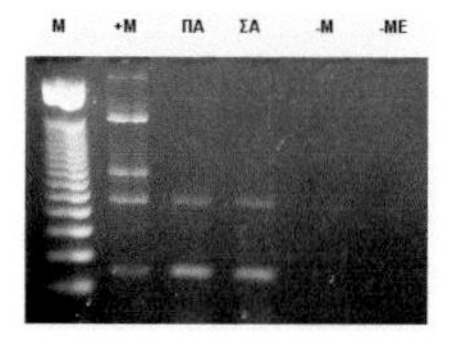

Figure 17.2. Polyacrylamide gel electrophoresis of multiplex PCR products. Lane M: Molecular weight marker (50 base pairs ladder); Lane +M: High molecular weight DNA multiplex PCR products; Lane ΠΑ: Parthenios sample multiplex PCR products; Lane ΣΑ: Putative brother multiplex PCR products; Lane -M: Negative control; Lane -ME: Negative extraction control.
Photo G. Zouganelis

The results from multiplex PCR determined the strategy for amplification and sequencing of the HVI mt-DNA region from both samples. Therefore, we used primers for the generation of products less than 300 base pairs. These fragments, named HVIA and HVIB, are overlapping and give complete information for the entire HVI region. The fragments of 287 base pairs corresponding to HVIA and 255 base pairs corresponding to HVIB that resulted from amplification are shown in Figure 17.3.

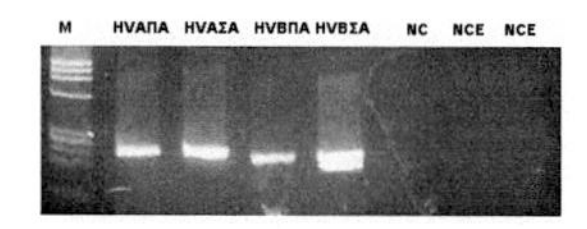

Figure 17.3. Polyacrylamide gel electrophoresis of mt-DNA PCR products. Lane M: Molecular weight marker (φX174); Lane HVAΠΑ: Parthenios HVIA products; Lane HVAΣΑ: Putative brother HVIA products; Lane HVBΠΑ: Parthenios HVIB products; Lane HVBΣΑ: Putative brother HVIB products; Lane NC: Negative control (reaction); Lanes NCE: Negative extraction controls. Photo G. Zouganelis

The PCR products were then purified from primer dimers and were sequenced with Big Dye Terminator chemistry (Applied Biosystems). In addition, DNA from the second author was amplified and sequenced. The complete HVI region from both bone samples and worker's samples is shown in Figure 17.4. Sample ΠΑ (Parthenios) presents a haplotype of 16126 T→C, 16236 C→A, 16237 A→G, whereas the sample ΣΑ (unknown man) displays a haplotype of 16256 C→T. Additionally, an author's sample (GZSEQ) presented a haplotype of 16104 C→T, 16384 G→A.

10. Pääbo 1989, pp. 1939–1943; Hagelberg and Clegg 1991, pp. 45–50; Parsons and Weedn 1997, pp. 109–138; Butler and Levin 1998, pp. 158–162; Seo et al. 2001, pp. 138–143; Gabriel et al. 2001, pp. 247–253.

	16040	*16050*	*16060*	*16070*	*16080*	*16090*	*16100*	*16110*	*16120*	*16130*
ΠA	GAAGCAGATT	TGGGTACCAC	CCAAGTATTG	ACTCACCCAT	CAACAACCGC	TATGTATTTC	GTACATTACT	GCCAGCCACC	ATGAATATTG	CACGGTACCA
ΣA	GAAGCAGATT	TGGGTACCAC	CCAAGTATTG	ACTCACCCAT	CAACAACCGC	TATGTATTTC	GTACATTACT	GCCAGCCACC	ATGAATATTG	TACGGTACCA
GZSEQ	GAAGCAGATT	TGGGTACCAC	CCAAGTATTG	ACTCACCCAT	CAACAACCGC	TATGTATTTC	GTACATTATT	GCCAGCCACC	ATGAATATTG	TACGGTACCA
CRS	GAAGCAGATT	TGGGTACCAC	CCAAGTATTG	ACTCACCCAT	CAACAACCGC	TATGTATTTC	GTACATTACT	GCCAGCCACC	ATGAATATTG	TACGGTACCA
	16140	*16150*	*16160*	*16170*	*16180*	*16190*	*16200*	*16210*	*16220*	*16230*
ΠA	TAAATACTTG	ACCACCTGTA	GTACATAAAA	ACCCAATCCA	CATCAAAACC	CCCTCCCCAT	GCTTACAAGC	AAGTACAGCA	ATCAACCCTC	AACTATCACA
ΣA	TAAATACTTG	ACCACCTGTA	GTACATAAAA	ACCCAATCCA	CATCAAAACC	CCCTCCCCAT	GCTTACAAGC	AAGTACAGCA	ATCAACCCTC	AACTATCACA
GZSEQ	TAAATACTTG	ACCACCTGTA	GTACATAAAA	ACCCAATCCA	CATCAAAACC	CCCTCCCCAT	GCTTACAAGC	AAGTACAGCA	ATCAACCCTC	AACTATCACA
CRS	TAAATACTTG	ACCACCTGTA	GTACATAAAA	ACCCAATCCA	CATCAAAACC	CCCTCCCCAT	GCTTACAAGC	AAGTACAGCA	ATCAACCCTC	AACTATCACA
	16240	*16250*	*16260*	*16270*	*16280*	*16290*	*16300*	*16310*	*16320*	*16330*
ΠA	AGTCAACTGC	AACTCCAAAG	CCACCCCTCA	CCCACTAGGA	TACCAACAAA	CCTACCCACC	CTTAACAGTA	CATAGTACAT	AAAGCCATTT	ACCGTACATA
ΣA	CATCAACTGC	AACTCCAAAG	TCACCCCTCA	CCCACTAGGA	TACCAACAAA	CCTACCCACC	CTTAACAGTA	CATAGTACAT	AAAGCCATTT	ACCGTACATA
GZSEQ	CATCAACTGC	AACTCCAAAG	CCACCCCTCA	CCCACTAGGA	TACCAACAAA	CCTACCCACC	CTTAACAGTA	CATAGTACAT	AAAGCCATTT	ACCGTACATA
CRS	CATCAACTGC	AACTCCAAAG	CCACCCCTCA	CCCACTAGGA	TACCAACAAA	CCTACCCACC	CTTAACAGTA	CATAGTACAT	AAAGCCATTT	ACCGTACATA
	16340	*16350*	*16360*	*16370*	*16380*	*16390*				
ΠA	GCACATTACA	GTCAAATCCC	TTCTCGTCCC	CATGGATGAC	CCCCCTCAGA	GTGCTACTCT				
ΣA	GCACATTACA	GTCAAATCCC	TTCTCGTCCC	CATGGATGAC	CCCCCTCAGA	GTGCTACTCT				
GZSEQ	GCACATTACA	GTCAAATCCC	TTCTCGTCCC	CATGGATGAC	CCCCCTCAAA	GTGCTACTCT				
CRS	GCACATTACA	GTCAAATCCC	TTCTCGTCCC	CATGGATGAC	CCCCCTCAGA	GTGCTACTCT				

Note: ΠA: Parthenios sequence; ΣA: Unknown individual sequence; CRS: HVI according to Andrews et al. 1999; GZSEQ: lab worker sequence. The complete sequences are between positions 16035 and 16395 of the HVI region. A = adenine, G = guanine, T = thymine, C = cytosine. The nucleotide differences in between sequences are highlighted.

Figure 17.4. Comparative representation of the HVI mt-DNA region from the bone samples

From the above, it is shown that the samples present different haplotypes from that derived from the worker. This fact excludes the possibility for intralaboratory contamination. In the same way, the two bone samples present four base pair differences between them. This difference excludes the possibility of the two samples being derived from a common maternal line and excludes the possibility of maternal kinship.

DISCUSSION

The above work is the first report involving mt-DNA analysis of skeletal remains with historical significance from Greece. Actually, there are few reports concerning kinship and DNA analysis of skeletal remains of historical importance published to date.[11] One of the reasons is that in many cases where skeletons are found (for example, during excavation or building renovation), the samples are not in good condition and the DNA is not always preserved sufficiently for molecular analysis. Fortunately, in the case of our project, both femora were in very good condition and the isolated DNA, although degraded, permitted the amplification of relatively large fragments (up to 287 base pairs). In fact, the state of preservation of DNA was similar to bones from up to 10 years since death examined previously in our laboratory.[12]

The good state of preservation allowed characterization of genomic DNA as shown from the amelogenin sexing assay. The amelogenin test is essential for the DNA characterization of forensic and archaeological specimens.[13] It is useful for DNA authentication[14] and indicates the DNA preservation in skeletal samples.[15]

The problem of contamination from external sources was tackled with the application of mt-DNA multiplex PCR. This assay showed that only relatively small DNA fragments could be amplified from bone extracts. If HMW DNA (contaminating DNA) was present, then in addition to the short amplicons, longer amplicons are formed, as has been shown by Alonso et al.[16] This is because there are enough intact DNA template molecules for an efficient amplification between contiguous forward and reverse primers of different pairs and between the outside forward and reverse primers. But if a highly degraded DNA sample is analyzed, only the short amplicons are generated.[17]

CONCLUSIONS

Four results can be drawn from this study. First, DNA was successfully isolated from a bone sample derived from Parthenios and another individual thought to be his brother. Second, the amylogenin sexing assay showed that both samples came from male individuals. Mt-DNA multiplex PCR was then employed to show that both samples contained degraded and authentic DNA. Finally, mt-DNA sequencing of the HVI region from both samples showed four base pair differences. This excludes the possibility that the two samples were derived from a common maternal line. Hence, the identification of Evmenios remains elusive.

11. Gill et al. 1994, pp. 130–135; Ivanov et al. 1996, pp. 417–420; Gill et al. 1995, pp. 9–10; Schultes, Hummel, and Herrmann 2000, pp. 37–44.

12. Zouganelis 2005.

13. Evison, Fieller, and Smille 1999, pp. 1–28; Manucci et al. 1994, pp. 190–193.

14. Meyer et al. 2000, pp. 87–90; Lassen, Hummel, and Herrmann 2000, pp. 1–8; Georgiou et al. 2002, pp. 165–172.

15. Evison, Fieller, and Smille 1999, pp. 1–28; Pääbo 1989, pp. 1939–1943.

16. Alonso et al. 2003, pp. 585–588.

17. Andrews et al. 1999, p. 147.

REFERENCES

Alonso, A., C. Albarrán, P. Martín, P. García, O. García, C. de la Rúa, A. Alzualde, L. F. de Simón, M. Sancho, and J. Piqueras. 2003. "Multiplex-PCR of Short Amplicons for mtDNA Sequencing from Ancient DNA," *International Congress Series* 1239, pp. 585–588.

Andrews, R. M., I. Kubacka, P. F. Chinnery, R. N. Lightowlers, D. M. Turnbull, and N. Howell. 1999. "Reanalysis and Revision of the Cambridge Reference Sequence for Human Mitochondrial DNA," *Nature Genetics* 23, p. 147.

Boles, T. C., C. C. Snow, and E. Stover. 1995. "Forensic DNA Testing on Skeletal Remains from Mass Graves: A Pilot Project in Guatemala," *Journal of Forensic Science* 40, pp. 349–355.

Butler, J. M., and B. C. Levin. 1998. "Forensic Applications of Mitochondrial DNA," *Tibtech* 6, pp. 158–162.

Evison, M. P., N. Fieller, and D. M. Smille. 1999. "Ancient HLA: A Preliminary Survey," *Ancient Biomolecules* 3, pp. 1–28.

Faerman, M., D. Filon, G. Kahila, C.L. Greenblatt, P. Smith, and A. Oppenheim. 1995. "Sex Identification of Archeological Human Remains Based on Amplification of the X and Y Amelogenin Alleles," *Gene* 167, pp. 327–332.

Gabriel, M. N., E. F. Huffine, J. H. Ryan, M. M. Holland, and T. J. Parsons. 2001. "Improved mt-DNA Sequence Analysis of Forensic Remains Using a 'Mini-Primer Set' Amplification Strategy," *Journal of Forensic Science* 46, pp. 247–253.

Georgiou, M., G. Zouganelis, M. Lambrinoudakis, and A. Koutselinis. 2002. "Sex Determination in Human Skeletal Remains of Forensic and Archeological Interest Using Molecular Techniques," *Iatriki* 82, pp. 165–172.

Gill, P., P. L. Ivanov, C. Kimpton, R. Piercy, N. Benson, G. Tully, I. Evett, E. Hagelberg, and K. Sullivan. 1994. "Identifications of the Remains of the Romanov Family by DNA Analysis," *Nature Genetics* 6, pp. 130–135.

Ginther, C., L. Issel-Tarver, and M.-C. King. 1992. "Identifying Individuals by Sequencing Mitochondrial DNA from Teeth," *Nature Genetics* 6, pp. 135–138.

Hagelberg, E., and J. N. Clegg. 1991. "Isolation and Characterization of DNA from Archaeological Bone," *Proceedings of the Royal Society London B* 244, pp. 45–50.

Holland, M. M., D. L. Fisher, L. G. Mitchell, W. C. Rodriguez, J. J. Canik, C. R. Merril, and V. W. Weedn. 1992. "Mitochondrial DNA Sequence Analysis of Human Skeletal Remains: Identification of Remains from the Vietnam War," *Journal of Forensic Science* 38, pp. 542–553.

Ivanov, P. L., M. J. Wadhams, R. K. Roby, M. M. Holland, V. W. Weedn, and T. J. Parsons. 1996. "Mitochondrial DNA Sequence Heteroplasmy in the Grand Duke of Russia Georgij Romanov Establishes the Authenticity of the Remains of Tsar Nicholas II," *Nature Genetics* 12, pp. 417–420.

Kolman, C. J., and N. Tuross. 2000. "Ancient DNA Analysis of Human Populations," *American Journal of Physical Anthropology* 111, pp. 5–23.

Krings, M., A. Stone, R. W. Schmitz, H. Krainitzki, M. Stoneking, and S. Pääbo. 1997. "Neandertal DNA Sequences and the Origin of Modern Humans," *Cell* 90, pp. 19–30.

Lassen, C., S. Hummel, and B. Herrmann. 2000. "Molecular Sex Identification of Stillborn Individuals (Traufkinder) from the Burial Site Aegerten," *Anthropologischer Anzeiger* 58, pp. 1–8.

Manucci, A., K. M. Sullivan, P. L. Ivanov, and P. Gill. 1994. "Forensic Application of a Rapid Amplification of the X-Y Homologous Gene Amelogenin," *International Journal of Legal Medicine* 106, pp. 190–193.

Meyer, E., M. Wiese, H. Bruchhaus, M. Claussen, and A. Klein. 2000. "Extraction and Amplification of Authentic DNA from Ancient Human Remains," *Forensic Science International* 113, pp. 87–90.

Pääbo, S. 1989. "Ancient DNA: Extraction, Characterization, Molecular Cloning, and Enzymatic Amplification," *Proceedings of the National Academy of Sciences* 86, pp. 1939–1943.

Parsons, T. J., and V. W. Weedn. 1997. "Preservation and Recovery of DNA in Postmortem Specimens and Trace Samples," in *Forensic Taphonomy: The Post Mortem Fate of Human Remains,* ed. W. D. Haglund and M. H. Sorg, Boca Raton, pp. 109–138.

Schultes, T., Hummel, S., and B. Herrmann. 2000. "Ancient DNA-typing Approaches for the Determination of Kinship in a Disturbed Collective Burial Site," *Anthropologischer Anzeiger* 58, pp. 37–44.

Seo Y., T. Uchiyama, K. Shimizu, and K. Takahama. 2000. "Identification of Remains by Sequencing of the Mitochondrial DNA Control Region," *American Journal of Forensic Medicine and Pathology* 21, pp. 138–143.

Stoneking, M., T. Melton, J. Nott, S. Barritt, R. Roby, M. Holland, V. Weedn, P. Gill, C. Kimpton, R. Aliston-Greiner, and K. Sullivan. 1995. "Establishing the Identity of Anna Anderson Manahan," *Nature Genetics* 9, pp. 9–10.

Zouganelis, G. 2005. "Isolation and Characterization of DNA in Skeletal Remains of Forensic and Archaeological Importance" (diss. Univ. of Athens).

INDEX